THE LIBRARY
ST. MARY'S COLLEGE OF MARYLAND
ST. MARY'S CITY, MARYLAND 20686

Evolution of Prokaryotes

Academic Press Rapid Manuscript Reproduction

The proceedings of a symposium held under the auspices of the
Federation of European Microbiological Societies and
the Deutsche Forschungsgemeinschaft
in Munich, Federal Republic of Germany, 16–18 September 1984

Evolution of Prokaryotes

Edited by

Karl H. Schleifer

Lehrstuhl für Mikrobiologie
Technische Universität München
Munich, Federal Republic of Germany

Erko Stackebrandt

Institut für Allgemeine Mikrobiologie
Christian-Albrechts-Universität
Kiel, Federal Republic of Germany

1985

ACADEMIC PRESS
(Harcourt Brace Jovanovich, Publishers)
London   Orlando   San Diego   New York
Toronto   Montreal   Sydney   Tokyo

COPYRIGHT © 1985, BY ACADEMIC PRESS INC. (LONDON) LTD.
ALL RIGHTS RESERVED.
NO PART OF THIS PUBLICATION MAY BE REPRODUCED OR
TRANSMITTED IN ANY FORM OR BY ANY MEANS, ELECTRONIC
OR MECHANICAL, INCLUDING PHOTOCOPY, RECORDING, OR
ANY INFORMATION STORAGE AND RETRIEVAL SYSTEM, WITHOUT
PERMISSION IN WRITING FROM THE PUBLISHER.

ACADEMIC PRESS INC. (LONDON) LTD.
24-28 Oval Road
LONDON NW1 7DX

United States Edition published by
ACADEMIC PRESS, INC.
Orlando, Florida 32887

BRITISH LIBRARY CATALOGUING IN PUBLICATION DATA
Evolution of Prokaryotes.—(FEMS symposia
　publications; v. 29)
　　1. Prokaryotes—Evolution
　　I. Schleifer, Karl H.　　II Stackebrandt, Erko
　　III. Series
　　589.9′038　　　QR18

LIBRARY OF CONGRESS CATALOGING IN PUBLICATION DATA
Main entry under title:

Evolution of prokaryotes.

　　(FEMS symposia publicátions ; 29)
　　Based on a symposium on the evolution of
prokaryotes, held in Munich Sept. 1984 and sponsored by
the Deutsche Forschungsgemeinschaft and the
Federation of European Microbiological Societies.
　　Includes index.
　　1. Bacteria—Evolution—Congresses.　　I. Schleifer,
Karl H.　　II. Stackebrandt, Erko.　　III. Deutsche
Forschungsgemeinschaft.　　IV. Federation of European
Microbiological Societies.　　V. Series: FEMS symposium ;
no. 29.
QR81.7.E96　　1985　　　589.9′038　　　85-47780
ISBN 0-12-625130-4 (alk. paper)
PRINTED IN THE UNITED STATES OF AMERICA

85 86 87 88　　　9 8 7 6 5 4 3 2 1

Contents

Contributors	vii
Preface	ix

Why Study Evolutionary Relationships among Bacteria?   1
    Carl R. Woese

Prokaryotic Genome Evolution: What We Might Learn from the Archaebacteria   31
    W. Ford Doolittle and Charles J. Daniels

The Evolution of the Transcription Apparatus   45
    W. Zillig, R. Schnabel, K. Stetter, M. Thomm, F. Gropp and W.D. Reiter

Evolution of Translation   73
    A. Böck, M. Jarsch, H. Hummel and G. Schmid

Transposable Elements and Evolution   91
    R. Schmitt, P. Rogowsky, S.E. Halford and J. Grinsted

The Role of IS- and Complex Transposable Elements in the Evolution of New Genes   105
    Heinz Saedler

5 S Ribosomal RNA as a Tool for Studying Evolution   115
    R. De Wachter, E. Huysmans and A. Vandenberghe

Evolution of Light Energy Conversion   143
    K. Krishna Rao, Richard Cammack and David O. Hall

The Evolution of Bacterial Respiration *C. W. Jones*	175
Evolution of Chemolithoautotrophy *Bärbel Friedrich*	205
Evolution of Autotrophic CO_2 Fixation *G. Fuchs and E. Stupperich*	235
Evolution in the Citric Acid Cycle *P. D. J. Weitzman*	253
Evolution of Arginine Metabolism *V. Stalon*	277
Phylogeny and Phylogenetic Classification of Prokaryotes *E. Stackebrandt*	309
Evolution of the Systematics of Bacteria *Otto Kandler*	335
Index	363

Contributors

Numbers in parentheses indicate the pages on which the authors' contributions begin.

A. Böck (73), *Lehrstuhl für Mikrobiologie, Universität München, D-8000 Munich 19, Federal Republic of Germany*

Richard Cammack (143), *Department of Plant Sciences, King's College, London SE24 9JF, England*

Charles J. Daniels (31), *Department of Biochemistry, Dalhousie University, Halifax, Nova Scotia, Canada B3H 4H7*

R. De Wachter (115), *Departement Biochemie, Universiteit Antwerpen (U.I.A.), B-2610 Antwerp, Belgium*

W. Ford Doolittle (31), *Department of Biochemistry, Dalhousie University, Halifax, Nova Scotia, Canada B3H 4H7*

Bärbel Friedrich (205), *Institut für Mikrobiologie, Universität Göttingen, D-3400 Göttingen, Federal Republic of Germany*

G. Fuchs (235), *Abteilung für Angewandte Mikrobiologie, Universität Ulm, D-7900 Ulm, Federal Republic of Germany*

J. Grinsted (91), *Departments of Biochemistry and Microbiology, University of Bristol, Bristol BS8 1TD, England*

F. Gropp (45), *Max-Planck-Institut für Biochemie, D-8033 Martinsried, Federal Republic of Germany*

S. E. Halford (91), *Departments of Biochemistry and Microbiology, University of Bristol, Bristol BS8 1TD, England*

David O. Hall (143), *Department of Plant Sciences, King's College, London SE24 9JF, England*

H. Hummel (73), *Lehrstuhl für Mikrobiologie, Universität München, D-8000 Munich 19, Federal Republic of Germany*

E. Huysmans (115), *Departement Biochemie, Universiteit Antwerpen (U.I.A), B-2610 Antwerp, Belgium*

M. Jarsch (73), *Lehrstuhl für Mikrobiologie, Universität München, D-8000 Munich 19, Federal Republic of Germany*

C. W. Jones (175), *Department of Biochemistry, University of Leicester, Leicester LE1 7RH, England*

Otto Kandler (335), *Botanisches Institut der Universität München, D-8000 Munich 19, Federal Republic of Germany*

K. Krishna Rao (143), *Department of Plant Sciences, King's College, London SE24 9JF, England*

W. D. Reiter (45), *Max-Planck-Institut für Biochemie, D-8033 Martinsried, Federal Republic of Germany*

P. Rogowsky (91), *Lehrstuhl für Genetik, Universität Regensburg, D-8400 Regensburg, Federal Republic of Germany*

Heinz Saedler (105), *Max-Planck-Institut für Züchtungsforschung, D-5000 Cologne 30, Federal Republic of Germany*

G. Schmid (73), *Lehrstuhl für Mikrobiologie, Universität München, D-8000 Munich 19, Federal Republic of Germany*

R. Schmitt (91), *Lehrstuhl für Genetik, Universität Regensburg, D-8400 Regensburg, Federal Republic of Germany*

R. Schnabel (45), *Max-Planck-Institut für Biochemie, D-8033 Martinsried, Federal Republic of Germany*

E. Stackebrandt (309), *Institut für Allgemeine Mikrobiologie, Christian-Albrechts-Universität, D-2300 Kiel, Federal Republic of Germany*

V. Stalon (277), *Laboratoire de Microbiologie, Faculté des Sciences, Université Libre de Bruxelles, B-1070 Brussels, Belgium, and Institut de Recherches du CERIA, Brussels, Belgium*

K. Stetter (45), *Institut für Biochemie, Genetik und Mikrobiologie, Universität Regensburg, D-8400 Regensburg, Federal Republic of Germany*

E. Stupperich (235), *Abteilung für Angewandte Mikrobiologie, Universität Ulm, D-7900 Ulm, Federal Republic of Germany*

M. Thomm (45), *Institut für Biochemie, Genetik und Mikrobiologie, Universität Regensburg, D-8400 Regensburg, Federal Republic of Germany*

A. Vandenberghe (115), *Departement Biochemie, Universiteit Antwerpen (U.I.A.), B-2610 Antwerp, Belgium*

P. D. J. Weitzman (253), *Department of Biochemistry, University of Bath, Bath BA2 7AY, England*

Carl R. Woese (1), *Department of Genetics and Development, University of Illinois, Urbana, Illinois 61801, USA*

W. Zillig (45), *Max-Planck-Institut für Biochemie, D-8033 Martinsried, Federal Republic of Germany*

Preface

A symposium on the "Evolution of Prokaryotes" was held in Munich 16–18 September 1984. The meeting was sponsored by the Deutsche Forschungsgemeinschaft and the Federation of European Microbiological Societies. The organizing committee of the symposium consisted of A. Böck, K. H. Schleifer, E. Stackebrandt and W. Zillig. The organizers would like to express their gratitude to the sponsors who enabled them to hold this meeting.

In contrast to higher evolved organisms (eukaryotes), bacteria lack the morphological, developmental and fossil evidence for tracing their phylogenetic relationships. Within the last decade, however, advances in molecular biology and the resulting information, increasing almost exponentially, have brought about new insights into and a new understanding of the evolution and phylogeny of prokaryotes. The organizers of the symposium were convinced that it was the right time for brainstorming the area of the evolution of prokaryotes by bringing together scientists of various biological disciplines (biochemistry, biophysics, genetics, microbiology and molecular biology). This allows us to present a series of articles in which the various aspects of microbial evolution are dealt with in depth and breadth.

The scene for the symposium and the book has been set by the contribution of C. R. Woese, focusing on the importance of studying evolution. The following contributions discuss genome evolution, the genetic heterogeneity of the archaebacterial kingdom and the role archaebacteria may have played in the evolution of the eukaryotic cell. The next three chapters deal with the influence of transposable and IS-elements on the stability and evolution of genes.

The following articles can be classified under the "Evolution of Metabolism". The evolution of complex processes (e.g. photosynthesis, respiration, CO_2 fixation and chemolithoautotrophy) is dealt with, as are the evolutionary changes within certain metabolic pathways. The final two contributions focus on an attempt to use the phylogenetic data for a future genealogically based classification.

We hope that the book may lay the basis for a vivid discussion between scientists of different disciplines and that it may stimulate student interest in this fascinating aspect of biology.

Munich and Kiel
November 1984

Karl Heinz Schleifer
Erko Stackebrandt

WHY STUDY EVOLUTIONARY RELATIONSHIPS AMONG BACTERIA?

Carl R. Woese

Department of Genetics and Development
University of Illinois
Urbana, Illinois, USA

INTRODUCTION

Microbiology is a field that has developed from its inception virtually untouched by Darwin's grand idea. We can not relate the bacteria in phylogenetic terms. We do not understand how one bacterial phenotype might have arisen from another. We tend not to take evolutionary considerations into account in any discussion, planning, evaluation, etc., of our experiments with bacteria. If we speak of bacterial evolution at all, it seems to be during those idle, expansive moments of after-dinner speculation. Microbiology is certainly among the least evolutionarily oriented of biological disciplines.

For several generations microbiologists have been taught that since phylogenetic considerations play no role in the day to day workings of the field, they are of no real value. C.B. van Niel was perhaps the last great microbiologist to take bacterial evolutionary relationships seriously, and even he, by the end of his career seems to have given the problem up as intractable -- focusing instead on the determinative classification of bacteria (1, 2). We have yet to begin to appreciate the effect such a protracted, rationalized, and ultimately institutionalized failure to introduce evolutionary concepts into the structure of microbiology has had, on the field itself, on its relationship to biology as a whole, and on the quality of our appreciation of biology.

One can, however, be assured that this effect has been significant. The evolutionary scope of the bacteria far transcends that of the eukaryotes (at least those eukaryotes of which we have knowledge); bacterial evolutionary history spans nearly the entire history of our planet, and the two histories, biological and physical, are closely intertwined. Moreover, bacteria have played an important role in the evolution of the eukaryotic cell. It is fair to say that many if not most of the deeper evolutionary questions and their answers lie locked in the evolutionary history of microorganisms (3). Yet our whole understanding of evolution has been formulated without considering bacterial evolution.

To say that the main reason for studying bacterial phylogenetic relationships, now that these can be determined, is to establish a proper bacterial classification seems gross understatement. To say that we study bacterial phylogeny to understand the evolution of bacteria -- their biochemistry, their ecological relationships, etc. -- is even too parochial. We need to study bacterial phylogenetic relationships in order to gain deep insights into the evolutionary process and the origin of cellular life -- and thereby to understand what biology is all about!

PAST PREJUDICES AND PRECONCEPTIONS

As we begin to see bacterial phylogeny for the first time (4), we do not do so with clear eyes; perception is clouded by conservatism and prejudice, largely embodied in the existing system of classification (*e.g.* Bergey's Manual). Borrowed wholesale from metazoan biology, this classification was forced upon the bacterial world without due consideration as to whether or not it is appropriate. Certainly the formalized categories, with their Latin-Greek names, command an authority and convey a finality that is totally inappropriate. Many microbiologists may believe that the phylogenetic relationships now being discovered will serve merely to clean up, improve, or otherwise refine the existing taxonomy. This is a misconception. The present categories are not mere inexact approximations. They are at fundamental variance with the true phylogeny of bacteria (4, 5). The microbiologist conforms to, thinks within, and tends to perpetuate this taxonomy. Instead he should be questioning not only the validity of families such as *Rhodospirillaceae* and genera such as *Pseudomonas*, but what relevance the conventional taxonomic hierarchy, the concepts of genus, family, order, etc., have in the world of microbial relationships.

We are also strongly influenced in our attitude toward bacterial phylogeny by the conventional wisdom regarding the origin of life and the relationship between "prokaryotes" and eukaryotes. Oparin told us that the first organisms to arise on this planet, in an ocean rich in organic compounds, were extremely heterotrophic anaerobes, and that autotrophy and phototrophy were then later evolutionary inventions in a certain few sublines (6). To the microbiologist this has meant that the first *bacteria* were anaerobic heterotrophs, and the phototrophic bacteria (which arose subsequently in some particular subline) are in essence phylogenetically separate from non-photosynthetic bacteria (2). Such a view is strongly imprinted on the present system of bacterial classification — *e.g.* the family *Rhodospirillaceae*, which comprises photosynthetic bacteria only.

"Prokaryote-eukaryote" is a simple, but subtly deceptive concept, one that has strongly influenced and distorted our understanding of bacteria. Over the years the eukaryote-prokaryote notion has undergone enormous change, whose significance the biologist yet has fully to recognize. Originally the distinction between the two was made on the basis of non-comparable properties; the prokaryote did not possess this or that feature characteristic of eukaryotic cells, *e.g.* nucleus circumscribed by a membrane, organelles, and so on (7-10). Since there were no analogous properties defined for prokaryotes (*i.e.*, ones that were not possessed by the eukaryotic cell), the initial definition of a prokaryote was, then, a negative one — a "prokaryote" is any cell that is not eukaryotic (and has no distinctive features of its own in the light microscope).

Amazingly, the classical biologist took "prokaryote" so defined to represent a phylogenetically coherent category (as, for good reason, "eukaryote" did). Thus, when prokaryote and eukaryote became redefined in terms of molecular phenotype — a definition framed in terms of *comparable* properties — the biologist saw no reason not to let *Escherichia coli* represent the entire world of bacteria (11)! He characterized almost no other bacteria in molecular terms, and when he did, it was without a phylogenetic question in mind. If the phylogenetic coherence of the category prokaryote had at any point been questioned, the archaebacteria would have been discovered then and there.

The pernicious nature of the prokaryote-eukaryote concept is embodied in the prefix "*pro*". The *pro*karyote arose before the eukaryote. The *pro*karyote is more primitive than the eukaryote. And the more primitive type is the aboriginal type of cell. The *eu*- has then, somehow arisen from the *pro*-: Eukaryotic organelles came from prokaryotic endosymbionts; the eukaryotic nucleus is a residue of genes from prokaryotic endosymbionts; the original "eukaryotic" cell that hosted all the endosymbionts was itself a prokaryote that had lost its cell wall (9, 12).

Probably most of these prejudices and just-so stories will prove untrue. In any case, we have no business accepting them as dogma -- this is not the time to interpret new findings (with which we are being deluged) in the light of old prejudices.

THE MEASUREMENT OF PHYLOGENETIC RELATIONSHIPS IN BACTERIA

The classically-recognized phenotypic characteristics of the bacteria tend to be too simple to serve as the basis upon which to construct a system of (phylogenetic) classification. Bacteria cannot be reliably classified by their phenotypes. [This, of course, is not strictly true, for many aspects of their molecular phenotypes are complex enough to serve such a function, but for the present, one needs additional criteria by which to assess which of the molecular criteria are the phylogenetically representative ones.] And the problem is compounded by the fact that some (but not all) bacterial genes are subject to interspecies transfer.

The measurement of phylogenetic relationships among bacteria must rely primarily upon genotypic criteria. An organism's genotype is more than complex enough to serve as the basis upon which to define a system of classification. Any (extensive) homology seen in molecular sequence comparisons cannot reasonably represent convergence; it must reflect common ancestry. The problem has always been, of course, one of how to measure the genotype, of how to determine genetic sequence or its equivalent. This problem exists no longer (13). The only major consideration today is which gene or genes (or their products) one should sequence, which genes are the appropriate molecular chronometers.

The first molecule used extensively for phylogenetic measurement was cytochrome c; the tree constructed on the basis of cytochrome c sequence comparisons confirmed and extended the classical eukaryotic tree, and was largely responsible for establishing in biology the concept of the molecular sequence clock (14-15). Cytochrome c, however, is not the proper molecular chronometer for the study of bacterial phylogeny. The reason for this is simply the tremendous scope of bacterial evolution. The phylogenetic tree for eukaryotes constructed from cytochrome c comparisons (16) forms only one branch of the tree defined by the α subdivision of the purple photosynthetic bacteria (4, 17-18), which in turn is only one of several phylogenetic clusters that constitute the greater purple bacterial unit, which

WHY STUDY EVOLUTIONARY RELATIONSHIPS? 5

then constitutes only one of approximately ten major "phyla" of the eubacteria (4-5, 18; Woese et al., mss in preparation). Thus, it is not surprising that cytochrome c is too fast a molecular chronometer to be accurate in the bacterial realm. The molecule also does not remain functionally constant across the phylogenetic distances encontered among the bacteria, and does not even occur in many bacterial groups.

The 5S ribosomal RNA sequence has become a popular molecular chronometer (19-20). However, it too, is not optimal. The 5S rRNA is too small and too constrained to be well suited for this purpose. The molecule comprises 4-5 separate double helical elements. Occasionally one of these will become "redesigned" during the course of evolution, which means that a number of non-clock-like changes become introduced into the sequence, which in turn lead to overestimates of phylogenetic distance and (more importantly) to falsely low estimates of phylogenetic branching order.

The Ribosomal RNA Chronometer

The two large ribosomal RNAs are excellent molecular chronometers for measuring phylogenetic relationships among bacteria. They are extremely constant in function, universal in distribution, (readily isolated), moderately to highly conserved in sequence across great phylogenetic distances, and are large (21). As a result they can measure both close (e.g. species within the same "genus") and distant relationships (i.e. between kingdoms), and measure them as or more accurately than any other molecular chronometer. [The molecule's size is quite important, for it distinguishes the large ribosomal RNAs as molecular chronometers from smaller molecules in at least two ways: Large molecules tend not to be so constrained throughout their sequence as are their smaller counterparts, and they constitute a larger number of functionally defined subunits, domains, than do the smaller molecules. Instead of the 4-5 separate helices (presumably semi-independent domains) seen in the 5S rRNA, the 16S rRNA contains about fifty such. For this reason the above-mentioned occasional "redesign" of a domain will affect a smaller portion of the molecule, and so introduce less uncertainty into phylogenetic measurement (21). Not surprisingly, the phylogenetic categories defined by the larger molecule appear more consistent, more "regular", than do those identified on the basis of 5S rRNA.]

We have to understand the ribosomal RNA chronometer in detail, for upon this understanding turns the interpretation of sequence comparisons. One of its important characteristics is that the changes in rRNA sequence one sees are to a first approximation *selectively neutral*. ["Changes" in this case are not necessarily the individual changes at a given position in the sequence, however. They are the combinations of these that result in viable lines of descent. For example, it is unlikely that a change in one member of a base pair in a helix is a selectively neutral event; however, altering both members of the pair (in a way that retains pairing) can be so.] The reasons for making this assertion are two: For one, ribosomal components appear interchangeable among species, at least within a primary kingdom (22), which would be possible only if the ribosomes in the species involved were extremely similar in structure. For another, the independently occurring variations one observes in rRNA sequence tend to be the same in all lines of descent, at least within a given primary kingdom (18; Woese, unpublished observation); such similarity in pattern could not occur unless underlying structure were extremely similar in all cases.

The second characteristic of the rRNA chronometer is that the occurrence of changes in rRNA sequence can vary in frequency by at least two orders of magnitude at different positions in the molecule. Some positions readily change from species to species within the same "genus", others are nearly invariant across an entire primary kingdom, and some appear not to vary at all (23). For this reason the rRNA chronometer is capable of measuring the full range of evolutionary relationships.

A third characteristic of the chronometer is suggested by this remarkable disparity in the rates at which mutations become fixed at different positions in the molecule. This great a differential would not seem to reflect a corresponding differential in underlying mutation rates. Rather, certain positions in the molecule must be parts of structurally and/or functionally defined units (comprising several bases); and to change a base therein requires the changing of others in the unit as well, if the overall change is to be selectively neutral. Such composite changes must be higher order functions of an organism's mutation rate, which makes them rare. This property means that ribosomal RNAs (and presumably other macromolecules as well) have a chronometric structure different from what we generally assume. This is best explained by analogy: Macromolecular chronometers are usually treated as though they are clocks with one hand, i.e. sequence comparisons are conventionally used merely to generate one number, the number of differences (or similarities) between two aligned sequences. The proper analogy for the rRNA chronometer, however, is a clock with *multiple* hands, a conventional clock. However, this RNA "clock" has the unusual characteristic that when it speeds up (due to an increase in the mutation rate in a line of descent), the various hands do not speed up in simple proportion. More specifically, the "slower hands" speed up (proportionately) more than do the "faster hands". This characteristic makes the macromolecule a more powerful chronometer than we now assume it to be, for it means that the relative clock speed within groups of organisms can be determined intrinsically, which then improves the determination of branching orders, and means that the roots of trees derived from sequence data can be intrinsically delimited, if not defined (which is not now considered to be possible).

The rRNA chronometer has one other recognizable property, i.e., linkage between adjacent positions as regards the frequency of sequence change. If one position changes infrequently, then adjacent positions are likely to change infrequently as well, and so on. [This is a more or less obvious consequence of secondary structure-function constraints.]

The proper quantitative model for the rRNA chronometer has yet to be developed, however.

The Measurement of Bacterial Phylogenetic Relationships by Ribosomal RNA

At present approximately 400 bacterial species have been chacterized in terms of the (16S) ribosomal RNA sequences. Full sequences exist for only a few of these, however. The remainder have been characterized by the method of oligonucleotide cataloging (4-5, 24), which involves digesting the rRNA with ribonuclease T1 followed by separation and sequencing of the resulting oligonucleotides. Such an oligonucleotide catalog contains about 500-600 nucleotides in oligomers of lengths greater than five (running up to lengths in the 15-20 range). The catalogs are analysed in terms of binary association coefficients, or Sab values (24). In such an analysis pair-wise comparisons of catalogs are made, and the number of nucleotides in oligomers of length greater than five common to the two catalogs is determined. [The Sab value is this number divided by half the total number of nucleotides in all oligomers of length greater than five in the two catalogs (24).] A table of such Sab values for a set of organisms is then subject to unweighted pair group cluster analysis (24). The important difference between this and a conventional analysis of full sequences is that the Sab method is biased for those portions of the rRNA sequence in which *runs* of contiguous nucleotides are conserved (in composition). It is, therefore, relatively more sensitive to the phylogenetic informa-

TABLE 1

Oligonucleotides Defining the Three Primary Kingdoms

Sequence	Position in 16S rRNA	% Occurrence among		
		Eubacteria	Archaebacteria	Eucaryotes
CYUAAYACAUG	50	83	0	0
AYUAAG	"	1	62	100
AUUYAG	"	1	29	0
..ACUCCUACG	340	97	0	0
CCCUACG	"	0	97	0
..ACNUCYANG	"	0	0	100
CUAAYUAYG	510	59	0	0
CUAAYUYYG	"	33	0	0
YYUAAAG	570	3	97	0
UUAAAA.	"	0	0	92
AUACCCYG	795	93	3	0
AUACCCG	"	1	88	0
AUACCG	"	0	0	92
..AAACUCAAAG	910	89	0	0
..AAACUCAAAUG	"	9	0	0
AAACUUAAAG	"	0	100	100
CACAAG	935	97	0	0
CACYACAA..	"	0	94	0
CACCACCAG	"	0	0	84
..UUAAUUCG	960	97	0	0
..UYAAUUUG	"	0	0	100
..UYAAUYG	"	1	100	0
CAACCYUYR	1110	91	0	0
ACCCNCR	"	0	100	0
ACYYYR.		0	0	84
..CCYUUAYR	1210	94	0	0
CCCCG	"	0	100	0
CCCYUAG	"	0	0	92
UUCCCG	1380	93	0	0
UCCCUG	"	0	97	100
CUCCUUG	1390	0	94	0
CCCUUUG	"	0	0	92
UACACACCG	1400	99	0	100
CACACACCG	"	1	100	0
AUCACCUC..	1535	91	100	0
AUCAYU....	"	0	0	100

Y=pyrimidine; R=purine; N=any base. See Table 2 caption.
Numbers are approximate position in E. coli sequence (23)

WHY STUDY EVOLUTIONARY RELATIONSHIPS?

tion contained in the more highly conserved areas of sequence.

Given a sufficient amount of sequence (catalog) information, "signature" analysis is also possible (18, 25). Individual positions or groups of contiguous positions in the molecule can have compositions that are characteristic of particular phylogenetic groupings. The *spectrum* of allowable substitutions at certain positions and their frequency of occurrence can both have phylogenetic significance. Signature analyses are particularly useful in resolving some of the uncertain branching orders or in detecting rapidly evolving lines of descent (18, 25).

OVERVIEW OF BACTERIAL PHYLOGENY

[It is not my intention to give a detailed presentation of bacterial phylogeny here. That is done elsewhere in this volume. The following is merely an overview.]

It came as a great surprise in the late 1970's to discover that life on this planet was not fundamentally of two types, prokaryotes and eukaryotes, but rather comprised *three* types, the eubacteria, the archaebacteria, and the eukaryotes (26). The two classes of prokaryotes appear no more related to one another than either is to the eukaryotes. The three major groupings can be defined both phenotypically and genotypically. [It is important to distinguish between genotypic and phenotypic criteria in a phylogenetic context. The number of genotypic configurations consistent with the exact same phenotype is enormous. Therefore, genotypic change tends to be selectively neutral and so, for the most part independent of phenotypic change. This means that genotypic change is basically a pure time measure (albeit not always easy to interpret). Phenotypic change, on the other hand, is more a measure of evolutionary progression.] Each of the three major groups (primary kingdoms) shows a number of distinctive phenotypic characteristics -- *e.g.* the peptidoglycan walls of eubacteria, the ether-linked lipids of archaebacteria -- by which the group is solidly defined and readily distinguished (21, 27). The three are also clearly defined and distinguished by all available genotypic (sequence) criteria. Ribosomal RNA sequences within each kingdom are more than 70% similar, whereas between two primary kingdoms sequence similarity is 60% or less (23, 28; Olsen *et al.*, in preparation). Table 1 is a collection of sequences at various positions in 16S rRNA that define and distinguish the three primary kingdoms.

TABLE 2

Oligonucleotide Signatures Defining the Eubacterial "Phyla" and Their Main Subdivisions

Sequence	posn rRNA	Gm pos [1A]	Gm pos [1B]	purple [2A]	purple [2B]	purple [2C]	spirochete [3A]	spirochete [3B]	spirochete [3C]	bac-cyt [4A]	bac-cyt [4B]	cyn [5]	grn [6]	cfx [7]	srd [8]	rad [9]	pln [10]
CYUAAYACAUG	050	97	100	100	0	96	0	0	100	100	100	100	100	0	94	0	0
CYUUACACAUG	"	0	0	0	100	0	0	0	0	0	0	0	0	0	0	0	0
CCUAAUG	"	0	0	0	0	0	0	0	0	0	0	0	0	100	0	0	0
CUUAAG	"	0	0	0	0	0	0	0	0	0	0	0	0	0	0	100	0
YCACAYYG	315	95	95	100	92	98	44	100	100	0	0	100	100	0	94	0	0
RCACAYUG	"	0	2	0	0	2	50	0	0	0	0	0	0	0	6	0	0
UCCCCACAUUG	"	0	0	0	0	0	0	0	0	100	0	0	0	0	0	0	0
..CCCCACACUG	"	0	0	0	0	0	0	0	0	0	75	0	0	0	0	0	0
YCACAG	"	0	0	0	0	0	0	0	0	0	0	0	0	0	0	100	0
AAUAUUG	365	24	93	52	0	90	0	0	0	100	100	0	100	0	38	0	0
AAUCUG	"	4	0	17	4	0	0	100	0	0	0	0	0	0	12	0	0
AAUUUG	"	0	0	0	93	0	0	0	0	0	0	0	0	0	50	0	0
AAUYUUYG	"	18	0	0	0	0	0	0	0	0	0	0	0	75	0	0	100
AAUCUUAG	"	0	0	17	0	0	0	0	0	0	0	0	0	0	0	0	0
AAUAUUCCG	"	0	0	0	0	0	44	0	0	0	0	0	0	0	0	0	0
AAUUUYCG	"	1	0	0	0	0	0	0	0	0	0	100	0	0	0	0	0
AAUCUCCG	"	24	0	0	0	0	44	0	75	0	0	0	0	25	0	0	0
AAUYUUYCACA.	"	23	0	0	0	0	6	0	25	0	0	0	0	0	0	0	0
AAUAUUCCACA.	"	0	0	0	0	0	6	0	0	0	0	0	0	0	0	0	0

Sequence	Pos																
CYAACUAYG	510	94	100	9	100	11	19	100	50	0	0	0	0	0	0	6	0
CYAAUUACG	"	0	0	0	0	0	81	0	50	0	0	0	0	0	0	0	0
CUAAAUACG	"	4	0	0	0	0	0	0	0	0	0	0	0	0	0	0	0
CUAACUYYG	"	0	0	91	0	83	0	0	0	64	100	86	100	0	50	88	0
CUAAUUCCG	"	0	0	0	0	0	0	0	0	36	0	14	0	0	50	0	0
AAAUUCG	700	0	0	0	0	0	0	0	0	0	0	0	0	0	0	50	0
AAAUCCG	"	0	0	0	0	0	0	0	0	0	0	0	0	0	0	31	0
AAACUCAAAG	910	91	0	4	26	8	100	100	100	0	100	0	100	0	0	6	100
CUAAAACUCAAAG	"	0	98	0	0	6	0	0	0	0	0	0	0	0	0	19	0
UUAAAACUCAAAG	"	1	0	22	11	9	0	0	0	0	0	0	0	0	0	0	0
AUAAAACUCAAAG	"	0	0	0	4	0	0	0	0	0	0	0	0	0	0	6	0
AUUAAAACUCAAAG	"	5	0	74	56	13	0	0	0	0	0	0	0	0	0	6	0
ACUAAAACUCAAAG	"	0	0	0	0	2	0	0	0	0	0	0	0	0	0	62	0
CUUAAAACUCAAAG	"	0	0	0	0	0	0	0	0	0	0	0	0	0	75	0	0
.UUAAAACUCAAAUG	"	0	0	0	0	62	0	0	0	0	0	0	0	0	0	0	0
UUUAAUUCG	960	95	2	100	30	100	100	100	100	100	100	86	100	100	100	100	0
AUUAAUUCG	"	0	74	0	70	0	0	0	0	0	0	7	0	0	0	0	0
CUUAAUUCG	"	2	25	0	0	0	0	0	0	0	0	0	0	0	0	0	100
UCCUCAUG	1200	0	0	96	96	0	6	50	25	0	0	0	100	0	0	75	75
UCAUCAYG	"	8	36	4	4	100	75	0	75	58	0	43	0	0	0	19	25
AUCAUCAUG	"	2	3	0	0	0	0	0	0	0	0	0	0	0	0	0	0
UCAAAUCAUCAUG	"	80	51	0	0	0	19	0	0	0	0	0	0	0	0	0	0
UCAAAUCCUCAUG	"	0	0	0	0	0	0	50	0	0	0	0	0	0	0	0	0
UCAAAUCAUCACG	"	0	0	0	0	0	0	0	0	42	0	0	0	0	0	0	0
UCAAAUCAG	"	1	0	0	0	0	0	0	0	0	100	0	0	0	0	0	0

(con't)

11

TABLE 2 (con't)

Sequence	posn rRNA	Gm pos [1A 1B]		purple [2A 2B 2C]			spirochete [3A 3B 3C]			bac-cyt [4A 4B]		cyn [5]	grn [6]	cfx [7]	srd [8]	rad [9]	pln [10]
CCCUUAUR.	1210	0	0	13	100	23	93	0	0	8	0	0	0	0	56	0	0
CCCUUACR.	"	0	0	87	0	75	6	0	0	92	75	0	100	50	6	0	0
CCUUUAUG	"	2	0	0	0	2	0	100	0	0	25	0	0	0	37	0	100
UCCUUACG	"	0	0	0	0	0	0	0	0	0	0	0	0	0	0	100	0
CCCUUAUR.	"	94	93	0	0	0	0	0	100	0	0	21	0	0	0	0	0
CCCUUACR.	"	0	5	0	0	0	0	0	0	0	0	71	0	0	0	0	0
CUACAAUG	1240	92	100	100	0	94	88	100	100	0	100	29	0	75	75	100	38
UUACAAUG	"	0	0	0	0	0	0	0	0	100	0	0	0	0	0	0	0
AUACAAUG	"	1	0	0	0	0	0	0	0	0	0	0	100	0	0	0	0
UACUACAAUG	"	7	0	0	0	6	0	0	0	0	0	57	0	0	13	0	0
UAAUACAAUG	"	0	0	0	25	0	6	0	0	0	0	7	0	0	13	0	0
UCAUACAAUG	"	0	0	0	75	0	0	0	0	0	0	0	0	0	0	0	0
UCCUACAAUG	"	0	0	0	0	0	0	0	0	0	0	0	0	0	0	0	63
AUCACCUCCU.	1335	96	100	96	92	98	94	100	100	0	8	93	100	100	88	100	100
AAYACCUCCU.	"	0	0	0	0	0	0	0	0	100	83	0	0	0	0	0	0

1A,B=low (high) G+C Gram positive; 2A,B,C= α, β, γ subdivisions of purple bacteria; 3A,B,C=spirochetes, leptospiras, anaerobic halophiles; 4A,B=bacteroides, cytophagas; 5=cyanobacteria; 6=chlorobium group; 7=chloroflexus group; 8=sulfate reducers and relatives; 9=deinococcus group; 10=planctomyces group. Occurrences given as percent. Implied G preceeds all sequences except those preceeded by dots. Y=pyrimidine; R=purine; N=any base. See text for references.

The Eubacteria

The phylogeny of the eubacteria accords poorly with our preconceptions (4-5). The morphological criteria previously used to define bacterial groupings tend to produce inexact to misleading taxa (4, 29-30). Groupings defined on the basis of the biochemical characteristics fare little better (4, 18). While the Gram positive criterion, as modified by cell wall studies (31), yields a phylogenetically valid grouping, the Gram negative one does not (as might be expected). Photosynthetic species do not all cluster together to the exclusion of most non-photosynthetic species (17, 18).

The eubacteria actually comprise of the order of ten major "phyla", the order of branching of which from the common line of eubacterial descent is close enough that it cannot be distinguished by the oligonucletoide cataloging method. Hopefully, full sequencing will do so. Partial oligonucleotide signatures defining the major eubacterial "phyla" and their major subdivisions are shown in Table 2.

1- *The Gram Positive Bacteria* Gram positive bacteria, as defined by wall structure (31), form a phylogenetically coherent grouping that comprises two distinct branches, those organisms of low G+C content DNAs and those of high G+C (4-5). The basic phenotype of the low G+C group is clostridial, in the sense that most of the primary sublines in this group contain clostridial representation. Assuming an ancestral clostridial phenotype for the group means that the other phenotypes in the group have arisen largely through loss of clostridial properties -- loss of spore formation (*e.g. Eubacterium*), loss of rod shape (*e.g. Sarcina*), etc. The most unusual and so interesting of the sublines of the low G+C Gram positive bacteria is that leading to the genera *Bacillus*, *Lactobacillus* and *Streptococcus* and to the mycoplasmas (25). It would seem that evolution in this subline more or less recapitulates the development of the oxygen atmosphere. The lactobacillus and streptococcus phenotypes ostensibly represent a microaerophilic era, the bacillus phenotype the transition to a truely aerobic atmosphere. Two specific clostridia, *C. ramosum* and *C. innocuum*, must represent the phenotype from which the mycoplasmas arose, for they are specifically related to the mycoplasmas, to the exclusion of the other clostridia (25). The mycoplasmas, however, have taken a bizarre evolutionary course. Their evolution feels qualitatively different from a phenotypic perspective -- hence their classification as a separate class, *Mollicutes* (32-33). As judged by ribosomal

RNA, the mycoplasmas are bizarre even at the level of genetic sequence (25). We shall return to these interesting organisms below.

The high G+C Gram positive eubacteria would seem to be derived from some general actinomycete phenotype (29, 34-35). They are unusual phenotypically in their unique, complex morphologies and the variability therein -- a variability that extends to the molecular level; their wall structures, for example, exhibit far more variety in peptidoglycan type than do those of any other eubacterial group (31). It is interesting that morphologies run from the simplest -- *e.g.* species of *Micrococcus* -- to the most complex within this group (5).

The true phylogenetic groupings among the high G+C Gram positive bacteria are for the most part an admixture of the classical genera -- reflecting the fact that classical morphological criteria, even among these morphologically complex bacteria, are generally not phylogenetically valid characters (29, 34-35).

2- *The Purple Bacteria and Relatives* The purple bacteria and their relatives in one sense are the largest group of eubacteria: More classically recognized genera are grouped therein than in any other eubacterial "phylum" (4, 17-18, 30, 36-37). To a first approximation this group corresponds (with important exceptions) to the "Gram negative" bacteria. The evolutionary basis for the group is clearly photosynthetic; all sublines contain one or more genera of purple photosynthetic bacteria. The "phylum" comprises three main subdivisions (17-18, 37) -- the α subdivision, containing many species of purple non-sulfur bacteria plus non-photosynthetic genera, the β subdivision, containing a few species of purple non-sulfur bacteria and many non-photosynthetic species, and the γ subdivision, containing purple sulfur bacteria and many non-photosynthetic species, all well defined by oligonucleotide signature; see Table 2.

The juxtaposition of numerous photosynthetic and non-photosynthetic genera, of oxidative and reductive biochemistries, etc. suggests unsuspected evolutionary relationships among the various physiologies. There can be little doubt that the non-photosynthetic phenotypes have arisen from the photosynthetic one, rather than the reverse. The juxtaposition of genera that oxidize nitrogen compounds (*Nitrobacter*) with those that reduce them (*Rhodopseudomonas*, *Rhizobium*) strongly implies that the two metabolisms are closely related. It would seem that the general pathways are first established in a few lines, and evolution later tunes them to run in the oxidative or the reductive direction, to utilize this portion of the pathway only, and so on.

3-*The Spirochetes and Relatives*   The spirochetes are one of the rare cases where morphological criteria properly define a major phylogenetic grouping (4, 38). Within the spirochete group, however, the classically defined divisions once again prove faulty. The conventional distinction between spirochetes and treponemes is particularly misleading, obscuring the true distinctions.

The grouping defined by the spirochetes, treponemes and borrelia is a deep one in terms of lowest Sab values, which are in the 0.25-0.30 range (38). The relationship between this unit and the leptospiras is still more distant, so much so that it cannot be established by the customary Sab analysis -- but does emerge when signature analysis is used (38). Also joining this spirochete "phylum" at a very deep level is a newly recognized and major grouping of obligate anaerobic halophiles (38-40). This "phylum" must be a very ancient one.

4-*The Bacteroides, Flavobacteria and Cytophagas*   Although a grouping of the bacteroides, obligate anaerobes, with the aerobic flavobacteria and cytophagas is unexpected on phenotypic grounds, the three types, together with other relatives, constitute a distinct eubacterial "phylum" (41). This unusual juxtaposition of phenotypes can to some extent be rationalized in terms of an organism of intermediate phenotype, a yet-to-be named obligately anaerobic flexible rod (41; Stetter *et al.*, in preparation).

5-*The Cyanobacteria and Chloroplasts*   This group is one that has been studied both in terms of the cataloging technique (42-45) and by full sequencing (of one cyanobacterial and four chloroplast examples) (46-50). The "phylum" is not a particularly deep one phylogenetically unless the chloroplast examples are included (4); however, too few cyanobacteria have been cataloged to claim that the apparent depth is representative of the true depth. The data are insufficient to determine whether the chloroplasts (or green plants) have arisen specifically from *Prochloron*, an organism that contains both chlorophylls a and b (51), or from the cyanobacteria *per se*. Failure to resolve the matter probably is due to the "clock speed problem" (see below). Resolution will require a few full sequences of appropriate bacterial rRNAs.

6-*The Green Sulfur Bacteria*   Perhaps the most remarkable phylogenetic characteristic of this "phylum" is its lack of phylogenetic depth — the more so because it is a major anaerobic photosynthetic phenotype. The lowest Sab values in the group are in the 0.40 range (Gibson et al., in preparation). No non-photosynthetic relatives of the green sulfur bacteria are yet known. A specific relationship between them and the other green photosynthetic bacteria, of the chloroflexus phenotype, cannot be established by the oligonucleotide cataloging method (4, 17); if such does exist, it would be at an extremely deep level.

7-*The Chloroflexus Group* The evolutionary structure of this "phylum" stands in sharp contrast to that seen for the green sulfur bacteria. The chloroflexus group appears to represent one of the deepest branchings from the common line of eubacterial descent (4), and also seems deep internally as well (Gibson et al., in preparation). Although relatively few species are known to belong to the group, impressive phenotypic diversity already exists; the group also contains the non-photosynthetic genera *Herpetosiphon* and *Thermomicrobium*.

The main, if not the only resemblances between *Chloroflexus* and *Chlorobium* are in the auxiliary light gathering system, the chlorobium vesicles (52). Significantly, the two types differ in the structure of their photosynthetic reaction centers (53). It has been suggested by Pierson (personal communication) that the auxiliary light gathering system may be plasmid encoded, which in turn suggests the possibility of its genes being transferred between species. If so, the two types of green photosynthetic bacteria may not actually be related. Interspecies gene transfer might also help to explain the facts that the chlorobium types constitute a phylogenetically shallow group and ostensibly have no non-photosynthetic relatives.

Because of its general diversity, phylogenetic depth and deep branching, the chloroflexus group is one of several (another being the planctomyces group) that possibly represents a primitive eubacterial phenotype.

8-*The Sulfate Reducing Bacteria and Relatives* Another somewhat unlikely grouping of species is that involving the sulfate reducing bacteria, the myxococci and the bdellovibrios (54-55); Stackebrandt et al., mss in preparation). The grouping is an ancient one as judged by lowest Sab values — which are in the 0.33 range (55; Stackebrandt et al., mss in preparation). It is not clear from catalog analysis whether the bdellovibrios alone constitute a coherent unit within this "phylum", or whether the bdellovibrio condition arose more than once from other phenotypes in the "phyum" (55).

9-*The Radioresistant Micrococci* Because the group is internally shallow — lowest Sab values above 0.50 — the *Deinococcus* group appears to be an incomplete phylogenetic unit (56). Nevertheless, by Sab measure and oligonucleotide signature, this group of eubacteria stands decidedly alone within the eubacteria (4), and so represents a separate eubacterial "phylum". Perhaps the most characteristic features of the Deinococci are their radiation resistance and unique and complex cell walls (56).

10-*The Planctomyces-Pasteuria Group* Species in this group are remarkable in that their walls are devoid of peptidoglycan (57). As defined by oligonucleotide cataloging the group comprises *Planctomyces* and *Pasteuria* species and certain wall-less isolates (58). By Sab measure the group is only remotely related to the other eubacteria (though specifically so) (58). Unfortunately, this does not prove that the planctomyces and their relatives are necessarily an earlier branching than the other eubacteria from their common line of descent. Falsely deep branchings are usually, if not always, deduced for species that come from rapidly evolving lines of descent. The mycoplasmas are an excellent example of this (25). If it were not for the fact that certain clostridia are close (specific) relatives, the true phylogenetic position of the mycoplasmas would be almost impossible to ascertain (25). Since they have no known normal relatives, this central question remains unanswered for the planctomyces group.

If the planctomyces line were evolving at normal rates (and so were, therefore, a true deep branching), then oligonucleotide catalogs for members of the group might show certain similarities to outgroup catalogs (in this case the archaebacteria) not seen in other eubacterial catalogs. Examination of the appropriate catalogs shows this perhaps to be case, but the extent of outgroup resemblance is not spectacular (Woese, unpublished analysis). This intriguing question should resolve when full sequences are known.

Conclusions : While approximately ten "phyla" of eubacteria can be defined by (partial) sequence comparisons, the order of their branching from the common eubacterial line (and from one another) remains completely unresolved -- so we have yet to create a framework in which to conceptualize the evolution of eubacterial biochemistries and various other properties. A few specific relationships between "phyla" are suggested by the catalog data -- e.g. the spirochetes with the Gram positive bacteria (Woese *et al.*, unpublished analysis) -- but these are definitey tentative. Hopefully,

TABLE 3

Signatures Defining Various Archaebacterial Groups

Sequence	position in rRNA	"Methanogen" group A	B	C	D	H	T	Sulfur Group
AYUAAG	50	100	100	100	67	0	100	0
AUUUAG	"	0	0	0	0	89	0	0
AAACCUCCG	365	89	100	0	0	0	0	0
AAACCUUUACA..	"	0	0	0	0	100	0	0
AAAACUUUACA..	"	0	0	100	100	0	0	0
AAAACUG	"	0	0	0	0	0	100	0
UAAYACCG	535	100	80	100	100	100	0	0
UAACACCCG	"	0	0	0	0	0	100	0
UAACACCAG	"	0	0	0	0	0	0	100
..UAAUCCY..	705	100	0	100	100	100	100	33
AUCCYUG	"	0	100	0	0	0	0	0
CACNACAAC.	935	100	100	100	100	100	100	0
CACCACAAG	"	0	0	0	0	0	0	100
UUUAAUYG	960	100	100	100	100	100	100	0
CUYAAUUG	"	0	0	0	0	0	0	100
AUuCAACG	970	100	100	0	0	0	100	0
ACuCAACG	"	0	0	100	100	100	0	0
ACAUCUCACCAG	980	89	0	0	0	78	0	0
ACCCR	1110	89	100	100	100	100	0	0
ACCCCCA..	"	0	0	0	0	0	100	100
CACUCAUR	1150	0	100	0	0	0	0	0
CUACAAUG	1240	100	100	100	100	100	0	0
CUACAAAG	"	0	0	0	0	0	100	0
UAAUCG	1350	100	100	100	100	89	100	0
AAUUAUCCG		0	0	100	0	0	0	0
AAAUCUUG		0	0	0	100	0	0	0

Symbols as in Tables 1 & 2; lower case = modified base
A=*Methanobacteriales*, C= *Methanomicrobiales*, H=halobacteria
B= *Methanococcales*, D= *Methanosarcinales*, T= *Thermoplasma*

more extensive sequencing will resolve these matters.

Already it is clear, however, that many of our prejudices regarding the various aspects of bacterial evolution are suspect, if not wrong. Thus, at very least, we have to stop teaching these as dogma. The conventional wisdom according to which the first bacteria were non-photosynthetic heterotrophs (2, 6) is now suspect. No indication exists in the emerging eubacterial phylogeny that heterotrophy preceeded autotrophy or that photosynthesis evolved later than anaerobic, fermentative metabolism. If anything, the structure of of the eubacterial tree suggests that the common eubacterial ancestor was photosynthetic (3-5, 23). Four of the eubacterial "phyla" contain (widespread) photosynthetic representation. And, it has been shown recently (Stackebrandt and Woese, unpublished) that the major remaining group devoid of photosynthetic species, the Gram positive eubacteria, peripherally includes *Heliobacterium chlorum*, a newly discovered photosynthetic anaerobe which possesses a unique variation of bacterial chlorophyll (59).

The Archaebacteria

Unlike the eubacteria, the archaebacteria appear to comprise only two major "phyla", the methanogens, extreme halopliles and the genus *Thermoplasma* on the one hand, the sulfur-dependent archaebacteria ("thermoacidophiles") on the other (4, 60-61). Separation between the two — Sab's in the 0.15-0.18 range (61) — is comparable to, but somewhat greater than that seen among most eubacterial "phyla". [For reference, Sab's between eubacteria and archaebacteria are in the 0.10 range (4, 26).] As mentioned above, all archaebacterial (16S) ribosomal RNAs are at least 70% similar. Comparisons with outgroup sequences reveal that the two major branches of the archaebacteria are isochronic, i.e. they appear to evolve at about the same rate (23, 28; Olsen *et al.*, mss in preparation). Consequently, the depth of the division between the two archaebactrial "phyla" would not seem to reflect rapid evolution in either of the archaebacterial lines of descent. Also all of the archaebacterial sequences are closer to the common ancestral rRNA sequence than are the sequences from one or both of the other kingdoms (28).

An abbreviated oligonucleotide signature defining and distinguishing among the various major archaebacterial subdivisions is shown in Table 3.

The Methanogens and Their Relatives. By Sab measure this archaebacterial "phylum" comprises six distinct groupings of organisms (4, 25, 61). *Thermoplasma acidophilum* is the sole representative of a group that forms the deepest branching in this line of descent, Sab about 0.20 (4). The remaining five groups -- the *Methanobacteriales*, *Methanomicrobiales*, *Methanosarcinales*, *Methanococcales* and the extreme halophiles-- branch subsequently in an order that is not clearly resolved by the customary analysis (4, 61). [It should be noted that the depth of division even among these groups, in the range of 0.25 by Sab measure (61), is comparable to that among most eubacterial "phyla" (4).] Each of these Orders is also relatively deep phylogenetically -- lowest Sab values in these various groupings are: *Methanobacteriales*- 0.35, *Methanomicrobiales*- 0.44, *Methanosarcinales*- 0.45, *Methanococcales*- 0.40 and the extreme halophiles- 0.41 (61; G. E. Fox, unpublished calculation).

The branching of *Thermoplasma* from the others is sufficiently deep that one should question whether the group it represents is specifically clustered with the methanogens and extreme halophiles, or is actually a separate, third major line of archaebacterial descent. The fact that its RNA polymerase subunit type resembles that seen in the sulfur-dependent archaebacteria, not that seen in the methanogens, further strengthens the latter possibility (62). However, in several other phenotypic respects (*e.g.* extent of modification of ribosomal and transfer RNAs) this wall-less thermoacidophile would seem to cluster with the methanogens and extreme halophiles (25).

The grouping of methanogens, extreme halophiles and the wall-less thermoacidophile is difficult to rationalize on phenotypic grounds. Their niches seem fundamentally different, as do their metabolisms. No common cofactors have been reported. Yet the brave prediction from the genotypic evidence is that specific biochemical similarities will ultimately be uncovered.

The Sulfur-Dependent Archaebacteria. Few species from this "phylum" have been characterized by the cataloging method. Even so, the phylogenetic depth of the group is already below the 0.30 level (60), suggesting that the group will prove a deep one. The group has also been characterized by DNA-rRNA hybridization, and shown to be phylogenetically coherent thereby (63). The sulfur-dependent archaebacteria are remarkable phenotypically in that all isolates so far are thermophilic (though not all are acidophilic), strongly suggesting the ancestral phenotype for the "phylum" to be so. Among these thermophiles are those that grow at the highest optimum termperatures so far reported, the *Pyrodictium* series -- growing optimally at 105 °C (64).

The phenotypic similarites between the archaebacterial "phyla" are relatively few (by eubacterial standards) but striking -- e.g. ether-linked lipids (65); lack of ribothymine in transfer RNAs (and its replacement, at postion 54 in the molecule, by modified or unmodified pseudouridine) (66); and, the reduction of molecular sulfur (67).

The phenotypic differences between the two "phyla" are equally striking. Perhaps the most striking of these is that ribosomal and transfer RNAs are much more highly modified in the sulfur-dependent archaebacteria (in which respect they resemble the eukaryotes) than in the methanogenic cluster of archaebacteria (56, 68). Differences in the shape of the 50S ribosome particle have been noted (69). The mode of cell division differs between the two groups, as do the kinds of viruses each possesses (70). Such phenotypic variety tends not to occur in the other two primary kingdoms. The general range of phenotypic variety among the archaebacteria tends to be greater than in the other kingdoms (21) -- e.g. 5S rRNA secondary structure (71-72) and RNA polymerase subunit pattern (62).

THE MODE OF BACTERIAL EVOLUTION

The only evolution we can lay any claim to understanding is metazoan evolution, and our understanding of it is indeed superficial. It is therefore important to understand the ways in which bacterial evolution resembles metazoan evolution. Given that strikingly different systems and settings are involved in the two cases, parallels between them should prove helpful in defining and understanding basic evolutionary principles.

Metazoan evolution has several salient general characteristics. Organisms in the major evolutionary groupings, e.g. the phyla, are quite distinct from one another phenotypically. What would be judged as intermediate forms are rarely encountered. Moreover, the type of evolution that has given rise to the major metazoan groupings, usually referred to as macroevolution, seems qualitatively different from that involved in "normal" evolutionary progression (73). Although it involves radical phenotypic change, the process, strangely, tends to occur over a much shorter time span than does "normal" evolution (73). Do these phenomena have counterparts on the bacterial level?

Although it is not always apparent in terms of their phenotypes (as we now define them), the eubacteria do appear to fall into distinct classes. Definite divisions can also be seen in terms of the phylogenies generated by rRNA sequence comparisons; each of the eubacterial "phyla" is well defined and well separated from the others by Sab values (4), and each has a distinctive oligonucleotide signature; see Table 2. At the lower taxonomic levels, however, this degree of distinction seems to disappear (4, 5).

Bacteria also show a form of macroevolution. The bacterial form can be clearly defined on the molecular level, and so, may be helpful not only in understanding the as yet inscrutable pattern of bacterial evolution, but also in understanding metazoan macroevolution.

Mycoplasmas are the primary example of macroevolution among bacteria, though others can be recognized. Phenotypically the mycoplasmas are unique enough to warrant classification as a separate class, *Mollicutes* (32). Yet on the genotypic level they can be recognized genealogically as members of the low G+C Gram positive bacteria, from the same subline that spawned lactobacilli, streptococci, etc.; and they have two specific relatives that are *bona fide* clostridia (25).

The mechanism underlying rapid evolution, discussed in detail elsewhere (Woese, in preparation), basically involves an increased mutation rate in a line of descent. As a result, the field of variants in which the line evolves is markedly augmented in both number and kind, which in turn means that the line can and (statistically speaking) will explore evolutionary avenues not open to "normal" lines of descent. This effect can be seen at the molecular level in the catalogs of mycoplasma rRNAs, which tend to lack a higher than normal fraction of the oligonucleotides that are kingdom-level invariant in sequence. These catalogs differ more than one would expect from one another, as well (25).

The reason for the elevated mutation rate in the mycoplasma lines of descent appears to be their small genome sizes (32) -- which enable them to sustain higher than normal mutation rates and still replicate their complement of genes with the same accuracy (per genome) as do normal bacteria (Woese, in preparation). For this reason, too, the mycoplasmas appear permanently locked into the rapid (macro-) evolution mode. Conceivably, there could be unusual environmental circumstances under which normal lines of descent could temporarily exist with elevated mutation rates, and so undergo macroevolutionary saltations.

The genotypic hallmarks of groups that have undergone saltatory (macro-) evolution would be: (i) relative coherence within the group, by Sab or other sequence measure, and (ii) disproportionate distance (by the same measure) from outgroups. One group that seems to fit this definition is the high G+C Gram positive bacteria (29). By Sab measure the group is not particularly deep -- lowest Sab's within the group above 0.40 (5). And, their rRNA catalogs are missing a higher proportion of the highly conserved (ancestral) oligonucleotides than are those of the low G+C subdivision of the Gram positive eubacteria and most other major eubacterial groups (29, 34-35; Woese, in preparation). Another group exhibiting the same characteristics is the *Bacteroides*; by Sab measure they are further removed from outgroups than are their specific relatives, the flavobacteria and cytophagas, and their catalogs lack a higher proportion of highly conserved oligonucleotides, as well (41).

My argument would demand that both of these groups appear unusual phenotypically. The high G+C Gram positive eubacteria are deemed so -- some microbiologists have tended to classify them as quite separate from other bacteria (74). So far, the bacterioides remain unspectacular phenotypically.

If this concept be valid, many other, more subtle, examples of bacterial rapid evolution will come to light, and with them the patterns that permit us to understand this mode of evolution. Are all major changes in an ancestral phenotype -- e.g. the evolution of non-photosynthetic lines from photosynthetic ones -- brought about by this saltatory type of evolution?

THE RELATIONSHIPS AMONG THE THREE PRIMARY KINGDOMS

The evolutionary relationships among the three primary kingdoms is a central biological question. These groupings represent the initial radiation from that state we call the universal ancestor, and so they are primary, defining of the speciation process. Do any of the three phenotypes resemble the universal ancestor more than the others (3, 21)? Have any one of the groups given rise to one or both of the others (3, 21, 72, 75)? While questions of this sort cannot be settled at present, I would like to consider one of them in particular, mainly for its heuristic value, *i.e.* have one or both of the other primary kingdoms arisen from within the archaebacteria?

If one or both of the other kingdoms had arisen from the archaebacteria, they would have had to do so through macroevolution, for both are quite distinct phenotypically from the archaebacteria (and from one another), right down to the molecular level -- which, then, makes the archaebacteria a more primitive phenotype than one or both of the others. This would be consistent with the above-mentioned fact that archaebacterial 16S rRNA sequences are closer to the ancestral 16S rRNA sequence than are the 16S rRNA sequences from one or both of the other two kingdoms.

Had either of the other two kingdoms arisen from the archaebacteria, one would expect to find some trace of this in the phenotypes and genotypes of that group. Genotypically (in terms of sequences) no evidence so far supports the thesis. The eubacterial 16S rRNA sequence is not significantly closer to its counterpart in one of the two archaebacterial subdivisions than in the other; the same holds for the eukaryotic 16S-like rRNA (Olsen *et al.*, in preparation). An insufficient number of tRNAs from the sulfur-dependent archaebacteria have been sequenced that a meaningful statement can be made regarding them in this context.

However, on the phenotypic level some suggestive evidence (albeit difficult to interpret clearly) does exist. As noted above, the two major subdivisions of the archaebacteria differ significantly in details of their molecular phenotypes. The mode of division characteristic of the methanogen branch (formation of cross walls) and their viruses (true phages) are shared specifically by the eubacteria, as is the degree of modification of ribosomal and transfer RNAs and certain of the idiosyncracies in ribosome morphology. On the other hand the sulfur-dependent archaebacteria share specifically with the eukaryotes features such as high level of modification of ribosomal and transfer RNAs, and perhaps other idiosyncracies in ribosome morphology (60, 68, 69). In addition recent studies show a specific resemblance between these two as regards genomic organization of ribosomal RNA genes (Pace and Woese, unpublished).

These few points of phenotypic similarity are exceptions to the generalizations that define the three primary kingdoms; they are few enough that the case is not a strong one, and one cannot discard convergence as an explanation for the specific resemblance. Also, since the root of the universal phylogenetic tree cannot be determined, one cannot tell whether some of the resemblances are residual ancestral properties rather than shared derived characters.

If the archaebacteria have given rise to the other two kingdoms, then they represent an heretofore unrecognized stage in evolution — a stage that immediately followed that of the universal ancestor, but preceeded the stages defined by the other two kingdoms. Regardless of how this matter resolves, it is evident that studies in bacterial phylogeny are opening a vast door to the recesses of evolution.

We need to understand the universal ancestor.

ACKNOWLEDGEMENTS

The author's work on eubacteria has been supported by grants from the National Science Foundation (USA), while that on archaebacteria has been supported by NASA.

REFERENCES

1. Stanier, R.Y. and van Niel, C.B. (1941). The main outlines of bacterial classification. *J. Bacteriol.* 42,437-466.
2. van Niel, C.B. (1946). The classification and natural relationships of bacteria. *Cold Spring Harbor Symp. Quant. Biol.* 11,285-301.
3. Woese, C.R. (1983). The primary lines of descent and the universal ancestor. *In* "Evolution from Molecules to Man" (Ed D.S. Bendall) pp. 209-233. Cambridge University Press, Cambridge.
4. Fox, G.E., Stackebrandt, E., Hespell, R.B., Gibson, J., Maniloff, J., Dyer, T.A. Wolfe, R.S. Balch, W.E., Tanner, R., Magrum, L.J, Zablen, L.B., Blakemore, R., Gupta, R., Bonen, L., Lewis, B.J., Stahl, D.A., Luehrsen, K.R., Chen, K.N. and Woese, C.R. (1980). The phylogeny of prokaryotes. *Science* 209,457-463.
5. Stackebrandt, E. and Woese, C.R. (1981). The evolution of prokaryotes. *In* "Molecular and Cellular Aspects of Microbial Evolution". (Eds M.J. Carlile, J.R. Collins, and B.E.B. Moseley) Society for General Microbiology Symposium Vol. 32 pp. 1-31. Cambridge University Press, Cambridge.
6. Oparin, A.I. (1957). "The Origin of Life on Earth". 3rd edn. Oliver and Boyd, Edinburgh-London.
7. Chatton, E. (1937). Titres et travoux scientifiques. Sete, Sottano.
8. Stanier, R.Y. and van Niel, C.B. (1962). The concept of a bacterium. *Arch. Mikrobiol.* 42,17-35.
9. Stanier, R.Y. (1970). Some aspects of the biology of cells and their possible evolutionary significance. *In* "Organization and control in prokaryotic and eukaryotic cells". (Eds H.P. Charles and B.C.J.G. Knight) Society for General Microbiology Symposium Vol. 20 pp. 1-38. Cambridge Univesity Press, Cambridge.
10. Murray, R.G.E. (1974). A place for bacteria in the living world. *In* "Bergey's manual of determinative bacteriology" (Eds. R.E. Buchanan and N.E. Gibbons) 8th edn, pp. 4-9. The Williams and Wilkins Co., Baltimore.

11. Lehninger, A.L. (1975) "Biochemistry". 2nd edn. Worth Publishers, Inc., New York.
12. Margulis, L. (1970). "Origin of Eucaryotic Cells". Yale University Press, New Haven.
13. Sanger, F. (1981). Determination of nucleotide sequences in DNA. *Science* 214,1205-1210.
14. Zuckerkandl, E. and Pauling, L. (1965). Molecules as documents of evolutionary history. *J. Theor. Biol.* 8,357-366.
15. Fitch, W.M. and Margoliash, E. (1967). Constrution of phylogenetic trees. *Science* 155,279-284.
16. Schwartz, R.M. and Dayhoff, M.O. (1978). Origins of prokaryotes, eukaryotes, mitochondria and chloroplasts. *Science* 199,395-403.
17. Gibson, J., Stackebrandt, E., Zablen, L.B., Gupta, R. and Woese, C.R. (1979). A phylogenetic analysis of the purple photosynthetic bacteria. *Curr. Microbiol.* 4,245-249.
18. Woese, C.R., Stackebrandt, E., Weisburg, W.G., Paster, B.J., Madigan, M.T., Fowler, V.J., Hahn, C.M., Blanz, P., Gupta, R., Nealson, K.H. and Fox, G.E. (1984). The phylogeny of purple bacteria: The alpha subdivision. *System. Appl. Microbiol.* 221 (in press).
19. Hori, H. and Osawa, S. (1979). Evolutionary change in 5S RNA secondary structure and a phylogenetic tree of 54 5S rRNA species. *Proc. Natl. Acad. Sci. USA* 76,381-385.
20. Kuntzel, H. (1982). Phylogenetic trees derived from mitochondrial, nuclear, eubacterial and archaebacterial rRNA sequences: Implications on the origin of eukaryotes *Zbl. Bakt. Hyg., I. Abt. Orig.* C3,31-39.
21. Woese, C.R. (1982). Archaebacteria and cellular origins: An overview. *Zbl. Bakt. Hyg., I. Abt. Orig.* C3,1-17.
22. Nomura, M., Mizushima, S., Ozaki, M., Traub, P. and Lowry, P.E. (1969). Structure and function of ribosomes and their molecular components. *Cold Spring Harbor Symp. Quant. Biol.* 34,49-61.
23. Woese, C.R., Gutell, R., Gupta, R. and Noller, H.F. (1983). Detailed analysis of the higher-order structure of 16S-like ribosomal ribonucleic acids. *Microbiol. Rev.* 47,621-669.
24. Fox, G.E., Pechman, K.R. and Woese, C.R. (1977). Comparative cataloging of 16S rRNA: Molecular approach to prokaryotic systematics. *Inter. J. System. Bacteriol.* 27,44-57.
25. Woese, C.R., Maniloff, J. and Zablen, L.B. (1980). Phylogenetic analysis of the mycoplasmas. *Proc. Natl. Acad. Sci. USA* 77,494-498.

26. Woese, C.R. and Fox, G.E. (1977). Phylogenetic structure of the prokaryotic domain. *Proc. Natl. Acad. Sci. USA* 74,5088-5090.
27. Woese, C.R., Magrum, L.J. and Fox, G.E. (1978). Archaebacteria. *J. Molec. Evol.* 11,245-252.
28. Gupta, R., Lanter, J. and Woese, C.R. (1983). Sequence of the 16S ribosomal RNA from Halobacterium volcanii, an archaebacterium. *Science* 221,656-659.
29. Stackebrandt, E. and Woese, C.R. (1980). The phylogenetic structure of the coryneform group of bacteria. *Zbl. Bakt. Hyg.*, I. *Abt. Orig.* Cl,137-149.
30. Woese, C.R., Blanz, P., Hespell, R.B. and Hahn, C.M. (1982). Phylogenetic relationships among various helical bacteria. *Curr. Microbiol.* 7,119-124.
31. Schleifer, K.H. and Kandler, O. (1972). Peptidoglycan types of bacterial cell walls and their taxonomic implications. *Bacteriol. Rev.* 36,407-477.
32. Razin, S. (1978). The mycoplasmas. *Microbiol. Rev.* 42,414-470.
33. Gibbons, N.E. and Murray, R.G.E. (1978). Proposals concerning the higher taxa of bacteria. *Int. J. Syst. Bacteriol.* 28,1-6.
34. Stackebrandt, E. and Woese, C.R. (1981). Towards a phylogeny of the actinomycetes and related organisms. *Curr. Microbiol.* 5,197-202.
35. Stackebrandt, E., Wummer-Fuessl, B., Fowler, V., and Schleifer, K.-H. (1981). DNA homologies and ribosomal RNA similarities among spore-forming members of the order Actinomycetales. *Int. J. System. Bacteriol.* 31,420-431.
36. Woese, C.R., Blanz, P. and Hahn, C.M. (1984). What isn't a pseudomonad: The importance of nomenclature in bacterial classification. *System. Appl. Microbiol.* 5,179-195.
37. Woese, C.R., Weisburg, W.G., Paster, B.J., Hahn, C.M., Tanner, R.S., Krieg, N.R., Koops, H.-P., Harms, H. and Stackebrandt, E. (1984). The phylogeny of purple bacteria: the beta subdivision. *System. Appl. Microbiol.* (in press).
38. Paster, B.J., Stackebrandt, E., Hespell, R.B., Hahn, C.M. and Woese, C.R. (1984). The phylogeny of spirochetes. *System. Appl. Microbiol.* 217 (in press).
39. Oren, A., Weisburg, W.G., Kessel, M. and Woese, C.R. (1984). Halobacteroides halobius gen. nov., sp. nov., a moderately halophilic anaerobic bacterium from the bottom sediments of the Dead Sea. *System. Appl. Microbiol.* 5,58-70.

40. Oren, A., Paster, B.J. and Woese, C.R. (1984). Haloanaerobiaceae: A new family of moderately halophilic, obligatory anaerobic bacteria. *Appl. System. Microbiol.* 5,71-80.
41. Paster, B.J., Ludwig, W., Weisburg, W.G., Stackebrandt, E., Hespell, R.B., Hahn, C.M., Reichenbach, H., Stetter, K.O. and Woese, C.R. (1984). A phylogenetic grouping of the bacteroides, cytophagas, and certain flavobacteria. *System. Appl. Microbiol.* (in press).
42. Doolittle, W.F., Woese, C.R., Sogin, M.L., Bonen, L. and Stahl, D.A (1975). Sequence studies on 16S ribosomal RNA from a blue-green alga. *J. Molec. Evol.* 4,307-315.
43. Bonen, L. and Doolittle, W.F. (1976). Partial sequences of 16S rRNA and the phylogeny of blue-green algae and chloroplasts. *Nature* 261,669-673.
44. Bonen, L. and Doolittle, W.F. (1978). Ribosomal RNA homologies and the evolution of the filamentous blue-green bacteria. *J. Molec. Evol.* 10,283-291.
45. Stackebrandt, E., Seewaldt, E., Fowler, V.J. and Schleifer, K.-H. (1982). The relatedness of Prochloron sp. isolated from different didemnid ascidian hosts. *Arch. Microbiol.* 132,216-217.
46. Tomioka, N., and Sugiura, M. (1983). The complete nucleotide sequence of a 16S ribosomal RNA gene from a blue-green alga, *Anacystis Nidulans*. *Mol. Gen. Genet.* 191,46-50.
47. Schwarz, Z. and Kossel, H. (1980). The primary structure of 16S rDNA from *Zea mays* chloroplast is homologous to *E. coli* 16S rRNA. *Nature* 283,739-742.
48. Tohdoh, N. and Sugiura, M. (1982). The complete nucleotide sequence of a 16S ribosomal RNA gene from tobacco chloroplasts. *Gene* 17,213-218.
49. Dron, M., Rahire, M., Rochaix, J.-D. (1982). Sequence of the chloroplast 16S rRNA gene and its surrounding regions of *Chlamydomonas reinhardii*. *Nucleic Acids Res.* 10,7609-7620.
50. Graf, L., Roux, E., Stutz, E., and Kossel, H. (1982). Nucleotide sequence of a *Euglena gracilis* chloroplast gene coding for the 16S rRNA: homologies to *E. coli* and *Zea mays* chloroplast 16S rRNA. *Nucleic Acids Res.* 10,6369-6381.
51. Lewin, R.A. (1981). *Prochloron* and the theory of symbiogenesis. *Annal. N. Y. Acad. Sci.* 361,325-328.
52. Madigan, M.T. and Brock, T.D. (1977). Single "chlorobium-type" vessicles of phototrophically grown *Chloroflexus aurantiacus* observed using negative staining techniques. *J. Gen. Microbiol.* 102,279-285.

53. Pierson, B.K. and Thornber, J.P. (1983). Isolation and spectral characterization of photochemical reaction centers from the thermophilic green bacterium *Chloroflexus aurantiacus* strain J-10-fl. *Proc. Natl. Acad. Sci. USA* 80,80-84.
54. Ludwig, W., Schleifer, K.H., Reichenbach, H. and Stackebrandt, E. (1983). A phylogenetic analysis of the myxobacteria *Myxococcus fulvus, Stigmatella aurantiaca, Cystobacter fuscus, Sorangium cellulosum* and *Nannocystis exedens*. *Arch. Microbiol.* 135,58-62.
55. Hespell, R.B., Paster, B.J., Macke, T.J. and Woese, C.R. (1984). The origin and phylogeny of the bdellovibrios. *System. Appl. Microbiol.* 5,196-203.
56. Brooks, B.W., Murray, R.G.E., Johnson, J.L., Stackebrandt, E., Woese, C.R. and Fox, G.E. (1980). Red-pigmented micrococci: a basis for taxonomy. *Inter. Jour. System. Bacteriol.* 30,627-646.
57. Konig, W., Schlesner, H. and Hirsch, P. (1984). Cell wall studies on budding bacteria of the planctomyces-pasteuria group and on a *Prosthecomicrobium* sp. *Arch. Microbiol.* 138,200-205.
58. Stackebrandt, E., Ludwig, W., Schubert, W., Klink, F., Schlesner, H., Roggentin, T. and Hirsch, P. (1984). Molecular genetic evidence for early evolutionary origin of budding peptidoglycan-less eubacteria. *Nature* 307,735-737.
59. Gest, H. and Favinger, J.L. (1983). *Heliobacterium chlorum*, an anoxygenic brownish-green photosynthetic bacterium containing a "new" form of bacteriochlorophyll. *Arch. Microbiol.* 136,11-16.
60. Woese, C.R., Gupta, R., Hahn, C.M., Zillig, W. and Tu, J. (1984). The phylogenetic relationships of three sulfur dependent archaebacteria. *System. Appl. Microbiol.* 5,97-106.
61. Balch, W.E., Fox, G.E., Magrum, L.J., Woese C.R. and Wolfe, R.S. (1979). Methanogens: Reevaluation of a unique biological group. *Microbiol. Rev.* 43,260-296.
62. Schnabel, R., Thomm, M., Gerardy-Schahn, R., Zillig, W., Stetter, K.O., and Huet, J. (1983). Structural homology between different archaebacterial DNA-dependent RNA polymerases analyzed by immunological comparison of their components. *EMBO J.* 2,751-755.
63. Tu, J., Prangishvilli, P., Huber, H., Wildgruber, G., Zillig, W., and Stetter, K.O. (1982). Taxonomic relations between archaebacteria including 6 novel genera examined by cross hybridization of DNAs and 16S rRNAs. *J. Mol. Evol.* 18,109-114.

64. Stetter, K.O. (1982). Ultrathin mycelia-forming organisms from submarine volcanic areas having an optimum growth temperature of 105 °C. *Nature* 300,258-260.
65. Langworthy, T.A., Tornabene, T.G., and Holzer, G. (1982). *Zbl. Bakt. Hyg.*, I. *Abt. Orig.* C3,228-244.
66. Pang, H., Ihara, M., Kuchino, Y., Nishimura, S., Gupta, R., Woese, C.R. and McCloskey, J.A. (1982). Structure of a modified nucleoside in archaebacterial tRNA which replaces ribosylthymine. *J. Biol. Chem.* 257,3589-3592.
67. Stetter, K.O. and Gaag, G. (1983). Reduction of molecular sulfur by methanogenic bacteria. *Nature* 305,309-311.
68. Woese, C.R., Magrum, L.J. and Fox, G.E. (1978). Archaebacteria. *J. Molec. Evol.* 11,245-252.
69. Henderson, E., Oakes, M., Clark, M.W., Lake, J.A., Matheson, A.T., and Zillig, W. (1984). A new ribosome structure. *Science* 225,510-512.
70. Zillig, W., Schnabel, R. and Stetter, K.O. (1984). Archaebacteria and the origin of the eukaryotic cytoplasm. *Curr. Topics in Microbiol. and Immunol.* (in press).
71. Stahl, D.A., Luehrsen, K.R., Woese, C.R. and Pace, N.R. (1981). An unusual 5S rRNA from *Sulfolobus acidocaldarius*, and its implications for a general 5S rRNA structure. *Nucleic Acids Res.* 9,6129-6137.
72. Fox, G.E., Luehrsen, K.R. and Woese, C.R. (1982). Archaebacterial 5S ribosomal RNA. *Zbl. Bakt. Hyg.*, I. *Abt. Orig.* C3,330-345.
73. Simpson, G.G. (1944). "Tempo and mode in evolution". Columbia University Press, New York.
74. Hopwood, D.A. (1978). Extrachromosomally determined antibiotic production. *Ann. Rev. Microbiol.* 32, 373-392.
75. Woese, C.R. and Gupta, R. (1981). Are archaebacteria merely derived "prokaryotes"? *Nature* 289,95-96.

PROKARYOTIC GENOME EVOLUTION: WHAT WE MIGHT LEARN FROM THE ARCHAEBACTERIA

W. Ford Doolittle and Charles J. Daniels

Department of Biochemistry
Dalhousie University
Halifax, Nova Scotia, Canada

INTRODUCTION

1977 was an important year in evolutionary molecular biology. It saw the discovery of intervening sequences in eukaryotic protein-coding genes, the formulation of the progenote concept, and the recognition of the strangeness of the archaebacteria (1-3). We would like (i) to review the effects of these events on our thinking about early genome evolution, both prokaryotic and eukaryotic, (ii) to reformulate some questions about prokaryotic and eukaryotic genome evolution as questions about the molecular biology of the progenote, examining available data bearing on these questions, and (iii) to suggest directions of future research. We will operate from the premise that many issues in early eubacterial and eukaryotic genome evolution are best resolved, together and indirectly, through a more thorough examination of the archaebacteria*.

---

* We shall use the term prokaryote to refer to both eubacteria and archaebacteria. Many of the properties assumed, prior to 1977, to be generally prokaryotic are now known to be only generally eubacterial and when discussing such pre-1977 assumptions, we will write 'prokaryotes (eubacteria)' in an attempt to dispell this historically hidden confusion. We shall use the term 'eukaryotic nuclear' in referring to that evolutionary and genomic lineage determining the properties of eukaryotic nuclei and cytoplasms which, together with plastid and mitochondrial genomes, comprises the total eukaryotic genome (4).

The events of 1977

For the decade-and-a-half preceding 1977, it was difficult to think about microbial evolution in terms other than those of the prokaryote (eubacterium)/eukaryote division - the "most profound single evolutionary discontinuity in the contemporary biological world (5)". By 1977, the basics of that style of prokaryotic molecular biology practised by the eubacteria were well understood, and we knew enough about eukaryotic molecular biology to be able to say that it was different in many of its details.

We also knew by this time that eukaryotic cells were evolutionary chimeras, being the descendants of two (in fungi and animals) or three (in plants) genomic lineages: the plastid lineage, which could then be shown by 16S ribosomal RNA oligonucleotide cataloguing (now by direct gene sequencing) to be of cyanobacterial origin; the mitochondrial lineage, which could be similarly if not so convincingly shown to be at least *generally* eubacterial; and the nuclear lineage, then and now without apparent surviving free-living relatives (4, 6-9). So it was in fact specifically differences between the fundamental properties of the *nuclear* genomic lineage and eubacterial genomic lineages which reinforced belief in the primacy of the prokaryotic/eukaryote evolutionary discontiniuty. Some of these differences are listed in Table 1.

A satisfactory theory for the origins of both prokaryotic (eubacterial) and eukaryotic genomes required explanations of how these differences came about. Until 1977, the common position was that the eukaryotic nucleus evolved from a eubacterial nucleoid, not perhaps unlike that of modern day *E. coli,* (6,10), and relatively recently at that (Fig. 1). If this were so, and if eukaryotes are seen as evolutionarily more advanced ("higher") than prokaryotes, then differences from prokaryotes which they show must be either direct improvements on prokaryotic ways of doing things or indirect accommodations to such improvements (the uncoupling of transcription and translation and the capping and tailing of messenger RNAs could be seen as accommodations to the interposition of the nuclear membrane, for instance).

This position, common though it was, could only be held comfortably if not examined too carefully. There are too many ways in which eukaryotes are no "better" than prokaryotes (eubacteria), only different. How, for example, can the abandonment by eukaryotes of the clustering of genes into operons be seen as an improvement? Why should eukaryotes have abandoned an apparently efficient ribosome-messenger RNA recognition system involving specific base pairing

TABLE 1
Molecular Biological Differences between Eukaryotes and Prokaryotes (Eubacteria)

Trait	Eukaryotes	Eubacteria	Archaebacteria
genome	several linear chromosomes	one circular chromosome	unknown (like eubacteria?)
operons	no	yes	unknown
rRNA genes	18S, 5.8S, 28S transcriptionally linked, 5S unlinked	16S, 23S, 5S transcriptionally linked	like eubacteria, with variations
introns	yes, in protein, rRNA and tRNA genes	not as far as we know	yes, in some tRNA genes
promoters	"TATA" boxes for RNA polymerase II, other for I and III	-35 region and "Pribnow box"	unknown, perhaps unique
RNA polymerases	three, for three gene classes	one, but developmentally variable	probably one
capped mRNAs	yes	no	probably no
5' mRNA leaders	usually present, no Shine Dalgarno sequence	present Shine Dalgarno sequence	maybe short or absent
3' poly A tails	usually present, and stable	absent or unstable	probably absent
ribosome size	80S	70S	70S
rRNA size	18S, 28S, 5.8S, 5S	16S, 23S, 5S	16S, 23S, 5S
ribosome antibiotic sensitivity	cycloheximide and others	chloramphenicol and others	generally insensitive
diphtheria toxin sensitivity	yes	no	yes
initiator tRNA	methionine	formylmethionine	methionine

for one which sees only the ends of messenger RNAs (11). And why should they have evolved a whole new set of promoters and polymerases for the transcription of ribosomal and transfer RNA genes, and fundamentally altered protein-coding gene promoters?

One might have been able to come to terms with each of these and other such conundra separately, but together they seemed a bit much. The discovery of intervening sequences in eukaryotic nuclear and viral genes only made the situation worse.

The origin of introns. Immediately after that discovery, Gilbert proposed that intervening sequences sped the evolutionary process, and arose because they did so (1). According to this view, the parcelling out of protein-coding

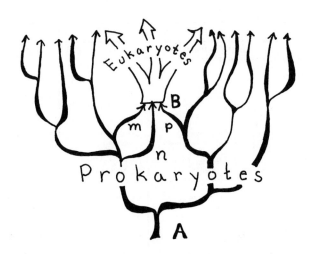

Fig. 1. Common understanding of prokaryotic/eukaryote evolutionary relationships before 1977. Points A and B are, respectively about 3.5 and 1.5 thousand million years ago. 'p' indicates the cyanobacterial lineage which gave rise to plastids, 'm' indicates the eubacterial lineage which gave rise to mitochondria, 'n' indicates the eubacterial lineage which gave rise to eukaryotic nuclei and cytoplasms.

information into small pieces (exons) facilities the chance assembly, either through mis-splicing or genetic recombination, of potentially functional novel proteins. Such a mechanism for the generation of evolutionary novelty might be required (in addition to normal gene duplication and mutation) to explain the greater evolutionary versatility (in terms of morphology and development, not metabolism) of eukaryotes.

This may all be true, and in fact data on several eukaryotic protein-coding genes do support the notion that exons are quasi-independent evolutionary units (12). However, the presence of introns only in eukaryotic genomes cannot be easily explained away as just one of those molecular biological improvements made at the time of a relatively recent prokaryotic/eukaryote transition. The introduction of an intervening sequence into an already functioning continuous gene poses an immediate selective disadvantage (unless splicing pre-exists and is cost-free) and no selective advantage until many more genes have been so interrupted. Even then, the advantage is not at the level of the individual but at the level of the group (13). All things considered, its simplest to suppose that the possession of introns is a primitive trait, one which eukaryotic nuclear genomes have "always" shown, and which eubacterial genomes (at least those we know) once had but have since lost - in the interest of efficiency (14,15).

The primitivity of eukaryotic nuclear molecular biology. If one supposes *that*, then it suddenly becomes very appealing to go on and suggest that many or all of the molecular biological idiosyncracies of the eukarytic nuclear genome are also essentially primitive in that lineage, that is, that the nuclear genome was "always" different from the modern eubacterial genome, at least in the ways in which it is different now; there never was an *E. coli*-like period in the evolutionary history of eukaryotic cells.

And if that is so, then the eukaryotic nuclear genomic lineage must have diverged from prokaryotic genomes a very long time ago indeed, by the following logic. (i) All eubacteria probably share a common molecular biology which is similar in (at least) those details in which eubacteria differ from eukaryotes listed in Table 1. One can make this assertion on the basis of similarities in (a) the structure of genes and transcription and translation signals, (b) ribosome structure and, as for as we know, function, (c) rRNA gene organization, (d) rRNA sequence and secondary structure, (e) RNA polymerase structure and (f) mRNA structure, between eubacteria such as *E. coli, Bacillus subtilis* and cyanobacteria. (ii) The cyanobacteria themselves are at least 3.5 thousand-million years old, and must themselves be derived from a common eubacterial ancestor older than that (19). (iii) The last common ancestor of eubacterial and eukaryotic nuclear genomes must be even older still (4,16).

The same conclusion could be drawn from quantitative data emanating from the laboratory of Carl Woese, and also first available to public scrutiny in 1977 (17). Those data were based on 16S ribosomal RNA T1-oligonucleotide catalogs and comprised the first comprehensive data set on homologous macromolecules encoded by both eubacterial and eukaryotic nuclear genomes from which phylogenetic conclusion could be drawn. (The largest other data set, that for cytochromes *c*, traces only mitochondrial and plastid phylogenies - there are precious few sequenced nuclear-encoded proteins [or genes] for which we also have sequenced prokaryotic homologs, and tRNAs provide, because of their smallness and con-

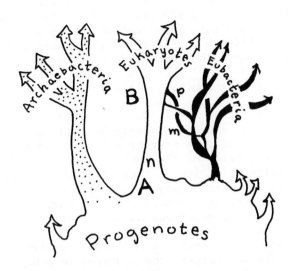

Fig. 2. Understanding of prokaryote/eukaryote evolutionary relationships after 1977. Symbols as in Fig. 1.

servation, little deep phylogenetic information).

And Woese's data quite clearly showed that, as far as 16S ribosomal RNA sequences are concerned, eubacteria are one kind of thing and eukaryotes are another. There is no specific branch of the eubacteria to which eukaryotic nuclei are especially close, and all eubacteria are closer to each other than any are to eukaryotic nuclei — not what one would expect if the situation were as shown in Figure 1. Figure 2 portrays the situation more accurately.

The concept of the progenate. It seems simplest, then, to assume that those differences in detail exhibited by eubacterial and eukaryotic nuclear molecular biologies represent independently evolved adaptations of the machinery of replication, transcription and translation. That is, that they are separately achieved solutions to problems of efficiency and accuracy of information transfer which had not been solved in the last common ancestor shared by eubacteria and eukaryotic nuclei.

Arguments of this sort were first advanced by Woese and Fox (2), just before the discovery of intervening sequences. The prokaryote/eukaryote distinction was, they reminded us, taxonomic, not phylogenetic. Although eukaryotes may indeed have evolved from cells without nuclei or organelles, to call these early cells 'prokaryotes' was misleading, since this term had, in the 15 years since receiving formal definition (18), taken on a phylogenetic meaning.

Progenotes were, in their view, "simpler, more rudimentary entities" than modern prokaryotes. In them, "molecular functions would not be of the complex, refined nature we associate with functions today", and thus "subsequent evolution would alter functions mainly in the sense of refining them" (19). Such evolution would differ, they claimed, both in tempo and mode from evolution within eubacterial and eukaryotic nuclear lineages since shortly after their foundations some 3.5 thousand million years ago. In tempo because selection pressures for increased efficiency and accuracy of the mechanisms of information transfer were intense. And in mode because, perhaps, "the world of progenotes may be more a world of semi-autonomous subcellular entities that somehow group to give 'loose' (ill-defined) cellular forms". A world in which "it is easy to picture a ready exchange, a flow, among subcellular entities - be they called genes, plasmids, viruses, selfish DNA, organelles or whatever" (19).

The discovery of the archaebacteria. The notion that eubacteria and eukaryotes arose separately from truly primitive ancestors, from progenotes in which elements of the phenotype-genotype linkage were still being forged, suggested that one might find still a third equally distinct 'kingdom' of ancient divergence. That one was found somehow eases acceptance of the progenote concept at the deepest (non-rational) level.

Archaebacteria give evidence of still other, and therefore independently achieved, solutions to problems of accuracy and efficiency of information transfer unsolved in the progenote. They also possess molecular biological traits which appear specifically either eukaryotic or eubacterial. These are mixed, in such a way that it may be possible to decide upon primitive and derived states in the evolution of the information transfer system in a way which would not be possible if there were only two primary kingdoms.

How we might use the archaebacteria

Primitive and advanced states. Suppose we had sufficient data to derive a rooted phylogenetic tree relating archaebacteria, eubacteria and eukaryotic nuclei, and that tree were as shown in Figure 3.

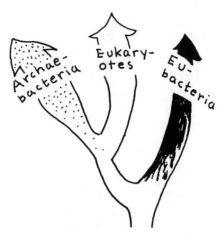

Fig. 3. *One of the three possible rooted phylogenetic trees relating the three kingdoms*

Then we might well conclude that traits shared by archaebacteria and eubacteria (70S ribosomes, ribosomal RNAs of 16S, 23S and 5S, single RNA polymerases, uncapped and largely untailed mRNAs) are truly primitive (present in progenotes) - the eukaryotic version thereof being derived (20) from them and thus properly considered advanced. Of traits shared by archaebacteria and eukaryotic nuclei we could not be certain, but we could at least eliminate the possibility that these were the consequences of the interposition of a nuclear membrane or the acquisition of plastids and mitochondria, and we could be reasonably certain that such traits were truly old ones. Arguments for the antiquity of intervening sequences (14,15) gained considerable support from the finding of introns in tRNA genes of *Sulfolobus* for instance (21,22).

Complications. There will of course be problems with such a simple approach, and we may never be able to complete such an analysis. Available molecular sequence data (for ribosomal proteins and ribosomal RNAs for instance) do not clearly support any of the three possible rooted trees relating the three kingdoms. Genetic exchange between separately evolving progenotes may have been sufficiently extensive that it is necessary to draw separate trees for different macromolecules - it may make no sense to speak of lineages of genomes in this context (19). There may be four primary kingdoms instead of three, with eukaryotes and eubacteria each being separately related to two different members of what we now call the archaebacteria (23). And

finally, it may not be sensible to hope to root the tree.
Without some independent way to assess differences in evolutionary rates between the three kindgoms, we may always have to resort to nonquantitative judgements about which states of a molecular biological trait are the most primitive. The latter is, as the history of the prokaryote/eukaryote concept itself clearly illustrates, a risky endeavor. However, the archaebacteria provide us with a vitally needed third perspective from which to make such judgements, and the academic effort involved is in the best traditions of evolutionary thought. We would hope that Carl Woese is not offended when we suggest that he has, almost single-handedly, driven evolutionary molecular biology back into the nineteenth century.

QUESTIONS AND THE BEGINNINGS OF SOME ANSWERS ABOUT PROKARYOTE EVOLUTION

To see if this works at all, it's necessary to look at some examples. The data are fragmentary, and largely unpublished. We'll proceed with the order suggested by Table 1, ommitting traits for which we have too little data to even guess.

Ribosomal RNA genes

In eubacteria, these are characteristically linked into single transcriptional units of the structure 5'-16S-tRNA(s)-23S-5S-(tRNA)-3' (36,37). In eukaryotes the order is 18S-5.8S-28S. 5S is unlinked transcriptionally, and there are no intergenic spacer tRNA genes. 5.8S rRNA is the homolog of the 5' end of 23S rRNA (38). In some fungi and protists, 5S genes are linked, physically, to 28S rRNA genes, and this situation has been described as an evolutionary intermediate (39,40). Such a description makes sense only in terms of the pre-1977 view of prokaryote/-eukaryote evolution and there is in fact no way of deciding, *a prior*, whether the eukaryote type of organization derives from the eubacterial one, or *vice versa*.

In archaebacteria, there is much variability, and transcriptional linkage has not been definitely shown. However, in *Halobacterium halobium*, *H. volcanii* and *H. cutirubrum* the physical order is unquestionably 5'-16S-tRNA-23S-5S-3', and the latter species shows distal gene for a tRNA^{cys} (37,41,42). The intergenic tRNA gene is for ala in *H. cutirubrum* and *Methanococcus vannielli* (37, 43). Although physical linkage of 16S and 23S is not demonstrable in some species, and some have an 'extra' gene for 5S (45), archaebacteria clearly exhibit the eubacterial type of organization. We could conclude this to be primitive if the rooted tree were as in Figure 3, but the most pertinent molecular sequence data, those for 16S rRNAs (28,30,31) suggest that, at least for rRNA genes, it is not.

Promoters

Eubacteria may in general possess promoters like those of *E. coli*, with conserved sequence centered around the 10th and 35th nucleotide 5' to the transcription initiation site (16,47). A general eukaryotic promoter for RNA polymerase II is hard to define, but the conserved TATAA sequence about 25 to 35 nucleotides 5' and other semiconserved sequences further upstream seem to be important. It is difficult to see how either situation could be derived from the other, and it seems crucial to know what kind of promoters the archaebacteria have. Relevant published halobacterial sequences include those for bacteriorhodopsin, a protein which somehow regulates the synthesis of bacteriorhodopsin, and the presumed genes (open reading frames) within a number of insertion sequence like elements (32,33, F. Pfeifler, personal communication). Although there are some common sequences, their presence is by no means universal and, in those few cases where we can identify transcription start sites, these common sequences are not in the same positions as the conserved sequences of *E. coli* promoters.

There are unpublished data of which we are unaware, but we'll hazard the guess that archaebacterial promoter sequences, if they can indeed be defined in this way, are like neither eukaryotic nor eubacterial promoter sequences. If we are right about that, then we'd suggest that progenotes did not have well-defined promoter sequences. Proteins could have been produced from large, effectively polycistronic, transcripts and transcription could have been initiated infrequently from a relatively small number of good promoters or nearly randomly, from a large number of quite poor ones. There would clearly have been selection for the introduction, on a gene by gene or operon by operon basis, of promoters recognizable by the RNA polymerase of the day. The sequence recognition properties of the polymerase would co-evolve with promoter sequences and, if those secondarily introduced gene promoters were not essential, would evolve differently in different lineages. Promoters in the three kingdoms would differ, as apparently they do. What seems not to be likely is that primitive cells had, for instance, all *E. coli*-like or all eukaryotic-like, promoters. Such a situation, once established, should be immutable.

Intervening sequences

The discussion of the antiquity of intervening sequences has centered around those in protein-coding genes (14,15), since it is for these for which the evolutionary proposal of Gilbert (1) is relevant. However, many of the same arguments can be used to defend an ancient origin for intervening sequences in genes for tRNAs. The two published sequences for intervening sequence-containing tRNA genes from *Sulfolobus solfataricus* (21) show these archaebacterial intervening sequences to be not unlike their eukaryotic nuclear counterparts in position (one nucleotide 3' to the anticodon), in size (15-25 nucleotides) and in the ability of intervening sequence-specific and exon-specific nucleotides to form an extended secondary structure in place of the anticodon stem of the mature tRNA. We have recently sequenced an intervening sequence-containing tRNA^{trp} gene from *Halobacterium volcanii*. It contains an intervening sequence which, except for its unusual length (105 base pairs) is generally like these others.

One can suppose that tRNA intervening sequences currently have functions (in processing or influencing rates of transcription, for instance) but it does not seem economical to suggest that they arose for such a purpose. One could also suppose that the uniformity of positioning of tRNA intervening sequences in different tRNA genes of a given organism reflects structural requirements of a single tRNA-splicing enzyme. The identical positioning of archaebacterial and eukaryotic nuclear tRNA intervening sequences can only bespeak evolutionary homology, however. The deeper question here is this: does the identity in positioning of introns in genes for different tRNAs reflect the presence of a similarly positioned intervening sequence in some ancestral gene from which all tRNA genes are derived, or does it reflect selection (by requirements of transcription and processing systems) from among a variety of kinds of early tRNA genes with intervening sequences occuring in a greater variety of positions. The archaebacteria hold the greatest promise of providing the diversity of intervening-sequence containing tRNA gene with which to address such questions.

5' mRNA leaders

We know the transcription initiation sites of only two archaebacterial protein coding genes - the *bop* gene and the linked *brp* gene (32, F. Pfeifler, personal communication). The first is only two nucleotides upstream from a functional translation initiation codon, and the second may be identical with the first position of the translation initiation codon. This suggests that, at least for these mRNAs, there is no Shine-Dalgarno-like ribosome-messenger RNA interactions, for all that the 16S ribosomal RNA bears the appropriate 3'-terminal pyrimidine rich stretch (28,30,31). Sequences 3' to the initiation codon might be involved, but is is equally tempting to suggest that only a 5' end is required, that initiation is thus more eukaryote-like, and archaebacteria will thus prove to lack operons (11). The speculation is premature, but the experimental approach to testing it is obvious.

RNA polymerases

Zillig has compared the properties of eubacterial, archaebacterial and eukaryotic RNA polymerases elsewhere in this volume, and the latter two classes appear to be the more closely related. It is important to know whether archaebacteria possess but a single major polymerase or whether there are multiple distinct activities for different gene classes (that is, protein, rRNA and tRNA) as in eukaryotes.

FUTURE DIRECTIONS

It is clear that we are just beginning to be able to ask questions about the progenote, and that we can only do so because we have the archaebacteria as a third point from which to triangulate. We hope that is also clear that this approach will prove fruitful - we have in hand a new way of looking at early evolution, new questions to ask, and new organisms with which to answer them. The archaebacteria are more than just an evolutionary novelty. They hold the key to an understanding of the evolution of all three kingdoms.

REFERENCES

1. Gilbert, W. (1977) Why genes in pieces? **Nature 271**, 501.
2. Woese, C.R. and Fox, G.E. (1977). The concept of cellular evolution. **J. Molec. Evol 10**, 1-6.
3. Fox, G.E., Magrum, L.J., Balch, W.E., Wolfe, R.S., and Woese, C.R. (1977). Classification of methanogenic bacteria by 16S ribosomal RNA characterization. **Proc. Natl. Acad. Sci. USA 74**, 4537-4541.

4. Gray, M.W. and Doolittle, W.F. (1982). Has the endosymbiont hypothesis been proven? Microbiol. Revs. 46. 1.
5. Stanier, R.Y., Adelberg, E.A. and Ingraham, J.L. (1976). "The Microbial World". Prentice-Hall, Inc. Englewood Cliffs, New Jersey.
6. Margulis, L. (1970). "Origin of Eukaryotic Cells". Yale University Press, New Haven.
7. Bonen, L. and Doolittle, W.F. (1976). Partial sequences of 16S rRNA and the phylogeny of blue-green algae and chloroplasts. Nature 261, 669.
8. Schwarz, Z., and Kossel, H. (1980). The primary structure of 16S rDNA from Zea mays chloroplast is homologous to E. coli 16S rRNA. Nature 283, 739..
9. Spencer, D.F., Schnare, M.N. and Gray, M.W. (1984). Pronounced structural similarities between the small subunit ribosomal RNA genes of wheat mitochondria and Escherchia coli. Proc. Natl. Acad. Sci. USA 81, 493.
10. Cavalier-Smith, T. (1975). The origin of nuclei and of eukaryotic cells. Nature 256, 563.
11. Kozak, M. (1984). Compilation and analysis of sequences upstream from the translational start site in eukaryotic mRNAs. Nucleic Acids Res. 12, 857.
12. Blake, C.C.F. (1983) Do genes-in-pieces imply proteins in pieces? Nature 273, 267.
13. Maynard Smith, J. (1978). "The Evolution of Sex", Cambridge University Press, Cambridge.
14. Doolittle, W.F. (1978). Genes-in-pieces, were they ever together? Nature 272, 581.
15. Darnell, J.E., Jr. (1978). Implications of RNA-RNA splicing in evolution of eukaryotic cells. Science 202, 1257.
16. Doolittle, W.F. (1982). Molecular evolution. In "The Biology of the Cyanobacteria" (Eds. N.G. Carr and B.A. Whitton), pp. 307-331. Blackwell's, Oxford.Woese, C.R.
17. and Fox, G.E. (1977). Phylogenetic structure of the prokaryotic domain: the primary kingdoms. Proc. Natl. Acad. Sci. U.S.A. 74, 5088.
18. Stanier, R.Y. and van Niel, C.B. (1962). The concept of a bacterium. Arch. Mikrobiol. 42, 17.
19. Woese, C.R. (1982). Archaebacteria and cellular origins. 261. Blakt. Hyg. I. Abt. Orig. C3, 1.
20. Uzzell, T. and Spolsky, C. (1974). Mitochondria and plastids as endosymbionts: a revival of special creation? Am. Sci. 62, 334.
21. Kaine, B.P., Gupta, R., and Woese, C.R. (1983). Putative introns in tRNA genes of prokaryotes. Proc. Natl. Acad. Sci. U.S.A. 80, 3309.

22. Rogers, J. (1983). Introns in archaebacteria Nature 304, 685.
23. Lake, J.A., Henderson, E., Oakes, M. and Clark, M.W. (1984). Eocytes: a new ribosome structure indicates a kingdom with a close relationship to eukaryotes. Proc. Natl. Acad. Sci. U.S.A. 81, 3786.
24. Nazar, R.N. (1980). A 5.8S-like sequence in prokaryotic 23S rRNA. FEBS Letts. 19, 212.
25. Maizels, N. (1976). Dictyostelium 17S, 25S and 5S rDNAs lie within a 38,000 bp repeated unit. Cell 9, 431-439.
26. Hofman, J.D., Lau, R.H., and Doolittle, W.F. (1979). The number, physical organization and transcription of ribosomal RNA cistrons in an archaebacterium: Halobacterium halobium. Nucleic Acis Res. 7, 1321.
27. Jarsch, M. and Bock, A. (1983). DNA sequence of the 16S rRNA/28S rRNA intercistronic spacer of two rDNA operons of the archaebacterium Methanococcus vannielli. Nucleic Acids Res. 11, 7537.
28. Hui, I. and Dennis, P.P. (1984). Characterization of the ribosomal RNA gene clusters in Halobacterium cutirobrum. J. Biol. Chem., in press.
29. Newmann, H., Gierl, A., Tu, J., Liebrock, J., Staiger, D., and Zillig, W. (1983). Organization of the genes for ribosomal RNA in archaebacteria. Mol. Gen. Genet. 192, 66.
30. Gupta, R., Lauter, J.M. and Woese, C.R. (1983). Sequence of the 16S ribosomal RNA from Halobacterium volcanii, an archaebacterium. Science 221, 656.
31. Leffers, H., and Garrett, R.A. (1984). The nucleotide sequence of the 16S ribosomal RNA gene of the archaebacterium Halococcus morrhua. EMBO J. 3, 1613.
32. DasSarma, S., RajBhandary, U.L., and Khorana, H.G. (1984). Bacterio-opsin mRNA in wild-type and bacterio-opsin-deficient Halobacterium halobium strains. Proc. Natl. Acad. Sci. U.S.A. 81, 125.
33. Xu, L. and Doolittle, W.F. (1983). Structure of the archaebacterial transposable element ISH50. Nucleic Acids Res. 11, 4195.

THE EVOLUTION OF THE TRANSCRIPTION APPARATUS

W. Zillig, R. Schnabel, F. Gropp and W. D. Reiter

Max-Planck-Institut für Biochemie
Martinsried, Federal Republic of Germany

K. Stetter and M. Thomm

Institut für Biochemie, Genetik und Mikrobiologie
Universität Regensburg
Regensburg, Federal Republic of Germany

1. INTRODUCTION

The division of the prokaryotic domain into the urkingdoms of the Eubacteria and the Archaebacteria by Carl Woese and his collaborators (Woese and Fox, 1977; Woese et al., 1978) and in consequence most of our present knowledge of the natural systems of the two novel urkingdoms (Fox et al., 1980) rest on the comparison of sequences of T1 RNAase oligonucleotide catalogs of 16S rRNAs of a large number of species (Fox et al., 1977). An increasing body of additional evidence including comparative analyses of other parts of the translation apparatus (Hori et al., 1982; Fox et al., 1982; Matheson and Yaguchi, 1982; Yaguchi et al., 1982; Böck et al., 1984; Gupta, 1984; Kessel and Klink, 1982), of the transcription machinery (Zillig et al., 1982; Schnabel et al., 1983; Huet et al., 1983), of cell envelopes (Kandler, 1982) and membranes (review by Langworthy et al., 1982), of responses to antibiotics (review by Böck and Kandler, in the press) and of certain aspects of metabolism prove these revolutionizing ideas beyond reasonable doubt.

The error margin of the original S_{AB} method, which led to the recognition of the three entirely separate urkingdoms of life, the Eubacteria, the Archaebacteria and the "eukaryotic cytoplasm" or Eucyta (defined as the truly eukaryotic compartments of eukaryotic organisms), does not allow a significant estimation of S_{AB} values and thus phylogenetic distance between the kingdoms, so that the early branching sequence in the evolution of the living world remains undetermined.

As more evidence becomes available, the problem acquires new dimensions.

(1) Comparative sequence analysis of homologous macromolecules measures the distances between the present ends of lineages, not, as would be required, between their origins i.e. the early branching points. Many more sequences have to be determined such that "ur-sequences" of lineages can be derived.

(2) Evolution might not solely proceed via the separation of lineages but, to an unknown, though possibly small extent, also via horizontal gene exchange or recombination between different lineages. This has for example certainly occurred between the urmitochondrial and the eucytic compartments in early eukaryotes. Examples of such processes between distant prokaryotes are as yet unknown.

(3) The evolution rates of different genes within a lineage are not necessarily parallel to each other. This is the reason why comparative analysis of different macromolecules has led to somewhat different phylogenetic trees. That derived from 5S rRNA sequence (Fox et al., 1982; Hori et al., 1982) is for example differing in details from that obtained by comparison of T1 RNAase oligonucleotide catalogs of 16S rRNAs. This review will give other examples.

In the following, we will show that a comparison of the component patterns (complexity, homologies) and antibiotic responses of the DNA dependent RNA polymerases of <u>Archaebacteria</u>, <u>Eubacteria</u> and eukaryotes

(1) is independent evidence for the large phylogenetic distance between <u>Eubacteria</u> and <u>Archaebacteria</u>;
(2) testifies for the deep division between the two major branches of the urkingdom of the <u>Archaebacteria</u>, together with other feature designs;
(3) suggests a specific relation of <u>Archaebacteria</u>, especially of the sulfur dependent branch, with <u>Eucyta</u> which remains to be defined;
(4) suggests a specific relation of Methanobacteriales and

EVOLUTION OF THE TRANSCRIPTION APPARATUS 47

Methanococcales to Eubacteria, which remains to be defined; (5) exemplifies differences in the rates of evolution of different genes.

2. FACTS

2.1 Signal structures and transcription mechanisms

Except for the presence of certain "boxes", e.g. the TATAAT box plus the -35 box in eubacterial promoters, the TATA box in eukaryotic RNA polymerase II promoters, and certain downstream structures in RNA polymerase III initiation sites, the sequences interacting with RNA polymerases in transcription initiaton are ill defined. RNA polymerases appear to recognize the specificity of promoters better than the human mind or the computer. Accordingly, the few known archaebacterial sequences containing transcription starts (DasSarma et al., 1984; Klein, H., personal communication) do not allow the derivation of general sequences or structures of archaebacterial promoters. However, mapping of transcription starts has revealed that sequences acting as promoters for E. coli RNA polymerase are inactive in certain Archaebacteria and vice versa (J. Konisky, personal communication). In vitro investigations strengthen the conclusion that archaebacterial promoters differ from their eubacterial counterparts (Prangishvilli et al., 1982).

Like in many cases in Eubacteria, two different genes, encoding distinct components of methyl CoM reductase, are linked within one operon in Methanococcus voltae (Konheiser et al., 1984).

Since in vitro systems for cyclic, asymmetric, signal specific transcription by archaebacterial RNA polymerases are not yet available, details of the transcription process in Archaebacteria e.g. binding of the polymerase to the DNA, specificity, role of components of the polymerases are unknown. Not much more is known about transcription by the eukaryotic RNA polymerases, which so closely resemble those of Archaebacteria.

So far, transcription by isolated archaebacterial polymerases differs from that by eubacterial enzymes mainly in that they are, by themselves, unable to transcribe free native DNA specifically.

Thus, the present status of ignorance of transcription signal structures and mechanisms not only in Archaebacteria, but also in eukaryotes and, less so, in Eubacteria, prohibits a meaningful comparative analysis aiming at understanding the phylogeny of the transcription systems. Therefore, we concentrate here on the much more lucid situation pertaining to the comparison of structures (compositions) and component homologies of the "normal" DNA dependent RNA polymerases.

2.2 The DNA dependent RNA polymerases of the Eubacteria

are all of rather low complexity, with only four types of "true" components (not considering "binding proteins" like ω and τ), are almost all, in their wild type versions, sensitive to the antibiotics rifampicin, which specifically inhibits initiation, and streptolydigin, which specifically inhibits elongation of transcription.

They are all of the type $\beta'\beta\sigma\alpha_2$ (see Fig. 1), in which β' and β are two large peptides with molecular weights above or close to 100.000, α is present twice per enzyme monomer and σ is an initiation factor involved in promoter recognition and, at least in some cases, released after initiation. β' is involved in DNA binding, β appears to harbour important active sites (reviews: Zillig et al., 1976; Burgess, 1976).

Differences between different groups concern the relative sizes of β', β and σ. In gramnegative Eubacteria, β' is usually larger than β, in grampositive Eubacteria, it is the other way round. Gramnegative Eubacteria possess, in average, large σ components. In grampositive Eubacteria like Bacillus σ is smaller, but an additional component, δ in Bacillus, y in Lactobacillus, appears to be functionally homologous to a portion of the large σ of the gramnegatives (Dickel et al., 1980; Achberger and Whitely, 1981; Gierl et al., 1982), suggesting the division of the large polyfunctional σ of the gramnegatives into two peptides of different function, one, σ, concerned with promoter recognition, the other, δ or y, with the suppression of unspecific binding and/or initiaton.

In some phyla like the blue green algae, σ is a stoichiometric component (Herzfeld and Kiper, 1976), in others, like E. coli, it is a substoichiometric factor only bound to free core enzyme. In Bacillus and Lactobacillus, it

EVOLUTION OF THE TRANSCRIPTION APPARATUS

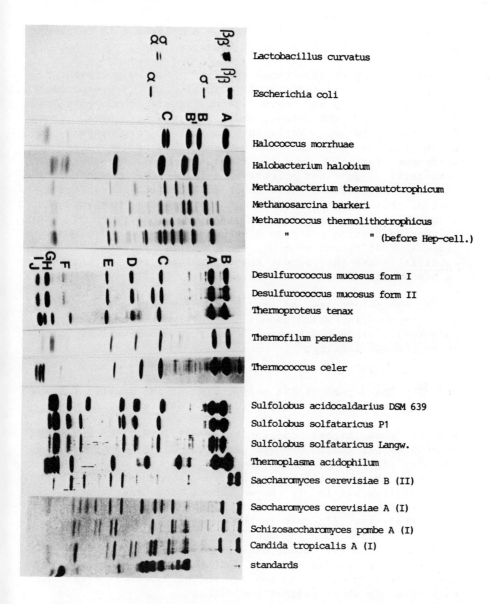

Fig. 1 SDS polyacrylamide gel electrophoresis patterns of DNA dependent RNA polymerase of eubacteria (top), methanogenic and halophic and sulfur dependent Archaebacteria (the latter including T. acidophilum) (middle) and Eucyta (bottom). Compiled and aligned from different runs.

replaces δ or y in the preinitiation complex and is released upon initiation (Gierl et al., 1982).

The homology of the components of RNA polymerases of different orders or families is evident from Western blot analysis employing antibodies against single components of one or the other polymerase, and also, in some cases, from functional tests, employing mutant components in reconstitution experiments. However, RNA polymerases of different families, e.g. grampositive and gramnegative Eubacteria, do not crossreact in the Ouchterlony in-gel-immunodiffusion test which is based on immunoprecipitation, thus allowing to determine phylogenetic distance in a qualitative manner. A discussion of the role of RNA polymerase binding proteins like ω and τ and of additional σ factors of different promoter specificities as found in B. subtilis, where they appear to be involved in sporulation, is beyond the scope of this review.

The RNA polymerases of organelles of eukaryotes, plastids and mitochondria, which should be derived from those of Eubacteria, are not discussed here because relevant evidence is completely lacking.

2.3 The DNA dependent RNA polymerases of the Eucytes i.e. of the nuclear compartment of the eukaryotes

are of much higher complexity, containing 10 or more components per enzyme monomer (see Fig. 1). A component is normally tightly and stoichiometrically bound to the enzyme particle, thus copurifying with other components and with the activity. Little is known about the requirement of certain components for enzymatic activity and even less about their specific role in the transcription process.

Even in primitive eukaryotes like yeast three types of nuclear RNA polymerases exist, differing in their functions and in their specific component patterns, but resembling each other in their general composition:

RNA polymerase I (or A) transcribes rRNA structural genes (including 5.8S rRNA) and is insensitive to the fungal poison α amanitin.

RNA polymerase II (or B) transcribes mRNA (protein) structural genes and is highly sensitive to α amanitin.

RNA polymerase III (or C) transcribes tRNA and 5S rRNA structural genes. Its sensitivity to α amanitin is intermediate.

Their component patterns each consist of two large subunits, with sizes around 100 kdaltons, and 8 to 12 smaller components (Roeder, 1976; Huet et al., 1982). However, only three of these, in Saccharomyces AB C27 (I,II,III 27), ABC 23 (I,II,III 23) and ABC 14.5 (I,II,III 14.5) are shared by all three enzymes. In addition, RNA polymerases I and III share AC 40 (I,III 40) and AC 19 (I,III 19) (Buhler et al., 1980).

2.4 The DNA dependent RNA polymerases of the Archaebacteria

are insensitive to rifampicin and α amanitin. Though forms lacking one or the other component (Sturm et al., 1980; Madon et al., 1983), in some cases inactive, and, sometimes forms with additional components (Prangishvilli et al., 1982) have been isolated, it appears that different types of enzymes as in eukaryotic nuclei do not exist.

The enzymes are of two types, one of high, the other of intermediate complexity. Those of the sulfur dependent Archaebacteria, which appear strikingly similar to yeast polymerase I, contain four heavy components, B,A,C and D, and more than six smaller components (E to J), those of the methanogens and extreme halophiles, which show a novel type of component pattern, contain five heavy components, A , B', B'', C and D, and few (about three) smaller components (see Fig. 1 and Table 1).

Their structures (and functions of which little is known) have been discussed in detail in previous reviews (e.g. Zillig et al., 1982).

2.5 Homologies of archaebacterial, eukaryotic and eubacterial DNA dependent RNA polymerases

2.5.1 The Western blotting technique

For the analysis of homologies between distinct components of corresponding multicomponent complexes of different origin, as for example the DNA dependent RNA polymerases of different organisms, the so-called Western blotting tech-

Table 1 Molecular weights (in kilodaltons) of components of archaebacterial DNA dependent RNA polymerases

Sub-unit	Methano-bacterium thermoauto-trophicum M	Methano-sarcina barkeri	Methano-coccus thermo-litho-trophicus	Methano-lobus volcani	Halo-bacterium halobium	Sulfo-lobus acido-caldarius	Sulfo-lobus solfa-taricus	Sulfo-lobus brierleyi	Thermo-plasma acido-philum	Thermo-proteus tenax	Thermo-filum pendens	Desul-furo-coccus mucosus	Thermo-coccus celer
A	119	97	109	102	148	101	101	106	108	104	102	106	105
B	-	-	-	-	-	122	121	130	135	130	132	135	125
B'	78	72	78	70	84	-	-	-	-	-	-	-	-
B''	56	66	53	73	69	-	-	-	-	-	-	-	-
C_1	42	46	49	49	49	44	41.5	42.5	56	42.5	45	44.5	44
C_2	-	-	-	42	46	-	-	-	-	-	-	(39.5)	-
D_1	33	30	25	31.5	23.5 (ϵ)	33	30	28	33	34	33	31.5	35
D_2	-	-	-	-	-	-	29.5	-	-	33	-	-	-
E	24.5	?	22.5	28	-	26	23	23	22	23	24	23	25
F	-	11	-	-	13	17.5	15.5	-	13.5	15.2	14	14.4	13
G	10.5	10.2	12	12	12.5	13.8	13.9	14.8	11.7	12.5	12	12.8	10.5
H	-	-	-	-	11.8	(4x) 11.8	(4x) 11.7	(4x) 12.9	11.4	(2x) 11.7	11.7	11.9	10.2
I	-	-	-	-	11.3	11.1	12.2	13.1	10.5	(2x) 11.3	11.0	11.1	10.0
J	-	-	-	-	-	10.8	11.0	10.8	-	10.8	-	10.7	-

The designations of the smaller components, especially of the RNA polymerases of Halobacterium and of methanogenic bacteria remain tentative until homology has been proved. C₂ in RNA polymerase from D. mucosus is an additional component of one form of this enzyme. D₁ and D₂ often form a doublet of unsharp bands with characteristic tinges.

nique of probing for immunochemical crossreactivity offers a short cut to nucleotide sequence comparison (Prager et al., 1980; Huet et al., 1982). Though only semiquantitative and by far not on the same high level of lucidity as sequence analysis, it yields valuable information speedily and with ease. The method consists of challenging blotted patterns of separated components of the complex, e.g. SDS polyacrylamide gel component patterns of RNA polymerases transferred to nitrocellulose sheets, with antibodies against single components. Crossreaction is visualized e.g. employing protein A of Staphylococcus aureus labelled with ^{125}J or anti-antibody covalently coupled to peroxidase, thus allowing detection by autoradiography or by a colour reaction. Positive reaction indicates that antigenic groups are shared between the immunogen and the reacting component. The strength of the reaction depends both on the number of shared antigenic determinants and on their affinity to the antibody which might be decreased by structural changes in evolution. For the purpose of phylogenetic studies, the antibodies employed for probing should be polyclonal i.e. directed against as many as possible of the antigenic determinants of the completely denatured polypeptides. The probability of crossreaction is determined by structural correspondence depending on phylogenetic distance but also by the size of the peptide such that the method is normally not suitable to prove homologies between small peptides (below 20 kdaltons) even if phylogenetically rather closely related.

2.5.2 Homologies

We collaborated with J. Huet and A. Sentenac in probing archaebacterial, eubacterial and eukaryotic RNA polymerases, both in spot tests and by Western blotting of SDS polyacrylamide gel electrophoresis component patterns, with antibodies against single components of yeast (Saccharomyces) RNA polymerases I (A) and II (B) (Huet et al., 1983). The homologies between different archaebacterial RNA polymerases were established employing antibodies against single components of the RNA polymerases of Sulfolobus acidocaldarius (R. Schnabel) and Methanobacterium thermoautotrophicum (M. Thomm), which were also used to challenge patterns of eukaryotic and eubacterial enzymes (Schnabel et al., 1983). Finally, antibodies against single subunits of E. coli RNA polymerase were used.

The RNA polymerases were from Escherichia coli, in few instances Bacillus cereus (Rexer et al., 1975) and Lactobacillus curvatus (Stetter and Zillig, 1974), representing the Eubacteria, Methanobacterium thermoautotrophicum (Stetter et al., 1980), Methanococcus thermolithotrophicus, Methanosarcina barkeri, Methanolobus volcani (all Thomm, 1983) and Halobacterium halobium (Zillig et al., 1978; Madon and Zillig, 1983), representing the methanogenic and extremely halophilic branch of Archaebacteria, Desulfurococcus mucosus, Thermoproteus tenax (both Prangishvilli et al., 1982), Thermofilum pendens (Zillig et al., 1983a), Thermococcus celer (Zillig et al., 1983b), and several Sulfolobus species (Zillig et al., 1980) representing the sulfur dependent branch of the Archaebacteria, Thermoplasma acidophilum (Sturm et al., 1980) as an isolated genus, and Saccharomyces cerevisiae, forms I(A) and II(B) Buhler et al., 1980) representing the Eucyta. The results are schematically shown in Fig. 2.

With exceptions, the crossreactions between corresponding components of Archaebacteria of both branches were strong. The crossreactions between corresponding components of archaebacterial and eucytic polymerases were of similar strength, often significantly stronger than between corresponding components of the eukaryotic polymerases I (A) and II (B) themselves. The observed crossreactions between homologous components of eubacterial RNA polymerases on the one hand and archaebacterial and eukaryotic RNA polymerases on the other were significant, but usually weak (Fig. 3).

For the purpose of discussion of the correspondence of the component patterns of different RNA polymerase types, we have proposed a nomenclature, in which the heaviest and the second components of the nuclear yeast RNA polymerases I and II are called A and B (Schnabel et al., 1983). In the RNA polymerases of all sulfur dependent Archaebacteria and of Thermoplasma acidophilum, the corresponding components appear in reversed succession, i.e. B with the largest apparent size followed by A. In contrast, in the enzymes of the methanogenic and extremely halophilic Archaebacteria, the heaviest component corresponds again to the heaviest of yeast and the second of Sulfolobus and has therefore to be termed A. The second and the third components of these enzymes both react with Sulfolobus B antibody but do not crossreact with each other. Thus they each correspond to a different portion of the eukaryotic or sulfur dependent component B. To account for their equivalence, we propose to

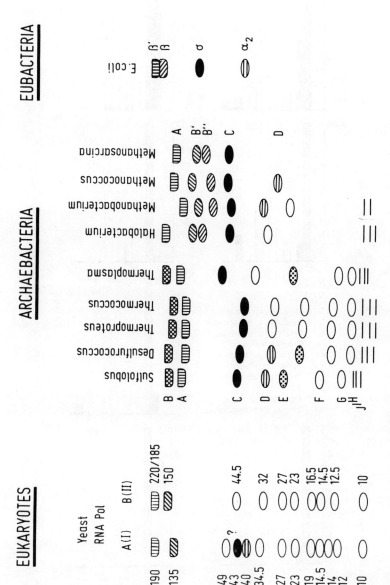

Fig. 2 Homologies between components in schematically drawn SDS polyacrylamide gel electrophoresis patterns as indicated by equal designs of corresponding bands.

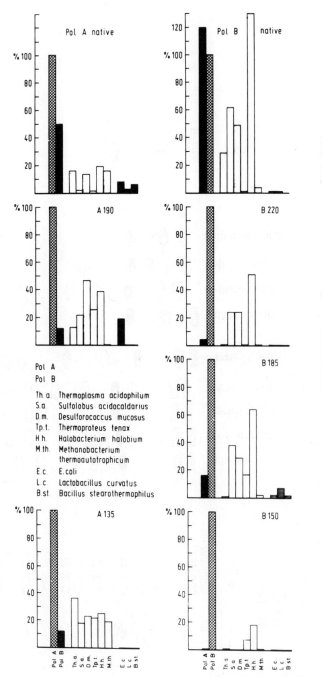

Fig. 3 Quantification of immunological crossreaction of different DNA dependent RNA polymerases with antibodies raised against native polymerase I (Pol A) and II (Pol B) of yeast and against the large components of these enzymes.

substitute their previous designations B and B' by B' and B'', respectively, in the order of decreasing apparent size. The large B subunit of the sulfur dependent Archaebacteria and Eucyta is either a fusion product of B' and B'', or B' and B'' are splitting products of B, depending on the direction of evolution, whether on the levels of transcription or translation, or posttranslationally, remains to be determined.

Antibodies against the B components of yeast RNA polymerases I and II recognize only the B', not the B'' subunit of H. halobium polymerase. The antibody against component B' of the RNA polymerase of Methanobacterium recognizes the B components of sulfur dependent Archaebacteria and the B' components of polymerases from methanogens and halophiles exclusively.

The antibody against component B'' of Methanobacterium recognizes the B components from sulfur dependent Archaebacteria and the B'', but not the B' components from Methanobacterium and Methanococcus. In the pattern from Methanolobus volcani, a representative of the Methanomicrobiales, it reacts strongly with B'', yet significantly though weakly also with component B'. In the pattern from H. halobium, it reacts almost as strongly with B' as with B'' (Fig. 4). Thus, in the Methanococcales, the split separates the antigenic sites in the same way as in the Methanobacteriales. In the Methanomicrobiales, it is in a slightly different position, in the Halobacteriales in a quite different position. The translocation of the B → B'+B'' split in the Methanomicrobiales might be the reason for the smaller size difference of B' and B'' as compared to that seen in other orders of this branch. In conclusion, the B' and the B'' components of the Methanococcales and Methanobacteriales represent different portions of the large B component from those of the Methanomicrobiales and the Halobacteriales.

The third components in size, C, of the enzymes of sulfur dependent Archaebacteria and Thermoplasma are homologous to the fourth components of methanogenic and extremely halophilic polymerases, which thus also have to be termed C. In the same manner, the homology has also been proved for the fourth component of the sulfur dependent Archaebacteria, D, and, in several cases, the fifth component, E. For smaller components the limitations of the method do not allow recognition of correspondence.

Though the crossreactions between components of eukaryotic and archaebacterial RNA polymerases on the one and eubacterial RNA polymerases on the other hand are weak, they suffice to elucidate interkingdom homologies: E. coli β' corresponds to component A of yeast and Archaebacteria. E. coli β antibody crossreacts with component B of yeast and Sulfolobus, exclusively with component B'' of Methanobacterium and Methanococcus, mainly with component B'' but weakly with B' of Methanolobus (Methanomicrobiales), and almost as strongly with component B' as with B'' of Halobacterium (fig. 4). In spite of the generally lower reactivity, it is particularly striking that the ratio of the responses of the B' and B'' components of various representatives of the methanogenic and extremely halophilic Archaebacteria to E. coli β antibody is very similar to the ratio of the responses of the same components to Methanobacterium B'' antibody, indicating correspondence specifically between Methanobacterium (and Methanococcus) B'' and E. coli β.

Antibody against component C from Sulfolobus reacts with either component A 43 or A 49 of yeast RNA polymerase I. A more precise assignment would require better resolution. In the E. coli polymerase pattern, it reacts strongly with the σ factor (Fig. 4). The other way round, E. coli σ -antibody reacts weakly but also significantly with the C components of both Methanolobus and Sulfolobus, i.e. of members of both branches of the archaebacterial kingdom (Fig. 4). Thus, the eubacterial σ corresponds to the archaebacterial C subunit. In yeast RNA polymerase I, C is probably component A43 or component A49.

E. coli α antibody reacts weakly with component A40. Since Sulfolobus D antibody also reacts with A40, it is highly probable that α corresponds to A40 and to the archaebacterial D component, though a direct crossreaction of α and D has not been observed.

Component E of Sulfolobus seems to correspond to yeast A27. A crossreaction with an E. coli subunit has not been observed. E is apparently involved in the interaction of the archaebacterial enzymes with the DNA (Schnabel et al., 1982). For smaller components, interkingdom homologies could not be established by immunochemical crossreaction.

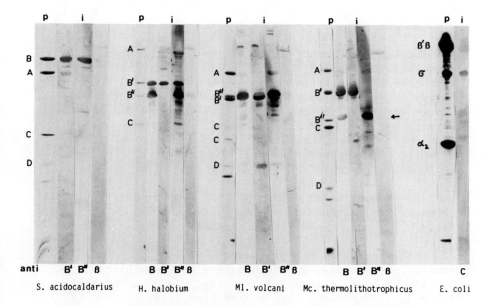

Fig. 4 Immunological crossreactions (Western blots) of components of archaebacterial and eubacterial RNA polymerases with antibodies raised against components B and C of S. acidocaldarius RNA polymerase, B' and B'' of Methanobacterium thermoautotrophicum RNA polymerase and β of E. coli RNA polymerase.

RNA polymerases, separated on SDS polyacrylamide gels and transferred to nitrocellulose sheets by electroblotting procedure, were challenged with antibodies.

Tracks labelled p show the component pattern visualized by Coomassie blue staining. Those labelled i show bound antibodies visualized with peroxidase-coupled antibodies.

3. Conclusions

(1) The large components A, B, C and D of the three types of eukaryotic nuclear RNA polymerases closely correspond to each other, to the components A, B, C and D of the polymerases of the sulfur dependent Archaebacteria and of Thermoplasma, to the components A, B', B'', C and D of the enzymes of the methanogenic and extremely halophilic Archaebacteria and, clearly though less closely to the subunits β', β, σ and α of the eubacterial RNA polymerases (Fig. 2).

This is convincing evidence for a common origin of all "normal" RNA polymerases and suggests a certain branching pattern in early evolution which shall be discussed below.

(2) Archaebacterial RNA polymerases differ from eukaryotic nuclear polymerases: The methanogens and halophiles by the B → B' + B'' division and the sulfur dependent Archaebacteria and Thermoplasma by a reversed size order of the components A and B. Both are insensitive to α amanitin. Archaebacterial differ from eubacterial RNA polymerases, as well in the considerably greater complexity of their component patterns, as in their insensitivity to rifampicin and streptolydigin.

(3) The more striking observation, however, is that they do not represent a uniform type, but rather clearly distinct types. The BACD... type realized in all RNA polymerases from sulfur dependent Archaebacteria and Thermoplasma differs from the ABCD... type represented by the three eukaryotic RNA polymerases only in the reversed size order of components A and B.

The AB'B''CD... types of the methanogenic and extremely halophilic Archaebacteria have apparently resulted from transcriptional, translational or posttranslational splits of the large component B of the sulfur dependent Archaebacteria and Thermoplasma.

In Methanolobus, representing the Methanomicrobiales, and more significantly in Halobacterium, this split leaves antigenic determinants linked in the respective B''s, which are separated in Methanobacterium B''. The split should therefore either have occurred independently (at least in its final consequence) several times in evolution, or it was specifically translocated by recombination processes after the separation of the respective lineages. In the case of the opposite direction of evolution, these conclusions would have to be reversed: the B B'+B'' split would have been a B'+B'' → B fusion. Recombination as a cause for the differences between the B's and B''s each of the Methanobacteriales and Methanococcales, the Methanomicrobiales and the Halobacteriales would in this case have been a necessity. This is a strong argument against this sequence of events.

(4) The homologies between corresponding components of Eucyta and Archaebacteria (of both branches) are quantitatively much stronger than between the latter and eubacterial components (Fig. 4). They appear even stronger than between different types of eukaryotic nuclear RNA polymerase components: evidence for the large phylogenetic distance, which separates the present eukaryotes from their one-polymerase-ancestors or, the other way round, for the primitive nature of the present Archaebacteria.

(5) The failure of the E. coli β antibody to react with the B'-components of Methanobacterium and Methanococcus and the striking resemblance of the reactivity ratios of the E. coli β and Methanobacterium B'' antibodies with the Methanolobus and Halobacterium B' and B'' components respectively suggests that E. coli β corresponds to B'' of Methanobacterium. Thus E. coli polymerase which exhibits the same B → B'+B'' split as Methanobacterium (at least as revealed by immunochemical homology) could have arisen from the same ancestor by loss of the B' component and additional streamlining (Fig. 5).

(6) The B'' component of Methanobacterium, which corresponds to only part of the B'' of Halobacterium, crossreacts with E. coli β, but not with the B component of yeast RNA polymerases I and II. In contrast, the B' component of Halobacterium, which corresponds to only part of B' of Methanobacterium, crossreacts with the B component of the yeast polymerases, but not with E. coli β. The B component of sulfur dependent Archaebacteria crossreacts with both Methanobacterium B'' and Halobacterium B'. This situation is visualized schematically in Fig. 5. Eubacteria and Eucyta have conserved different antigenic determinants of the B component of the sulfur dependent Archaebacteria, besides a core of common determinants. This opposite trimming is a strong argument for the ancestral nature of the sulfur dependent B and more generally of the Archaebacteria.

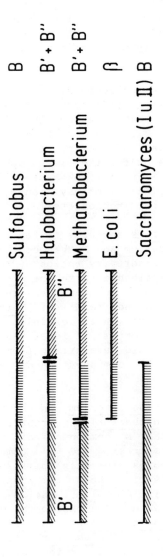

Fig. 5 Schematic representation of correspondence of E. coli β and Saccharomyces POL I B with B components of Halobacterium, Methanobacterium and Sulfolobus. Vertical lines connect corresponding structural elements i.e. antigenic sites. Horizontal lines do not give size of peptides but alignments of determinants.

(7) These data suggest that the original (ancestral) polymerase was that conserved in the sulfur dependent Archaebacteria and Thermoplasma (see Fig. 6).

The BACD type polymerase would then have given rise to the urkaryotic ABCD enzyme and to the (different!) AB'B''CD polymerases of the Halobacteriales and the methanogens, the first directly, the latter by a number of successive independent B → B'+B'' splits. Alternatively, different lineages could have separated after one original B → B'+B'' split, followed by its lineage specific recombinative translocation as discussed above. In any case, the eubacterial RNA polymerases appear derived from the same ancestral type as those of the Methanobacteriales and the Methanococcales, which are particularly close to each other, whereas the Methanomicrobiales are signifanctly more distant (Fig. 6). The big difference in the B split in H. halobium could be a consequence of the extreme genome instability of this organism but separates the Halobacteriales clearly from the other orders of the branch.

The events leading from an RNA polymerase of the Methanobacterium design to a typically eubacterial enzyme then were the loss of B' and additional streamlining by the loss of all (or most) components smaller than D (which corresponds to α).

It is not impossible though less probable that the eukaryotic polymerases were derived from the eubacterial enzymes by the opposite order of events: first acquisition of B' and of additional small components, then fusion of B'' and B'. A final decision between these two extreme trees has not been reached. The sequence of events, in one or the other direction, appears, however, logical and even cogent (Fig. 6).

(8) The RNA polymerases of Archaebacteria, especially of the sulfur dependent branch, appear strikingly "eukaryotic" like a number of other feature designs in Archaebacteria, e.g. the ADP ribosylation of EF2 of both branches by diphtheria toxin (Kessel and Klink, 1982), the existence, again in both branches, of replicating DNA polymerases sensitive to aphidicolin (Forterre et al., 1984; M. Nakayama and M. Kohiyama, personal communication; R. Schinzel and K.J. Burger, personal communication; H.P. Zabel, J. Winter, H. Fischer, E. Holler, personal communication; D. Prangishvilli and Zillig, 1984), the sequences of ribosomal "A-proteins"

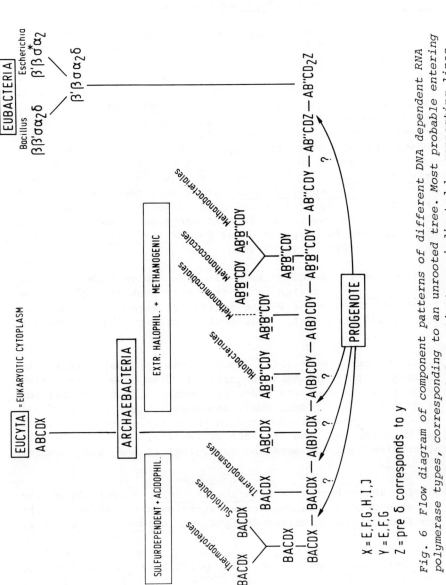

Fig. 6 Flow diagram of component patterns of different DNA dependent RNA polymerase types, corresponding to an unrooted tree. Most probable entering points for conversion to phylogenetic trees indicated by connecting lines from "progenote".

(Matheson and Yaguchi, 1982), the existence of introns in tRNA structural genes (Kaine et al., 1982; Daniels et al., 1984), the absence of formylation of methionyl initiator tRNA, the absence of the 3' terminal CCA sequence in the transcript and certain tRNA modifications (Gupta and Woese, 1980) (Table 2).

In a number of other feature designs, the sulfur dependent Archaebacteria appear closer to eukaryotes than the methanogens and extreme halophiles do (and, vice versa, the methanogens and halophiles closer to the Eubacteria). These comprise the shape of the large ribosomal subunit (Henderson et al., 1984; Lake et al., 1984), the 5S rRNA secondary structure as revealed by Fox et al. (1982), initiator tRNA structure (Kuchino et al., 1982), the complexity of ribosomal protein patterns (Schmid et al., 1982), the extent of rRNA modification (Woese et al., 1984) and cell division mechanisms (no septum formation). RNA polymerase composition appears a particularly valid feature of this type (Table 2).

However, Archaebacteria also have common designs considered typical for Eubacteria: they contain Shine Dalgarno sequences in mRNA and their complements in 16S rRNA, and restriction enzymes. The total sequences of 16S rRNA (Gupta et al., 1983; Leffers and Garrett, 1984; M. Jarsch and A. Böck, total sequence of 16S rRNA gene of Methanococcus vannielii, personal communication) also of the sulfur dependent Sulfolobus (R. Gupta and C. Woese, personal communication) appear more related to that of E. coli than to that of the eukaryote Dictyostelium. Archaebacteria are prokaryotes with all consequences. But this term refers rather to organisation than to phylogenetic status (Table 2).

(9) In summary, Archaebacteria appear as residues of an ancient primitive layer. One of its two surviving branches, comprising the sulfur dependent Archaebacteria, appears close to the ancestral Eucyta, the urkaryotes of Carl Woese, the other, methanogens and extreme halophiles, to the origin of the Eubacteria. An incomplete account of the basis of these views has been given previously (Zillig et al., in the press).

(10) Certain observations seem to contradict this picture, e.g. the exclusive existence of ether lipids in all known Archaebacteria, but of ester lipids in both modern kingdoms

Designs of Features of Archaebacteria

1. Eukaryotic, general

2. Eubacterial, general

3. More eukaryotic in sulfur dependent than in halophilic and methanogenic archaebacteria (and vice versa)

a) ADP ribosylation of EF2 by diphtheria toxin

b) replicating DNA polymerases aphidicolin sensitive

c) ribosomal A protein sequence

d) occurrence of introns

e) some tRNA modifications

f) no formylation of methionyl initiator tRNA

g) ACC 3' terminus of tRNA not encoded

a) Shine Dalgarno sequences

b) occurrence of restriction enzymes

c) 16S rRNA total sequences

(d) prokaryotic organisation)

a) RNA polymerase structure

b) shape of large ribosomal subunit

c) 5S rRNA secondary structure

d) initiator tRNA structure

e) complexity of ribosomal protein pattern

f) cell division mechanisms

(with one abnormal exception). Another conflicting observation is the complete lack of any sensitivity to ribosome specific antibiotics in sulfur dependent Archaebacteria, whereas methanogens share the sensitivity to a few inhibitors with eukaryotes, that to several others with Eubacteria, and Eubacteria and eukaryotes share responses not exhibited by Archaebacteria (Böck and Kandler, in the press).

Such inconsistencies and the non-congruence of the evolution rates of different genes prohibit the understanding of early divergence of and in the three kingdoms until further comparative studies yield complete agreement of all facts.

References

(1) Achberger, E.C. and Whitely, H.R. (1981). The role of the Delta peptide of the Bacillus subtilis RNA polymerase in promoter selection. J. Biol. Chem. 256, 7424-7432.

(2) Böck, A. and Kandler, O. Antibiotic Sensitivity of Archaebacteria. In "The Bacteria", Vol. 8, in the press.

(3) Böck, A., Jarsch, M., Hummel, H. and Schmid, G. (1984). Evolution of Translation. This volume.

(4) Buhler, J.M. Huet, J., Davies, K.E., Sentenac, A. and Fromageot, P. (1980). Immunological studies of yeast nuclear RNA polymerases at the subunit level. J. Biochem. 225, 9949-9954.

(5) Burgess, R.R. (1976). Purification and Physical Properties of E. coli RNA Polymerase. In "RNA Polymerase". Losick, R., and Chamberlin, M., eds., pp. 69-100. Cold Spring Harbor Laboratory, Cold Spring Harbor, New York.

(6) Daniels, C.J., Gupta, R. and Doolittle, W.F. (1984). Transcription and excision of a large intron in the tRNA^{Trp} gene of an archaebacterium, Halobacterium volcanii. Submitted to J. Biol. Chem.

(7) DasSarma, S., RajBhandary, U.L. and Khorana, H.G. (1984). Bacterio-opsin mRNA in wild-type and bacterio-opsin deficient Halobacterium halobium strains. Proc. Natl. Acad. Sci. USA 81, 125-129.

(8) Dickel,C.D., Burtis, K.C. and Doi, R.H. (1980). Delta factor increases promoter selectivity by Bacillus subtilis vegetative cell RNA polymerase. Biochem. Biophys. Res. Commun. 95, 1789-1795.

(9) Forterre, P., Elie, C. and Kohiyama, M. (1984). Aphidicolin inhibits growth and DNA synthesis in halophilic archaebacteria. J. Bact. 159, 800-802.

(10) Fox, G.E., Pechmann, K.J., Woese, C.R. (1977). Comparative cataloging of 16S ribosomal RNA: molecular approach to procaryotic systematics. Int. J. System. Bact. 27, 44-57.

(11) Fox, G.E., Stackebrandt, E., Hespell, R.B., Gibson, J., Maniloff, J., Dyer, T.A., Wolfe, R.S., Balch, W.E., Tanner, R.S., Magrum, L.J., Zablen, L.B., Blakemore, R., Gupta, R., Bonen, L., Lewis, B.J., Stahl, D.A., Luehrsen, K.R., Chen, K.N., Woese, C.R. (1980). The phylogeny of prokaryotes. Science 209, 457-463.

(12) Fox, G.E., Luehrsen, K.R. and Woese, C.R. (1982). Archaebacterial 5S ribosomal RNA. Zbl. Bact. Hyg., Orig. C 3, 330-345.

(13) Gierl, A., Zillig, W. and Stetter, K.O. (1982). The role of the components σ and y of the DNA-dependent RNA polymerase of Lactobacillus curvatus in promotor selection. Eur. J. Biochem. 125, 41-47.

(14) Gupta, R. and Woese, C.R. (1980). Unusual modification patterns in the transfer ribonucleic acids of archaebacteria. Current Microbiology 4, 245-249.

(15) Gupta, R., Lanter, J.M. and Woese, C.R. (1983). Sequence of the 16S ribosomal RNA from Halobacterium volcanii, an archaebacterium. Science 221, 656-659.

(16) Gupta, R. (1984). Halobacterium volcanii tRNAs: Identification of 41 tRNAs covering all amino acids, and the sequences of 33 class I tRNAs. J. Biol. Chem., in the press.

(17) Henderson, E., Oakes, M., Clark M.W., Lake, J.A., Matheson, A.T. and Zillig, W. (1984). A new ribosome structure. Science, 225, 510-512.

(18) Herzfeld, F. and Kiper, M. (1976). The reconstitution of Anacystis nidulans DNA-dependent RNA polymerase from its isolated subunits. Eur. J. Biochem. 62, 198.

(19) Hori, T., Itoh, T. and Osawa, S. (1982). The phylogenic structure of the metabacteria. Zbl. Bact. Hyg., I. Abt. Orig. C 3, 18-30.

(20) Huet, J., Sentenac, A. and Fromageot, P. (1982). Spot-immunodetection of conserved determinants in eukaryotic RNA polymerases. J. Biol. Chem. 257, 2613-2618.

(21) Huet, J., Schnabel, R., Sentenac, A. and Zillig, W. (1983). Archaebacteria and eukaryotes possess DNA-dependent RNA polymerases of a common type. The EMBO J. 2, 1291-1294.

(22) Jarsch, M. and Böck, A. (1984). Sequence of the 16S ribosomal RNA gene from Methanococcus vannielii. Submitted to The EMBO J.

(23) Kaine, B.P., Gupta, R. and Woese, C.R. (1983). Putative introns in tRNA genes of prokaryotes. Proc. Natl. Acad. Sci. USA 80, 3309-3312.

(24) Kandler, O. (1982). Cell wall structures and their phylogenetic implications. Zbl. Bakt. Hyg. I. Abt. Orig. C 3, 149-160.

(25) Kessel, M. and Klink, F. (1982). Identification and comparison of eighteen archaebacteria by means of the diphtheria toxin reaction. Zbl. Bakt. Hyg., I. Abt. Orig. C 3, 140-148.

(26) Konheiser, U., Pasti, G., Bollschweiler, C. and Klein, H. (1984). Physical mapping of genes coding for few subunits of methyl CoM reductase component C of Methanococcus voltae. Mol. Gen. Genet., in the press.

(27) Kuchino, Y., Hiara, M., Yabusaki, Y. and Nishimura, S. (1982). Initiator tRNAs from archaebacteria show common unique sequence characteristics. Nature 298, 684-685.

(28) Lake, J.A., Henderson, E., Oakes, M. and Clark, M.W. (1984). Eocytes: A new ribosome structure indicates a kingdom with a close relationship to eukaryotes. Proc. Natl. Acad. Sci. USA 81, 3786-3790.

(29) Langworthy, T.A., Tornabene, T.G. and Holzer, G. (1982). Lipids of archaebacteria. Zbl. Bakt. Hyg., I. Abt. Orig. C 3, 228-244.

(30) Leffers, H. and Garrett R.A. (1984). The nucleotide sequence of the 16S ribosomal RNA gene of the archaebacterium Halococcus morrhuae. The EMBO J. 3, 1613-1619.

(31) Madon, J., Leser, U. and Zillig, W. (1983). DNA-dependent RNA polymerase from the extremely halophilic archaebacterium Halococcus morrhuae. Eur. J. Biochem. 135, 279-283.

(32) Madon, J. and Zillig, W. (1983). A form of the DNA-dependent RNA polymerase of Halobacterium halobium, containing an additional component, is able to transcribe native DNA. Eur. J. Biochem. 133, 471-474 (1983).

(33) Matheson, A.T. and Yaguchi, M. (1982). The evolution of the archaebacterial ribosome. Zbl. Bakt. Hyg., I. Abt. Orig. C 3, 192-199.

(34) Prager, E.M., Wilson, A.C., Lowenstein, J.M. and Sarich, V.M. (1980). Mammoth albumin. Science 209, 287-289.

(35) Prangishvilli, D., Zillig, W., Gierl, A., Biesert, L. and Holz, I. (1982). DNA-dependent RNA polymerases of thermoacidophilic archaebacteria. Eur. J. Biochem. 122, 471-477.

(36) Prangishvilli, D. and Zillig, W. (1984). DNA-Dependent DNA-Polymerases of Thermoacidophilic Archaebacterium Sulfolobus acidocaldarius. Poster presented at the FEMS Symposium "Evolution of Prokaryotes", held in Munich, September 16-18, 1984.

(37) Rexer, B. (1976). Struktur und Aktivität der DNA-abhängigen RNA-Polymerase aus vegetativen und sporulierenden Zellen von B. cereus T. Thesis, Ludwig-Maximilians-Universität München.

(38) Roeder, R.G. (1976). Eukaryotic nuclear RNA polymerases. In "RNA Polymerase". Losick, R. and Chamberlin, M., eds. pp. 285-329, Cold Spring Harbor Laboratory, Cold Spring Harbor, New York.

(39) Schmid, G., Pecher, Th. and Böck, A. (1982). Properties of the translational apparatus of the archaebacteria. Zbl. Bakt. Hyg., I. Abt. Orig. C 3, 209-217.

(40) Schnabel, R., Zillig, W. and Schnabel, H. (1982). Component E of the DNA-dependent RNA polymerase of the archaebacterium Thermoplasma acidophilum is required for the transcription of native DNA. Eur. J. Biochem. 129, 473-477.

(41) Schnabel, R., Thomm, M., Gerardy-Schahn, R., Zillig, W., Stetter, K.O. and Huet, J. (1983). Structural homology between different archaebacterial DNA-dependent RNA polymerases analyzed by immunological comparison of their components. The EMBO J. 2, 751-755.

(42) Stetter, K.O. and Zillig, W. (1974). Transcription in Lactobacillaceae. DNA-dependent RNA polymerase from Lactobacillus curvatus. Eur. J. Biochem. 48, 527-540.

(43) Stetter, K.O., Winter, J. and Hartlieb, R. (1980). DNA-dependent RNA polymerase of the archaebacterium Methanobacterium thermoautotrophicum. Zbl. Bakt. Hyg., I. Abt. Orig. C 1, 201-218.

(44) Sturm, S., Schönefeld, V., Zillig. W., Janekovic, D., and Stetter, K.O. (1980). Structure and function of the DNA-dependent RNA polymerase of the archaebacterium Thermoplasma acidophilum. Zbl. Bakt. Hyg., I. Abt. Orig. C 1, 12-25.

(45) Thomm, M., Altenbuchner, J. and Stetter, K.O. (1983). Evidence for a plasmid in a methanogenic bacterium. J. Bacteriol. 153, 1060-1062.

(46) Woese, C.R. and Fox, G.E. (1977). The concept of cellular evolution. J. Mol. Evol. 10, 1-6.

(47) Woese, C.R., Magrum, L.J. and Fox, G.E. (1978). Archaebacteria. J. Mol. Evol. 11, 245-252.

(48) Woese, C.R., Gupta, R., Hahn, C.M., Zillig, W. and Tu, J. (1984). The phylogenetic relationships of three

sulfur dependent archaebacteria. System. Appl. Microbiol. 5, 97-105.

(49) Yaguchi, M., Visentin, L.P., Zukur, M., Matheson, A.T., Roy, C. and Strom, A.R. (1982). Amino-terminal sequences of ribosomal proteins from the 30S subunit of archaebacterium Halobacterium cutirubrum. Zbl. Bakt. Hyg., I. Abt. Orig. C 3, 200-208.

(50) Zillig, W., Palm, P. and Heil, A. (1976). Function and Reassembly of Subunits of DNA-Dependent RNA Polymerase. In "RNA Polymerase". R. Losick and M. Chamberlin, eds., pp. 101-125. Cold Spring Harbor Laboratory, Cold Spring Harbor, New York.

(51) Zillig, W., Stetter, K.O. and Tobien, M. (1978). DNA-dependent RNA polymerase from Halobacterium halobium. Eur. J. Biochem. 91, 193-199.

(52) Zillig, W., Stetter, K.O., Wunderl, S., Schulz, W., Priess, H. and Scholz, I. (1980). The Sulfolobus-"Caldariella" group: Taxonomy on the basis of the structure of DNA-dependent polymerases. Arch. Microbiol. 125, 259-269.

(53) Zillig, W., Stetter, K.O., Schnabel, R., Madon, J. and Gierl, A. (1982) Transcription in archaebacteria. Zbl. Bakt. Hyg., I. Abt. Orig. C 3, 218-227.

(54) Zillig, W., Gierl, A., Schreiber, G., Wunderl, S., Janekovic, D., Stetter, K.O. and Klenk, H.P. (1983a). The archaebacterium Thermofilum pendens represents a novel genus of the thermophilic, anaerobic sulfur respiring Thermoproteales. System. Appl. Microbiol. 4, 79-87.

(55) Zillig, W., Holz, I., Janekovic, D., Schäfer, W. and Reiter, W.D. (1983b). The archaebacterium Thermococcus celer represents a novel genus within the thermophilic branch of the archaebacteria. System. Appl. Microbiol. 4, 88-94.

(56) Zillig, W., Schnabel, R. and Stetter, K.O. (1984). Archaebacteria and the Origin of the Eukaryotic Cytoplasm. In "Current Topics of Microbiology and Immunology", in the press.

EVOLUTION OF TRANSLATION

A. Böck, M. Jarsch, H. Hummel and G. Schmid

Lehrstuhl für Mikrobiologie
Universität München
Munich, Federal Republic of Germany

INTRODUCTION

The concept of a second bacterial kingdom, that of the archaebacteria in addition to that of the eubacteria (1) proves to be of tremendous heuristic value for the understanding of the evolution of metabolism, of cellular components and of organisms themselves. The previous view of the dichotomous division of cell organization into the prokaryotic structure as opposed to the eukaryotic one has been supplemented by a third type, that of the archaebacteria; although basically prokaryotic in its organization it exhibits properties which are either unique or which were hitherto accepted as being characteristic solely for eukaryotic cells (1).

The translational apparatus is one of the distinguishing features between the prokaryotic and eukaryotic cell type. Recent analysis of the structure and function of the archaebacterial protein synthesis machinery revealed the existence of an intermediate type, concerning ribosome morphology (2), primary sequences of macromolecular components (3-5) and action of inhibitors (6, 7). Its biochemical and biological analysis, whose present state this review covers, promises a wealth of information on the path of evolution of the 70S eubacterial and the 80S eukaryotic protein synthesis systems.

PHYSICAL PROPERTIES OF ARCHAEBACTERIAL RIBOSOMES

At appropriate ionic conditions ribosomes from methanogenic bacteria sediment at 70S in sucrose gradients (1, 8). They

73

dissociate into subunits of 30S and 50S upon lowering the Mg^{++} concentration and increasing the concentration of monovalent cations (9).

Ribosomes from several sulfur-dependent archaebacteria (e. g. *Sulfolobus*) at any ionic condition tested (i.e. also at high Mg^{++}) can only be isolated in the dissociated state (10, 11). Their sedimentation positions in sucrose gradients match those of eubacterial 30S and 50S subunits (10, 11).

The relative ratio of RNA and protein has until now only been determined for ribosomal subunits from *Sulfolobus solfataricus* (11). Isodensity-equilibrium centrifugation in CsCl revealed an appreciably higher protein content for both the 30S (52 %) and 50S (37 %) subunit in comparison to those of the *E. coli* ribosome (38 % and 30 %, respectively) (11).

In the electron microscope characteristic structural features are displayed: 30S subunits from archaebacteria share with 40S subunits from eukaryotic cytoplasmic ribosomes the existence of an asymmetric protuberance, the "archaebacterial bill" (12). The structure of the 50S subunits as revealed by the electron microscope, on the other hand, is heterogeneous. In the sulfur-dependent archaebacteria it is characterized by two separated bulged extensions at the opposite site of the stalk (2). In large ribosomal subunits from an extreme halophilic organism these bulges were observed to be either much less extensive or completely lacking (2). Based solely on these morphological features a classification of the sulfur--dependent archaebacteria separate from the methanogen/halophile branch into a fourth primary kingdom was proposed (2). Recent analysis of the morphology of ribosomal subunits from a methanogenic organism, *Methanococcus vannielii*, however, demonstrated that there is a considerable overlap of these morphological features between the two major groups of archaebacteria (55). *M. vannielii* has 50S subunits with separated bulges at least as extensive in their size than those from *Sulfolobus* 50S subunits. Ribosome morphology, therefore, does not appear to be a concise differentiating criterion.

THE RIBOSOMAL RNA MOIETY

Archaebacteria possess the rRNA complement typical for eubacteria: 5S, 16S and 23S rRNA (1, 4). Since the evolutionary implications of primary and secondary structural features of 5S and 16S rRNA have been covered already in detail in recent reviews (3, 4, 13) we shall concentrate here on two aspects: (i) the genomic organization of genes for rRNA in archaebacteria, and (ii) the discussion of recently determined sequences.

rDNA Gene Organization

The organization of the genes for rRNA in archaebacteria is very heterogeneous. In organisms belonging to the methanogen/halophile branch genes for 16S, 23S, 5S rRNA are closely linked (14-16). The number of these apparent rRNA transcriptional units ranges from one per genome for *H. halobium* (14, 16) to four for *Methanococcus vannielii* (15). In the sulfur-dependent archaebacteria the organization of rDNA cistrons is more disperse: The genes for 16S and 23S are normally linked but with spacers of greatly varying lengths in between. The genes for 5S rRNA either are situated in the immediate 3' neighborship of the 23S rRNA cistron or they are separated from it by spacers of several kilobases length (16, 17).

In eubacteria the intercistronic spacer between the 16S and 23S rRNA genes may contain genes for tRNA. This is also the case for the archaebacteria analyzed in detail: the spacer region of *M. vannielii* (18), of *H. volcanii* (C.R. Woese, personal communication) and *H. cutirubrum* (19) codes for a tRNA^{Ala} gene. It is interesting to note that in eubacterial and plastidal intercistronic spacers the tRNA^{Ala} gene is always encoded together with a gene for tRNA^{Ile}. This suggests that the gene for tRNA^{Ala} was assembled first into the rRNA transcriptional unit and that the tRNA^{Ile} was acquired afterwards by some ancestor of eubacteria and chloroplasts.

Eukaryotic cytoplasmic ribosomes contain a fourth additional RNA molecule in stoichiometric amounts, 5.8S rRNA. On the basis of extensive sequence homology of 5.8S rRNA with the 5'-terminal fragment of prokaryotic 23S rRNA it was suggested that eukaryotes contain a processing mechanism which cleaves the prokaryotic 23S sequence into the eukaryotic equivalents (20). In archaebacteria, this sequence homology between 23S rRNA and eukaryotic 5.8S rRNA also exists. This may indicate that during evolution eukaryotic cells have acquired this processing step rather than that prokaryotes have lost it.

A striking feature of several archaebacteria, e.g. *Sulfolobus*, *Thermococcus* and *Methanococcus*, is the existence of additional genes for 5S rRNA unlinked to the presumptive 16S-23S-5S transcriptional units (15, 16). Since unlinked 5S rRNA genes are the normal situation for eukaryotic cells the single 5S gene from *Methanococcus* was analyzed in more detail. DNA sequence analysis revealed that there is a presumptive transcriptional unit for this single 5S rRNA gene and for seven tRNA genes, namely 5'-tRNA^{Thr}-tRNA^{Pro}-tRNA^{Tyr}-tRNA^{Lys} - 5S rRNA-tRNA^{Asp}-tRNA^{Lys}-tRNA^{Asp} (21). It demonstrates that linkage of genes coding for products with similar physiological function which is a wide-spread phenomenon in eubacteria (cf. 22) also exists in the genome of archaebacteria.

The sequence of the extra 5S gene from *Methanococcus* differs in as much as 13 positions from that encoded by one of the 16S/23S linked 5S genes (Fig. 1) of the same organism.

```
         10        20        30        40        50        60        70        80
TGACATAACGGTCAAAGCGGAGGTGTAACATCCGATCCCATCCCGATCTCGGAAATTAAGCCCTCCAGCGATTCTTTAAG
*** *  ******  *****  ***********************************************************  *****
TGATACGGCGGTCATAGCGGGGGTGTAACATCCGATCCCATCCCGATCTCGGAAATTAAGCCCTCCAGCGATTCCTTAAG
         10        20        30        40        50        60        70        80

         90       100       110       120
TACTGCTATCTAGTGGGAACAAAGTGACGCCGTTAGTCAC
************************ *******  *  ****
TACTGCTATCTAGTGGGAACAAGGTGACGCTGCCGATCAC
         90       100       110       120
```

Fig. 1 Sequence of the "extra" 5S rRNA gene (lower sequence) compared to that of a gene (upper sequence) linked to one of the rRNA transcriptional units.

Since most of the divergence is caused by compensatory base exchanges one has to conclude that the RNA transcribed from it has a biological function. In any case, however, it is clear that the extra 5S gene and the operon encoded gene evolved with greatly differing rates.

16S rRNA Sequences

Total sequences of 16S rRNA have been determined now for three extreme halophiles, *Halobacterium volcanii* (23), *Halococcus morrhuae* (24), *Halobacterium cutirubrum* (19) and for one methanogen, *Methanococcus vannielii* (25). Table 1 shows the similarity matrix of two archaebacterial sequences (*H. volcanii* and *M. vannielii*) and the *E. coli* and *Dictyostelium discoideum* (26) sequences.

TABLE 1

Primary structure homologies between 16S-like rRNA sequences from M. vannielii, H. volcanii, E. coli and D. discoideum

	1	2	3
1. *M. vannielii*	–		
2. *H. volcanii*	75 %	–	
3. *E. coli*	61 %	59 %	–
4. *D. discoideum*	55 %	54 %	49 %

The homology values indicate a specific relationship between the methanogen and the halophile. Moreover, both archaebacterial sequences show a distinctly higher homology to the eubacterial sequence than to the eukaryotic one suggesting a common ancestor of eubacteria and the methanogen/halophile branch of archaebacteria. The similarity matrix also indicates a slightly higher homology of the *Methanococcus* sequence - in comparison to that of the halophile - with both the *E. coli* and the *D. discoideum* sequences. This points to a slower rate of genetic drift for the 16S gene from the methanogen.

When the 16S rRNA sequences from the three extreme halophiles were compared with that of the methanogen equal values of homology, between 75 and 76 %, were obtained. The three halophiles amongst themselves share homologies between 88 and 89 %. This points to (i) an equal phylogenetic distance of *H. volcanii*, *H. cutirubrum* and *H. morrhuae*, and (ii) the early branching of the line leading to *Methanococcus* from an ancestor common to the three halophiles. Determination of more sequences, especially from methanogens, will reveal whether it is possible to trace the descent of extreme halophiles to one special group of methanogens.

The secondary structure model of the 16S rRNA from the methanogen (25) formed following the rules of the comparative approach (23) is essentially identical to that delivered by the sequences of the three halophiles (19, 23, 24). It exhibits all the unique features discussed by Gupta *et al.* (23) on the basis of the model for *H. volcanii* 16S rRNA. This fact provides strong evidence that "adaptation" to high salinity has not exerted strong selective pressure on the 16S rRNA molecule from the halophiles. The peculiar structural features thus reflect the phylogenetic state rather than adaptation to extreme environments.

THE RIBOSOMAL PROTEIN COMPLEMENT

Protein Number

70S eubacterial ribosomes contain 52 proteins, eukaryotic 80S organelles between 70 and 85, depending on the organism (27). The protein complexity of archaebacterial ribosomes has been analyzed until now only by gel electrophoretic techniques; extensive protein-chemical data necessary to provide definite proof for the individuality of the proteins separated on two-dimensional electropherograms are only available for *H. cutirubrum* (28, 29).

The results obtained for the ribosomes of several organisms from the methanogen/halophile group of archaebacteria are summarized in Table 2.

TABLE 2

Numbers of ribosomal proteins from methanogens and halophiles resolved by gel electrophoresis

Order	Organism	Proteins present at subunits			Total	Ref.
		30S	50S	Both		
M.-bacteriales	Mb. thermoauto- trophicum	22	32	–	54	(9)
	Mb. bryantii	23	32	–	55	(9)
	Mbr. arboriphilus	22	31	1	53-54	(9)
M.-microbiales	Msp. hungatei	21	34	1	55-56	(9)
M.-coccales	Mc. vannielii	25	35	2	60-62	(9)
Halobacteriales	H. cutirubrum	21	32	–	53	(30)

Table 2 shows that the numbers of proteins separated by the techniques employed are in the range of those reported for typically eubacterial ribosomes (27). An exception appears to be *Mc. vannielii* with a distinctly increased number of 30S subunit proteins.

The protein composition of ribosomes from *Sulfolobus* has been determined independently by two groups (10, 11). Employing different electrophoretic techniques there was excellent agreement in the result that this thermophilic organism possesses a 30S subunit with a higher protein complexity than eubacteria (see Table 3). To analyse whether this is a general property of thermophilic sulfur-dependent archaebacteria the protein composition of ribosomes was also analyzed for other members of this group (31). Table 3 summarizes the results.

TABLE 3

Ribosomal proteins from sulfur-dependent archaebacteria resolved by gel electrophoresis

Order	Organism	Proteins present at subunit			Total	Ref.
		30S	50S	Both		
Sulfolobales	S. solfataricus	27	34	3	61-64	(10,11)
Thermoproteales	D. mobilis	24	34	-	58	(31)
	Th. tenax	24	32	-	56	(31)
	Tc. celer	24	32	-	56	(31)

Again, the 50S subunits possess the typical eubacterial number of proteins, the 30S subunit may be somewhat more complex.

It should be emphasized that it is difficult to define whether a protein is an integral part of the ribosome or not. To this end, control experiments have been carried out by washing *Sulfolobus* (11) and *Methanococcus* ribosomes (our unpublished results) successively with increasing concentrations of NH_4Cl; it was found that in both cases the ribosomal protein pattern stayed constant and the poly(U) dependent polyphenylalanine formation was unimpaired up to NH_4Cl concentrations of 1.5 M. Above that concentration a few proteins were lost and elongation activity decreased. Although this finding supports the notion that the proteins resolved in the electropherograms are integral parts of the ribosome more detailed proteinchemical and also functional information is required for the definite proof.

Immunological Comparison and Sequence Analysis

The immunological relatedness of ribosomal proteins from archaebacteria with those from eubacteria and eukaryotes has been analyzed employing a variety of techniques like immunodiffusion, quantitative immunoprecipitation and immunoblotting (32, 33). The results from these semiquantitative analyses may be summarized as follows: (i) Antisera directed against ribosomal proteins from methanogens (*Mc. vannielii* and *Mb. bryantii*) cross-reacted extensively with ribosomal proteins from halophiles and sulfur-dependent thermophiles. (ii) Cross-reaction between even taxonomically distant organisms of eubacteria was much more extensive than between eubacteria and either archaebacteria or yeast. (iii) Ribosomal proteins differ greatly in their evolutionary rates; determinants of a few proteins are conserved in ribosomes from eubacteria, eukaryotes and archaebacteria; protein L2 (*E. coli* nomenclature) is the most striking example (34). Other proteins are more variable.

Although only qualitative, this immunological comparison supports the concept of archaebacteria as a distinct group separate phylogenetically from those of eubacteria and eukaryotes. They also indicate that different structural and functional components of the ribosome have evolved with quite divergent rates. Constraints imposing evolutionary conservativity could be essential functional or structural roles e.g. binding to ribosomal RNA; the latter would force a co-evolution of the protein together with its binding site at the RNA. It is of interest in this connection that several ribosomal proteins from eubacteria were found to bind to rRNA from archaebacteria (35).

Sequences of ribosomal proteins from archaebacteria have not yet been determined in a number sufficient for phylogenetic conclusions. Most of the efforts concentrated on the alanine-rich, acidic A-protein, the equivalent of *E.coli* L7/L12. Complete and partial A-protein sequences are known for *H. cutirubrum* and *M. thermoautotrophicum*, respectively (36), as well as for seven eubacteria (cf. 27, 37) and four eukaryotes (cf. 27). Their comparison revealed: (i) an extensive homology of the two archaebacterial sequences, (ii) a much higher homology of the archaebacterial sequences with the eukaryotic than with the eubacterial type ones (5). In fact, from several primary structural characteristics it could be concluded that archaebacterial and eukaryotic A-proteins are derived from a common ancestor (5). It is important, however, to collect more comparative sequence data for this statement, especially in view of a recently reported A-protein sequence of *Rhodopseudomonas sphaeroides* which contains a tyrosine in its N-terminal fragment, a feature previously thought to solely be characteristic of eukaryotic and archaebacterial A-proteins (37).

Other proteins which have been studied in more detail are the 5S rRNA binding proteins from *H. cutirubrum*. The partial sequences determined point to some homology both with the *E. coli* and the yeast 5S binding proteins, with a tendency for a somewhat greater resemblance to the protein from latter organism (5).

Acidity of Ribosomal Proteins

Two-dimensional separations of ribosomal proteins from methanogens had revealed (8) that a higher percentage of proteins migrated to the acidic side of Kaltschmidt and Wittmann gels (38) compared to those from any eubacterium analyzed. This was reminiscent of the electrophoretic behaviour of halobacterial ribosomal proteins which almost exclusively are strongly acidic (30). It was noted (8) that the number of proteins of the ribosome from methanogens being "acidic" correlated with their relationship to the extreme halophiles, determined by 16S rRNA oligonucleotide analysis (39). Jarrell *et al.* (40) have extended this analysis to additional organisms and also determined the intracellular potassium concentration. A perfect correlation was obtained between the two parameters; organisms possessing the highest internal K^+ concentration (as high as 1.2 M for *Methanobrevibacter arboriphilus*) contained the highest number of acidic proteins and <u>vice versa</u>. This correlation induced the speculation that the methanogens are derived from a halophilic ancestor and that the gradient of internal K^+ concentration and of the acidity of the proteins reflects the evolutionary distance to this common ancestor (8, 40).

SENSITIVITY TO PROTEIN SYNTHESIS INHIBITORS

The determination of in vivo susceptibility of archaebacteria to protein synthesis inhibitors had revealed a quite unusual spectrum (41, 42). A general conclusion which could be drawn was that (i) a great number of anti-70S targeted inhibitors is completely ineffective, e.g. many aminoglycosides or macrolides, and (ii) sensitivity can be detected to a few selected anti-70S directed drugs and also to the 80S inhibitor anisomycin.

The evaluation of these in vivo data, of course, immediately raised questions on whether the drug reaches the target, whether it acts at its classical site and whether this is the sole inhibitory site.

To answer some of these questions in vitro polypeptide synthesis systems for three methanogenic organisms: *M. thermoautotrophicum*, *M. formicicum* and *M. vannielii* were set up and used to determine the effects of about 50 protein synthesis inhibitors, both anti-70S and anti-80S targeted compounds (7, 43, 44; Sanz, Hummel, Amils and Böck, unpublished). The results are summarized in Figure 2.

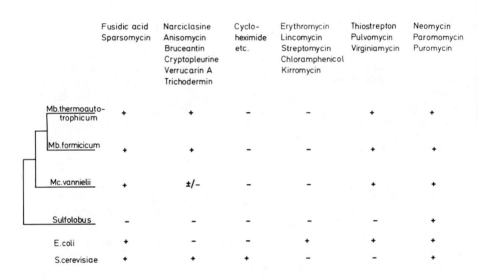

Fig. 2 *Sensitivity of poly(U) dependent polyphenylalanine synthesis to protein synthesis inhibitors. (+) strong inhibition; (+) inhibition only at concentrations above 0.1 mM; (-) no effect.*

The data for *Sulfolobus* are taken from work presented by the groups of Amils (45) and of Londei (46). The left part gives the relationship of these organisms in form of a qualitative dendrogram taken from the 16S S_{AB} data (1). The effects in the *E. coli* and *Saccharomyces* systems were determined as controls.

Several groups of compounds can be differentiated:
(i) One group is represented by fusidic acid - an inhibitor of elongation factor II - and sparsomycin; both act on 70S and 80S ribosomes; they are active on methanogens but not on *Sulfolobus*.
(ii) More interesting is a second class of compounds; they have been known until now as specific 80S protein synthesis inhibitors. Narciclasin, anisomycin and several trichothecene antibiotics are examples. They are potent inhibitors of ribosomes from the two *Methanobacterium* species, less of that from *Methanococcus* and inactive on *Sulfolobus*.
(iii) A third group represented by cycloheximide and comprising many other compounds (not shown) specifically block peptide bond formation at 80S ribosomes but not on archaebacterial and eubacterial ones.
(iv) The fourth group comprises antibiotics which are classical inhibitors of 70S ribosomes, but not of 80S eukaryotic ones. The macrolides, streptomycin, chloramphenicol and kirromycin belong to it. They do not inhibit poly(U) or poly(UG) directed polypeptide synthesis; for chloramphenicol, erythromycin and streptomycin it has been shown by equilibrium dialysis experiments that binding sites either are non-existent or that they have an extremely low affinity (44; 47).
(v) There are, however, 70S ribosome directed compounds which also act on methanogens; thiostrepton which binds to the GTPase center of the 50S subunit, pulvomycin which inhibits EF-Tu activity and the virginiamycins are examples.
(vi) The last group represented by neomycin and puromycin gives those compounds which act in any, i.e. also in the *Sulfolobus* system.

The in vitro inhibition of polypeptide formation by anisomycin, narciclasine and related compounds suggested the existence of a "eukaryotic" domain at the 70S ribosomes from these organisms. This assumption was strongly supported by the isolation of anisomycin resistant mutants from M. *formicicum* (48). Anisomycin resistance in these strains was demonstrated to be due to an altered 50S subunit. This suggests, that like in eukaryotic ribosomes (49, 50) anisomycin blocks protein synthesis by interfering with the large ribosomal subunit. Similar to anisomycin resistant mutants from yeast (50-52) those from M. *formicicum* were highly resistant in vitro to narciclasine, bruceantin, verrucarin A and trichodermin and they were hypersensitive to sparsomycin. This cross-resistance pattern indicates that the topology of the anisomycin interaction site at the 50S subunit is similar to that present at yeast ribosomes. It is therefore plausible to assume that this site did not arise by convergent evolution but rather by inheritance from a common ancestor. Since it is known that anisomycin and its related compounds block peptidyl-transfer and translocation the view might not be unreasonable that this domain of the 50S subunit is identical or close to the site where elongation factor II interacts. Elongation factor II from archaebacteria shares with eukaryotic elongation factor the diphtheria toxin sensitivity, as the work from Klink and coworkers (6) showed. Elongation factor II and its interaction site at the ribosome might therefore have evolved together.

Altogether, therefore, the antibiotic sensitivity pattern has shown that
(i) archaebacteria, and also the methanogens amongst themselves, do not display a uniform sensitivity pattern;
(ii) related organisms give an identical response. Particularily, the mesophilic and the thermophilic species of *Methanobacterium* behave identical. The sensitivity pattern, therefore, seems to reflect relationship;
(iii) Methanobacteriales and also halophiles (53) seem to possess a "eukaryotic" domain at their 50S subunits.

The fact that ribosomes from *Methanobacterium* simultaneously contain sites for anti-80S targeted drugs and for anti-70S directed ones (like the GTPase inhibitor thiostrepton) again suggests that different functional domains at the ribosome have evolved with different rates. It also points to a certain "hybrid" character of the translational machinery of archaebacteria.

An interesting question will be whether antibiotic sensitivity of protein synthesis is a reflection of the particular primary structure of ribosomal RNA. In other words, can one correlate the existence of a certain primary or secondary structure element of archaebacterial 16S or 23S rRNA with the activity or inactivity of an antibiotic. This in turn could give information on the involvement of rRNA partial structures on partial functions of the protein synthesis cycle.

CONCLUSIONS

Fig. 3 summarizes the data on several aspects of the archaebacterial translation apparatus. The three kingdoms are listed in a linear manner with the methanogen/halophile branch of archaebacteria situated somewhat closer to the eubacteria than to eukaryotes. The thermophiles, on the other hand, are tentatively placed closer to the eukaryotes mainly on the basis of the RNA polymerase homology (54). Quantitative data, however, are needed to support this order.

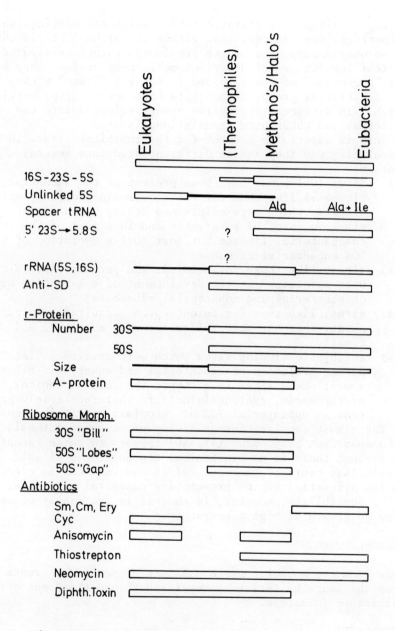

Fig. 3 Diagram of properties of the translational apparatus which archaebacteria share with either eubacteria or eukaryotes. Bars indicate the existence of common features; the degree of homologous features is symbolically indicated by the intensity of the bars. Anti-SD: Anti-Shine-Dalgarno.

Fig. 3 lists only characteristics which are distinguishing properties. One can see, that within the archaebacteria there is an overlapping of features previously either considered as typical for 70S or for 80S ribosomes. In an evolutionary sense the "phenotype" of the ribosome is not yet "fixed" within the archaebacterial group. It consists of a mosaic of properties which - in different assemblies - were selected into the eubacterial and eukaryotic translational system.

This is especially evident for the inhibitor sites. In a speculative way they can be differentiated into several classes:
(i) Sites (which may have been present at the ancestral ribosome) like that for the anisomycin group of antibiotics, which were maintained during evolution of ribosome into the eukaryotic and into part of the archaebacterial lineage but lost during evolution of the 70S eubacterial ribosome;
(ii) sites, like those for neomycin and paromomycin, which were maintained in the development of eukaryotic, archaebacterial and eubacterial ribosomes;
(iii) sites, like that for thiostrepton acquired by a common putative archaebacterial/eubacterial ancestor, and finally
(iv) antibiotic binding sites which were developed "late" during evolution of eubacterial and eukaryotic ribosomes; these sites (e.g. those for chloramphenicol, erythromycin, cycloheximide) are therefore solely present at eubacterial 70S or cytoplasmic 80S ribosomes.

The almost complete insensitivity of protein synthesis at ribosomes from *Sulfolobus* (45; 46) appears to be an exception. At present there is no biochemical explanation available. One possibility would be that the *Sulfolobus* ribosome is closest in its properties to the presumptive ancestral organelle; another possibility, however, is that it is the result of selection to growth at high temperature.

ACKNOWLEDGMENTS

Work from the authors' group has been supported by grants from the Deutsche Forschungsgemeinschaft and the Fonds der Chemischen Industrie.

REFERENCES

1. Woese, C.R., Magrum, L.J. and Fox, G.E. (1978). Archaebacteria, *J. Mol. Evol.* **11**, 245-252.

2. Lake, J.A., Henderson, E., Oakes, M. and Clark, M.W. (1984). Eocytes: A new ribosome structure indicates a kingdom with a close relationship to eukaryotes, *Proc. Natl. Acad. Sci. USA* 81, 3786-3790.
3. Fox, G.E., Luehrsen, K.R. and Woese, C.R. (1982). Archaebacterial 5S ribosomal RNA, *Zbl. Bakt. Hyg. I. Abt. Orig.* C3, 330-345.
4. Woese, C.R., Gutell, R., Gupta, R. and Noller, H.F. (1983). Detailed Analysis of the higher-order structure of 16S-like ribosomal ribonucleic acids, *Microbiol. Rev.* 47, 621-669.
5. Matheson, A.T. and Yaguchi, M. (1982). The evolution of the archaebacterial ribosome, *Zbl. Bakt. Hyg. I. Abt. Orig.* C3, 192-199.
6. Kessel, M. and Klink, F. (1980). Archaebacterial elongation factor is ADP-ribosylated by diphtheria toxin, *Nature* 287, 250-251.
7. Elhardt, D. and Böck, A. (1982). An in vitro polypeptide synthesizing system from methanogenic bacteria: Sensitivity to antibiotics, *Molec. Gen. Genet.* 188, 128-134.
8. Douglas, C., Achatz, F. and Böck, A. (1980). Electrophoretic characterization of ribosomal proteins from archaebacteria, *Zbl. Bakt. Hyg. I. Abt. Orig.* C1, 1-11.
9. Schmid, G. and Böck, A. (1982). The ribosomal protein composition of five methanogenic bacteria, *Zbl. Bakt. Hyg. I. Abt. Orig.* C3, 347-353.
10. Schmid, G. and Böck, A. (1982). The ribosomal protein composition of the archaebacterium *Sulfolobus*, *Molec. Gen. Genet.* 185, 498-501.
11. Londei, P., Teichner, A. and Cammarano, P. (1982). Particle weights and protein composition of the ribosomal subunits of the extremely thermoacidophilic archaebacterium *Caldariella acidophila*, *Biochem. J.* 209, 461-470.
12. Lake, J.A., Henderson, E., Clark, M.W. and Matheson, A.T. (1982). Mapping evolution with ribosome structure: Intralineage constancy and interlineage variation, *Proc. Natl. Acad. Sci. USA* 79, 5948-5952.
13. Noller, H.F. (1984). Structure of ribosomal RNA, *Ann. Rev. Biochem.* 53, 119-169.
14. Hofman, J.D., Lau, R.H. and Doolittle, W.F. (1979). The number, physical organization and transcription of ribosomal RNA cistrons in an archaebacterium: *Halobacterium halobium*, *Nucl. Acids Res.* 7, 1321-1333.
15. Jarsch, M., Altenbuchner, J. and Böck, A. (1983). Physical Organization of the genes for ribosomal RNA in *Methanococcus vannielii*, *Molec. Gen. Genet.* 189, 41-47.

16. Neumann, H., Gierl, A., Tu, J., Leibrock, J., Steiger, D. and Zillig, W. (1983). Organization of the genes for ribosomal RNA in archaebacteria, *Molec. Gen. Genet.* 192, 66-72.
17. Tu, J. and Zillig, W. (1982). Organization of rRNA structural genes in the archaebacterium *Thermoplasma acidophilum*, *Nucleic Acids Res.* 10, 7231-7245.
18. Jarsch, M. and Böck, A. (1983). DNA sequence of the 16S rRNA/23S rRNA intercistronic spacer of two rDNA operons of the archaebacterium *Methanococcus vannielii*, *Nucleic Acids Res.* 11, 7537-7543.
19. Hui, J. and Dennis, P.P. (1984). Characterization of the ribosomal RNA gene clusters in *Halobacterium cutirubrum*, *J. Biol. Chem.*, in press.
20. Nazar, R.N. (1983). The ribosomal 5.8S rRNA: eukaryotic adaptation or processing variant, *Canadian J. Biochem. Cell Biol.* 62, 311-320.
21. Wich, G., Jarsch, M. and Böck, A. (1984). Apparent operon for a 5S ribosomal RNA gene and for tRNA genes in the archaebacterium *Methanococcus vannielii*, *Molec. Gen. Genet.* 196, 146-151.
22. Nomura, M., Morgan, E.A. and Jaskunas, S.R. (1977). Genetics of bacterial ribosomes, *Ann. Rev. Genet.* 11, 297-347.
23. Gupta, R., Lanter, J.M. and Woese, C.R. (1983). Sequence of the 16S ribosomal RNA from *Halobacterium volcanii*, an archaebacterium, *Science* 221, 656-659.
24. Leffers, H. and Garrett, R.A. (1984). The nucleotide sequence of the 16S ribosomal RNA gene of the archaebacterium *Halococcus morrhuae*, *EMBO J.* 3, 1613-1619.
25. Jarsch, M. and Böck, A. (1984). Sequence of the 16S ribosomal RNA Gene from *Methanococcus vannielii*, *Nucleic Acids Res.*, submitted.
26. McCarroll, R., Olsen, G.J., Stahl, Y.D., Woese, C.R. and Sogin, M.L. (1983). Nucleotide Sequence of the *Dictyostelium discoideum* small-subunit rRNA inferred from the gene sequence. Evolutionary implications, *Biochemistry* 22, 5858-5868.
27. Wittmann, H.G. (1982). Components of bacterial ribosomes, *Ann. Rev. Biochem.* 51, 155-184.
28. Yaguchi, M., Visentin, L.P., Zuker, M., Matheson, A.T., Roy, C. and Strøm, A.R. (1982). Amino-terminal sequences of ribosomal proteins from the 30S subunit of Archaebacterium *Halobacterium cutirubrum*, *Zbl. Bakt. Hyg. I. Abt. Orig.* C3, 200-208.

29. Matheson, A.T., Yaguchi, M., Christensen, P., Rollin, C.F. and Hasnain, S. (1984). Purification, properties and N-terminal amino acid sequence of certain 50S subunit proteins from the archaebacterium *Halobacterium cutirubrum*, *Canadian J. Biochem. Cell Biol.* 62, 426-433.
30. Strøm, A.R. and Visentin, L.P. (1973). Acidic ribosomal proteins from the extreme halophile *Halobacterium cutirubrum*, *FEBS Letters* 37, 274-280.
31. Schmid, G. (1984). Untersuchungen zur Evolution ribosomaler Proteine von Archaebakterien. Dissertation Universität München.
32. Schmid, G. and Böck, A. (1981). Immunological comparison of ribosomal proteins from archaebacteria, *J. Bacteriol.* 147, 282-288.
33. Schmid, G. and Böck, A. (1984). Immunoblotting analysis of ribosomal proteins from archaebacteria, *System. Appl. Microbiol.* 5, 1-10.
34. Schmid, G., Strobel, O., Stöffler-Meilicke, M., Stöffler, G. and Böck, A. (1984). A ribosomal protein immunologically conserved in archaebacteria, eubacteria and eukaryotes, *FEBS Letters*, in press.
35. Thurlow, D.L. and Zimmermann, R.A. (1982). Evolution of protein binding regions of archaebacterial, eubacterial and eukaryotic ribosomal RNAs. In "Archaebacteria" (Ed. O. Kandler), Gustav Fischer Verlag, Stuttgart, New York.
36. Matheson, A.T., Yaguchi, M., Balch, W.E. and Wolfe, R.S. (1980). Sequence homologies in the N-terminal region of the ribosomal 'A' protein from *Methanobacterium thermoautotrophicum* and *Halobacterium cutirubrum*, *Biochim. Biophys. Acta* 626, 162-169.
37. Itoh, T. and Higo, K. (1983). Complete amino acid sequence of an L7/L12-type ribosomal protein from *Rhodopseudomonas spheroides*, *Biochim. Biophys. Acta* 744, 105-109.
38. Kaltschmidt, E. and Wittmann, H.G. (1970). Ribosomal proteins. VII. Two-dimensional polyacrylamide gel electrophoresis for finger-printing of ribosomal proteins, *Analyt. Biochem.* 36, 401-412.
39. Magrum, L.J., Luehrsen, K.R. and Woese, C.R. (1978). Are extreme halophiles actually "bacteria", *J. Mol. Evol.* 11, 1-8.
40. Jarrell, K.F., Sprott, G.D. and Matheson, A.T. (1984). Intracellular potassium concentration and relative acidity of the ribosomal proteins of methanogenic bacteria, *Canadian J. Microbiol.* 30, 663-668.
41. Pecher, T. and Böck, A. (1981). In vivo susceptibility of halophilic and methanogenic organisms to protein synthesis inhibitors, *FEMS Microbiol. Letters* 10, 295-297.

42. Hilpert, R., Winter, J., Hammes, W. and Kandler, O. (1981). The sensitivity of archaebacteria to antibiotics, *Zbl. Bakt. Hyg. I. Abt. Orig.* C2, 11-20.
43. Böck, A., Bär, U., Schmid, G. and Hummel, H. (1983). Aminoglycoside sensitivity of ribosomes from the archaebacterium *Methanococcus vannielii:* structure-activity relationship, *FEMS Microbiol. Letters* 20, 435-438.
44. Di Giambattista, M., Hummel, H., Böck, A. and Cocito, C. (1984). Action of synergimycins and macrolides on in vivo and in vitro protein synthesis in archaebacteria, *Molec. Gen. Genet.*, in press.
45. Sanz, J.L. and Amils, R. (1983). Sensitivity of thermoacidophilic archaebacteria to protein synthesis inhibitors. Abstract Meeting "Antibiotics 83", Jarandilla, Spain.
46. Londei, P. (1983). Action of inhibitors of protein synthesis on archaebacterial (*Caldariella acidophila*) ribosomes. Abstract Meeting "Antibiotics 83", Jarandilla, Spain.
47. Schmid, G., Pecher, T. and Böck, A. (1982). Properties of the translational apparatus of archaebacteria, *Zbl. Bakt. Hyg. I. Abt. Orig.* C3, 209-217.
48. Hummel, H. and Böck, A. (1985). Mutations in *Methanobacterium formicicum* conferring resistance to anti-80S ribosome targeted antibiotics, submitted for publication.
49. Barbacid, M. and Vazquez, D. (1974). (^{3}H)Anisomycin binding to eukaryotic ribosomes, *J. Mol. Biol.* 84, 603-623.
50. Jimenez, A. and Vazquez, D. (1975). Quantitative binding of antibiotics to ribosomes from a yeast mutant altered on the peptidyltransferase center, *Eur. J. Biochem.* 54, 483-492.
51. Schindler, D., Grant, P. and Davies, J. (1974). Trichodermin resistance-mutation affecting eukaryotic ribosomes, *Nature (London)* 248, 535-536.
52. Jimenez, A., Sanchez, L. and Vazquez, D. (1975). Simultaneous ribosomal resistance to trichodermin and anisomycin in *Saccharomyces cerevisiae* mutants, *Biochim. Biophys. Acta* 383, 427-434.
53. Kessel, M. and Klink, F. (1981). Elongation factors of the extremely halophilic archaebacterium *Halobacterium cutirubrum*, *Eur. J. Biochem.* 114, 481-486.
54. Zillig, W., Stetter, K.O., Schnabel, R., Madon, J. and Gierl, A. (1982). Transcription in archaebacteria, *Zbl. Bakt. Hyg. I. Abt. Orig.* C3, 218-227.
55. Stöffler-Meilicke, M., Breitenreuter, G., Strobel, O., Böck, A. and Stöffler, G. (1984). The structure of ribosomal subunits from archaebacteria as determined by (immuno) electron microscopy. Abstract FEMS Symposium on Evolution of prokaryotes, Munich.

TRANSPOSABLE ELEMENTS AND EVOLUTION

R. Schmitt and P. Rogowsky

Lehrstuhl für Genetik
Universität Regensburg
Regensburg, Federal Republic of Germany

S. E. Halford and J. Grinsted

Departments of Biochemistry and Microbiology
University of Bristol
Bristol, England

INTRODUCTION

Transposable elements are discrete genetic entities capable of translocation within and between chromosomes and plasmids and of random insertion without requiring obvious homology between the original and the target sequence. In addition, these elements promote DNA rearrangements such as deletions, inversions, and duplications, features essential for considering their potential role in the evolution of genomes and organisms. Transposable elements have been found in prokaryotic and eukaryotic organisms. In bacteria we distinguish between insertion elements (IS) and transposons (Tn). The former are relatively small (200 to 1400 base pairs) and contain mainly the genetic information for their transposition. The latter are larger movable entities (2.5 to 40 kilo base pairs) that in addition to transposition genes encode antibiotic and metal resistances, enterotoxin synthesis or novel metabolic enzymes favourable for the host under certain environmental conditions. Transposons have mostly been found as components of bacterial plasmids and their role in the rapid evolution of multiresistance factors is well-documented (1). Less is known about their role as part of the bacterial chromosome; however, we assume that they behave in an analogous manner with respect to rearranging chromosomal genetic material. The majority of these alterations remains unseen as they are lost due to deleterious effects on the cell. As will be seen, it requires special conditions to also detect the beneficial effects of chromosomal transposon insertions.

In this article we will focus on the members of a family of related eubacterial transposons (Tn3 family), notably on

Tn*1721*, that may serve as paradigm for illuminating the
following points:
(a) Transposon structure and gene organization.
(b) Transposition mechanisms.
(c) Transposon-mediated genome rearrangements and evolution.
(d) Evolution of a transposon family.

Transposon Structure and Gene Organization

To better understand DNA rearrangements generated by movable
genetic elements we will first consider the known structural
features and mechanisms of transposition. Three structural
features are common to essentially all transposable elements.
(i) They have inversely repeated ends (IR) of up to 40 base
pairs; (ii) these IRs flank a central region containing
transposition genes and - for transposons - additional determinants; (iii) upon transposition they generate a short
duplication of target DNA that flanks the inserted element
as direct repeats (2).
  Transposon Tn*1721* is particular in consisting of two
discrete portions, namely, a minor transposon (Tn*1722*) containing the genes required for transposition and an amplifi-
able region encoding resistance to tetracycline (*tet*; Fig.1)
(3). Three 38-base pair (bp) repeats, one separating the two
regions, render transposition of either the whole element or
the minor transposon alone possible (4). The element generates 5-bp direct repeats at the target site of insertion (5).
The *tet* region can undergo *recA*-dependent tandem duplication,
owing to homology regions (1.9 kb) that flank the resistance
determinant. Amplification generates a positive gene dosage
effect on tetracycline resistance, a condition favourable under high concentrations of the drug (6). We have shown that

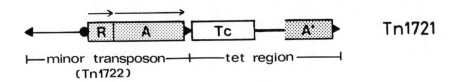

|—minor transposon—|——tet region——|
     (Tn1722)

*Fig. 1 Diagram of transposon Tn1721 showing central and ter-
minal 38-bp IRs (◀ ▶) flanking the minor transposon (Tn1722)
and the repetitious* tet *region. Genetic loci responsible for
transposition (●: resolution site; R: resolvase; A: transposase; A': truncated A) and tetracycline resistance (Tc)
and transcriptional polarities are shown. Shading indicates
homology with transposon Tn501.*

an "unchallenged" bacterial population contains approximately $1/10^8$ cells habouring Tn1721 in an amplified state, sufficient to ensure survival of the species under tetracycline "stress".

Tn1721 contains three adjacent genetic loci responsible for transposition, namely, *tnpA* the structural gene for the transposase enzyme, *tnpR* for the resolvase enzyme, and *res* the specific DNA site for resolution of cointegrates catalyzed by resolvase (Fig.1) (7). The genes *tnpR* and *tnpA* have the same transcriptional polarity, but independent promoters, since a deletion extending into the 5' end of *tnpR* is still transposase-proficient. Expression of *tnpR* is self-regulatory; by binding to the *res* site that overlaps the *tnpR* promoter resolvase represses the transcription of *tnpR* (7).

Transposition Mechanisms

Transposons move as discrete entities. There is no permutation of internal markers suggesting that both inverted ends are directly involved in the process of transposition. Two mechanisms have been postulated to interpret *in vivo* data on transposition (Fig.2). In conservative transposition the

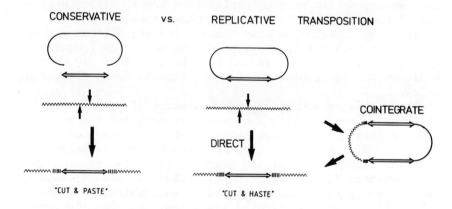

Fig. 2 Current models of transposition. Symbols denote transposable element (⇔), donor DNA (———), target DNA (∧∧∧∧∧), direct repeats (⦀⦀⦀) and single-strand cuts (↓↑). Note that the double-stranded donor and target DNAs are symbolized by single lines.

element is excised from the donor replicon, and staggered single-strand cuts mark the target site into which the transposon is ligated ("cut and paste"). This transposition mechanism appears to be operative with transposons Tn5 and Tn10, with IS elements, and with a number of movable elements of eukaryotes (8; N. Kleckner, pers. communication). Replicative transposition requires cutting at one or both ends and replication of the element ("cut and haste"). Routinely, this leads to a cointegrate intermediate that contains donor and target replicons fused by the two copies of the transposon (2). In a subsequent step, the cointegrate is resolved by element-encoded site-specific recombination to give donor and target both with a copy of the transposon (9). A possible by-pass of the cointegrate intermediate is known as "direct transposition "(Fig.2).

Replicative transposition is operative in Tn3-like elements which are described here. Two alternative models, symmetric and asymmetric, have been proposed for replicative transposition, depending on the use of both IRs simultaneously or on a processive mode starting at one IR (10-13). We have recently reported that truncated, "single-ended" derivatives of Tn1721 and Tn21 with a deletion of one IR are capable of transposing to a target replicon, if transposase is provided in cis or in trans (14,15). Restriction endonuclease and DNA-sequence analyses of the products suggested that an asymmetric mode of replicative transposition starting at the intact IR is operative under these conditions (16). These experiments also demonstrate the minimum prerequisites needed for transposition (though at low frequency), namely, a single intact IR and transposase. This may be significant for the evolution of such elements to be discussed below.

The mechanism governing the resolution of cointegrates is relevant to genome evolution, since it explains the generation of deletions and inversions. Cointegrates are efficiently resolved by site-specific, recA-independent recombination (9). The resolvase-catalysed reaction shows two interesting restrictions: It requires two res sites that are (i) located on the same replicon and (ii) that have identical orientation (17). Inversely oriented res sites, that lead to inversion of intervening DNA, are only used with low efficiency, if excessive resolvase is available (7). This strong restriction to the specific topology of "normal" cointegrates has been explained by a "tracking model", that requires binding of a resolvase dimer to one res site and diffusion along the intervening DNA until the two res sites are synapsed in proper orientation. Subsequently the enzyme catalyses breakage and reunion at a specific sequence (crossover point) to yield two catenated rings. These have to be separated by host-encoded DNA-gyrase (17).

This highly specific process requires a *res*-DNA sequence with the following properties: it should contain the crossover site and embody features that constitute the polarity of the locus. In addition, transcription signals for *tnpR* should be contained in *res* to account for autoregulation by resolvase.

The functional structure of *res* as shown in Fig.3 has been elucidated in three ways. (i) A 200-bp DNA sequence upstream of *tnpR* has been determined and shown to be essential for cointegrate resolution (7,18). (ii) *In vivo* resolution of a recombinant plasmid containing *res* sites of the related (but non-identical) Tn*501* and Tn*1721* led to a hybrid fused at the crossover point. DNA sequencing across the fusion indicated that recombination occurs in an 11-bp region of exact homology between Tn*501* and Tn*1721*, located 161-172 bp upstream of *tnpR* (19). This region also defines the transition from homology to non-homology between the two elements suggesting that site-specific recombination may have been an essential in their evolution. (iii) *In vitro* binding of purified resolvase to *res*-DNA results in three sites protected against DNase digestion (foot-printing). Site I (38 bp) most distal to *tnpR* contains the crossover point; site II (44 bp) overlaps the -35 promoter box and site III (31 bp) overlaps the -10 promoter box of the *tnpR* gene. Each site has features of partial dyad symmetry, and a consensus sequence of six "half-sites" (containing the motif GTC) can be established. The observed polarity of *res* may be explained by differential binding of resolvase to the three sites: the enzyme exhibits high affinity to site I (full protection at low concentrations of enzyme) and binds less efficiently to sites II and III (full protection only at high concentrations of enzyme). We are presently investigating, if this effect is due to interacting binding sites.

The resolvases of Tn*21* and Tn*1721* (79% homology) are interchangeable suggesting that the resolvase binding sites of the two elements must have similar sequences (18). As shown by asterisks (Fig.3), the three sites constituting *res* are located in regions of sequence homology between Tn*21* and Tn*1721*, thus lending independent support to our binding studies. Analogies to the structure of *res* in Tn*3* and $\gamma\delta$ concerning the principal arrangement of three sites (20) are apparent, but differences relate to their extent, to their spacing and to their DNA sequences (except for the GTC motif). In correspondence to these differences in *res* the resolvases of Tn*3*/$\gamma\delta$ on the one hand and of Tn*21*/Tn*1721* on the other have considerably diverged. There is only 32% homology in their primary sequences, which excludes an interchange of these functions (16,18). Abdel-Meguid et al. (21) have shown,

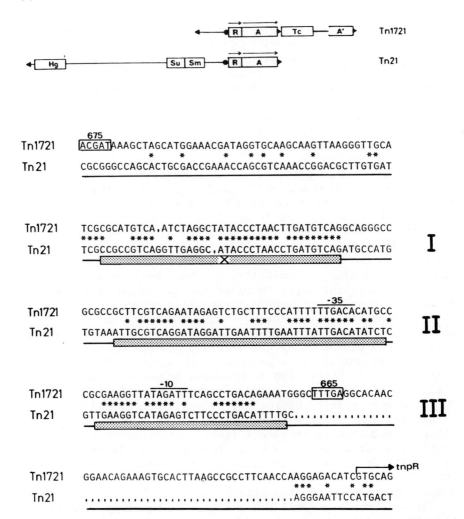

Fig. 3 Top: Genetic and physical maps of transposons Tn1721 and Tn21 superimposed with respect to their homologous transposition genes symbolised as in Fig.1. In addition, the determinants for mercuric (Hg), streptomycin (Sm) and sulfonamide (Su) resistances are shown. Bottom: Aligned DNA sequences of Tn1721 and Tn21 containing the res sites upstream of the resolvase gene (tnpR). Identical nucleotides are marked by asterisks between sequences, the extent of sites I to III defined by resolvase binding (▧▧▧), the crossover point (X), the tnpR promoter (-35, -10) and target sites of two transposon insertions (665, 675) used for sequencing are indicated.

that the N-terminal portion of γδ resolvase is responsible
for DNA binding, whereas a C-terminal fragment contains the
catalytic activity. A comparison of the res and resolvase
structures of Tn3/γδ and Tn21/Tn1721 is now feasible. It is
expected to yield interesting insights into the evolution of
specificity of protein-DNA interaction.

Transposon-Mediated DNA Rearrangements and Evolution

Transposable elements are constituents of most bacterial genomes and of many plasmids and bacteriophages. They can
alter both the organisation and the expression of prokaryotic genes at frequencies equal to or greater than mutation
events affecting a single or a few base pairs. Different
from point mutations, transposon-mediated events mostly involve larger segments of DNA and can therefore function as
agents of sudden jumps in an otherwise slow and continuous
evolution of DNA sequences. Transposon-induced alterations
can occur in several ways (Fig.4) (2). The <u>insertion</u> of a

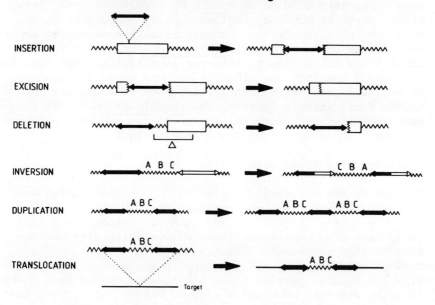

Fig. 4 Types of transposon-promoted rearrangements. Symbols: transposable elements (⟷ , ⟺), structural gene (▭), donor (∿∿∿) and target (——) DNA, deletion (Δ) and order of genes (A B C).

transposable element into a gene is an immediate event, frequently leading to gene inactivation. There are at least two exceptions of this rule. (i) Transposable elements can produce new regulatory signals leading to the expression of cryptic genes (22). (ii) It has recently been shown that cells of *Escherichia coli* harbouring transposon Tn5 or Tn10 have a growth advantage in chemostat competition over otherwise isogenic strains without these elements (23). It has been argued that these transposable elements enhance the mutation rate of the host bacterium, thus increasing the probability of a favourable mutation. The possibility that transposable elements have an evolutionary role as mutator genes is quite appealing and to us more satisfying than hypotheses that allocate to them the role of parasitic DNA (24). The effects of insertions can be reversed by rare precise excision of the element (Fig.4). More important for evolution are deletions that extend from one end of the element into adjacent DNA. These are either *recB,C*-dependent or may be the result of resolvase-catalysed deletion of DNA intervening between two neighbouring insertions in the genome. Two elements inserted in opposite orientation into a chromosome frequently lead to inversion of intervening DNA. They either serve as homology regions for *recA*-dependent recombination or as substrates for rare resolvase-catalysed inversions as detailed above. Duplications occur, if two elements flank the region of interest in direct orientation. As regions of homology the elements promote unequal crossovers between the arms of the replication fork thus leading to tandem duplications. Finally, the translocation of DNA fragments requires the concerted action of two flanking elements forming a large "compound transposon" (Fig.4).

Evolution of a Transposon Family

Members of the Tn3 family differ in the relative arrangement of their transposition genes and in the resistance determinants they carry (Fig.5). In Tn3 (and γδ) the *tnpR* and *tnpA* genes have divergent transcriptional polarity with the *res* site located in between (25). In contrast, Tn1721, Tn501, Tn21 and Tn4 contain *tnpR* and *tnpA* in the same orientation with *res* located outside (18). All these elements have 38-bp IRs with sequence homologies ranging between 50% (Tn3-Tn501) and 95% (Tn1721-Tn501).

Whereas the evolution of an ancestral element consisting of transposition genes (*tnpA*, *tnpR*) and sites (*res*, IR) is thought to be a distant event, the acquisition of resistance determinants must have occurred more recently. Consequently,

TRANSPOSABLE ELEMENTS AND EVOLUTION

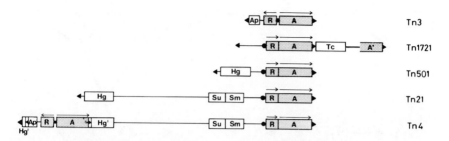

Fig. 5 Tn3-like transposable elements superimposed relative to their transposition genes (shaded). Tn4 diverged from Tn21 by an insertion of Tn3 into the Hg determinant (Ap: ampicillin resistance). Symbols as in Figs.1 and 3.

we will consider two phases of evolution: one concerning the ancestral elements and another dealing with the incorporation of various resistance markers.

Replicon fusion by single-ended transposons suggests that one IR plus transposase (*tnpA* gene product) are sufficient for transposition. Generation of a second IR, e.g. by hairpin loop replication, would yield a transposable element consisting of two IRs flanking *tnpA*. We have previously postulated (16) that specific recombination (*tnpR, res*) has been derived from an invertible element carrying a gene for flip-flop recombination analogous to *hin* of *Salmonella* (26). Subsequent deletion of one of the recombination sites flanking the gene would fix the invertible segment in one orientation or the other giving rise to the γδ line (opposite polarity of *tnpR* and *tnpA*) or the Tn*1722* line (identical polarity of *tnpR* and *tnpA*) as illustrated in Fig.6. Note that in each case *res* is located adjacent to the 5' end of the *tnpR* gene suggesting their common evolution and their incorporation as an entity into the element.

Our knowledge of the relationships between Tn*3*-like elements relies on the physical maps, on sequence homologies and on complementation studies (Table 1). The two branches represented by γδ and Tn*1722* (Fig.6), respectively, differ in their genetic organisation and cannot complement each other. In contrast, within each branch the TnpR and (in part) the TnpA functions are interchangeable (18,27). Tn*4* originating from the Tn*1722* line represents a special case of converging evolution between the two branches: this element was generated from Tn*21* by the insertion of Tn*3* into the mercuric resistance operon (Fig.5) (28).

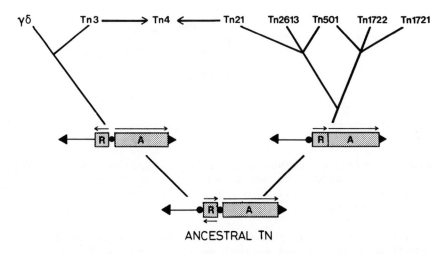

Fig. 6 Postulated pedigree of the Tn3 family. The two major lines represented by γδ- and Tn1722-like transposable elements diverged early when the invertible segment (●R●) was fixed in one or the other orientation relative to A. Subsequent diversification occurred by the incorporation of various resistance determinants shown in Fig.5 (Tn2613 is a Tn21-like Hg transposon) (16,29).

TABLE 1

Complementation of transposition functions (16,18)

Transposon probed	Tn1721		Tn501		Tn21		Tn3	
	A[a]	R	A	R	A	R	A	R
Tn1721	+	+	+	+	(+)[b]	+	−	−
Tn501	+	+	+	+	(+)	+	−	−
Tn21	−	+	−	+	+	+	−	−
Tn3	−	−	−	−	−	−	+	+

[a] Abbreviations: A = tnpA, R = tnpR.
[b] (+) denotes low efficiency of complementation.

The acquisition of resistance determinants (possibly originating in antibiotic producing soil bacteria such as *Streptomyces*) must have occurred some time after the divergence of the γδ and the Tn*1722* lines as diagrammed in Fig.6. The time scale during which these steps that shaped present-day transposons have occurred is not known. But in view of highly efficient recombination mechanisms that govern the incorporation and rearrangement of resistance markers these processes can be rapid, and it is conceivable that such events may have occurred lately and that they are still operative to engineer new transposons. We have previously described three distinct ways by which these genes have been incorporated into transposable elements involving illegitimate and homologous recombination with subsequent deletions to shape present-day transposons (16). These pathways have been deduced by following the genealogy of related elements and by probing their evolution in the test tube.

Conclusions

Transposable elements are motors for rapid alterations of genome organisation and expression. The mechanisms of illegitimate, site-specific and homologous recombination function separately or in concert to generate DNA rearrangements such as insertions, deletions, inversions, duplications or translocations. The same mechanisms are operative in shaping modern antibiotic or heavy metal resistance transposons. It should be noted that most of these processes operate by shuffling and reshuffling of DNA modules, thus creating a certain instability and plasticity of the prokaryotic genome. Given a limited genome capacity, this instability may be a necessary prerequisite for the adaptation to an ever changing environment. Transposable elements may thus have evolved as natural tools for genetic engineering to constantly rearrange the genome thus ensuring the survival of cells or entire populations that harbour such elements.

ACKNOWLEDGEMENTS

We are indebted to Helene Beier for excellent illustrations and to Simon Silver for critical revision of the manuscript. This investigation was supported by grants from the Deutsche Forschungsgemeinschaft to R.S. and from the Medical Research Council to J.G. and S.H.

REFERENCES

1. Cohen, S.N. (1976). Transposable genetic elements and plasmid evolution, *Nature* 263, 731-738.
2. Kleckner, N. (1981). Transposable elements in prokaryotes, *Ann. Rev. Genet.* 15, 341-404.
3. Schmitt, R., Bernhard, E. and Mattes, R. (1979). Characterisation of Tn1721, a new transposon containing tetracycline resistance genes capable of amplification, *Mol. Gen. Genet.* 172, 53-65.
4. Schmitt, R., Altenbuchner, J., Wiebauer, K., Arnold, W., Pühler, A. and Schöffl, F. (1981). Basis of transposition and gene amplification by Tn1721 and related tetracycline-resistance transposons, *Cold Spring Harbor Symp. Quant. Biol.* 45, 59-65.
5. Schöffl,F., Arnold, W., Pühler, A., Altenbuchner, J. and Schmitt, R. (1981). The tetracycline resistance transposons Tn1721 and Tn1771 have three 38-base pair repeats and generate five-base pair direct repeats, *Mol. Gen. Genet.* 181, 87-94.
6. Wiebauer, K., Schraml, S., Shales, S.W. and Schmitt, R. (1981). Tetracycline resistance transposon Tn1721: *recA*-dependent gene amplification and expression of tetracycline resistance, *J. Bacteriol.* 147, 851-859.
7. Altenbuchner, J. and Schmitt, R. (1981). Transposon Tn1721: site-specific recombination generates deletions and inversions, *Mol. Gen. Genet.* 190, 300-308.
8. Berg, D.E. (1983). Structural requirements for IS50-mediated gene transposition, *Proc. Natl. Acad. Sci. USA* 80, 792-796.
9. Reed, R.R. (1981). Resolution of cointegrates between transposons γδ and Tn3 defines the recombination site, *Proc. Natl. Acad. Sci. USA* 78, 3428-3432.
10. Shapiro, J.A. (1979). Molecular model for the transposition and replication of bacteriophage Mu and other transposable elements, *Proc. Natl. Acad. Sci. USA* 76, 1933-1937.
11. Arthur, A. and Sherratt, D. (1979). Dissection of the transposition process: a transposon-encoded site-specific recombination system, *Mol. Gen. Genet.* 175, 267-274.
12. Harshey, R.M. and Bukhari, A.J. (1981). A mechanism of DNA transposition, *Proc. Natl. Acad. Sci. USA* 78, 1090-1094.
13. Galas, D.J. and Chandler, M. (1981). On the molecular mechanism of transposition, *Proc. Natl. Acad. Sci. USA* 78, 4858-4862.

14. Mötsch, S. and Schmitt, R. (1984). Replicon fusion mediated by a single-ended derivative of transposon Tn1721, Mol. Gen. Genet. 195, 281-287.
15. Avila, P., de la Cruz, F., Ward, E. and Grinsted, J. (1984). Plasmids containing one inverted repeat of Tn21 can fuse with other plasmids in the presence of Tn21 transposase, Mol. Gen. Genet. 195, 288-293.
16. Schmitt, R., Mötsch, S., Rogowsky, P., de la Cruz, F. and Grinsted, J. (1984). On the transposition and evolution of Tn1721 and its relatives. In "Plasmids in Bacteria" (Eds. D. Helinski, S. Cohen and D. Clewell). Plenum Press, New York (in press).
17. Krasnow, M.A. and Cozzarelli, N.R. (1983). Site-specific relaxation and recombination by the Tn3 resolvase: recognition of the DNA path between oriented res sites, Cell 32, 1313-1324.
18. Diver, W.P., Grinsted, J., Fritzinger, D.C., Brown, N.L., Altenbuchner, J., Rogowsky, P. and Schmitt, R. (1983). DNA sequences of and complementation by the tnpR genes of Tn21, Tn501 and Tn1721, Mol. Gen. Genet. 191, 189-193.
19. Rogowsky, P. and Schmitt, R. (1984). Resolution of a hybrid cointegrate between transposons Tn501 and Tn1721 defines the recombination site, Mol. Gen. Genet. 193, 162-166.
20. Grindley, N.D.F., Lauth, M.R., Wells, R.G., Wityk, R.J., Salvo, J.J. and Reed, R.R. (1982). Transposon-mediated site-specific recombination: identification of three binding sites for resolvase at the res sites of γδ and Tn3, Cell 30, 19-27.
21. Abdel-Meguid, S.S., Grindley, N.D.F., Templeton, N.S. and Steitz, T.A. (1984). Cleavage of the site-specific recombination protein γδ resolvase: The smaller of two fragments binds DNA specifically, Proc. Natl. Acad. Sci. USA 81, 2001-2005.
22. Saedler, H., Reif, H.J., Hu, S. and Davidson, N. (1974). IS2, a genetic element for turn-off and turn-on of gene activity in E.coli, Mol. Gen. Genet. 132, 265-289.
23. Chao, L., Vargas, C., Spear, B.B. and Cox, E.C. (1983). Transposable elements as mutator genes in evolution, Nature 303, 633-635.
24. Doolittle, W.F. and Sapienza, C. (1980). Selfish genes, the phenotype paradigm and genome evolution, Nature 284, 601-603.
25. Heffron, F., McCarthy, B.J., Ohtsubo, H. and Ohtsubo, E. (1979). DNA sequence analysis of the transposon Tn3: three genes and three sites involved in transposition of Tn3, Cell 18, 1153-1164.

26. Zieg, J. and Simon, M. (1980). Analysis of the nucleotide sequence of an invertible controlling element, *Proc. Natl. Acad. Sci. USA* 77, 4196-4200.
27. Grinsted, J., de la Cruz, F., Altenbuchner, J. and Schmitt, R. (1982). Complementation of transposition of *tnpA* mutants of Tn*3*, Tn*21*, Tn*501* and Tn*1721*, *Plasmid* 8, 276-286.
28. Foster, T.J., Nakahara, H., Weiss, A.A. and Silver, S. (1979). TransposonA-generated mutations in the mercuric resistance genes of plasmid R100.1, *J. Bacteriol.* 140, 176-181.
29. Tanaka, F., Yamamoto, T. and Sawai, T. (1983). Evolution of complex resistance transposons from an ancestral mercury transposon, *J. Bacteriol.* 153, 1432-1438.

THE ROLE OF IS- AND COMPLEX TRANSPOSABLE ELEMENTS IN THE EVOLUTION OF NEW GENES

Heinz Saedler

Max-Planck-Institut für Züchtungsforschung
Cologne, Federal Republic of Germany

The role of IS-elements in the evolution of the E. coli chromosome and some Plasmids has been addressed at various times in the past (1, 2, 3).
It is the aim of this article to describe a scenario in which transposable elements play a key role in the evolution of a new gene.
By way of introduction a historical overview is given on the important steps of IS-mediated rearrangements, most of which will be needed for the scenario mentioned above.

1. DETECTION AND PROPERTIES OF IS-ELEMENTS

In the mid 60's I was studying, in P. Starlinger's laboratory strange mutations in the galactose operon of E. coli. These polar mutations, although reverting spontaneously, did not respond to any mutagen (4). Later on we found out that they were due to small DNA insertions (5). These always integrated as units (6, 7) and hence they were considered elements.
In subsequent years I, with various collaborators concentrated mostly on IS1 and IS2. Together with B. Heiss I showed that both of these elements were constituents of the E. coli chromosome where they occured in multiple copies (8). In collaboration with S. Hu, E. Ohtsubo and N. Davidson their presence on F and R plasmids was noted. Here they occupied strategic positions for interaction with the host chromosome or separating the r-determined from the RTF unit (9). In the same years we discovered that IS2 seemed to have interesting properties in that it caused turn-off of gene expression

if integrated in one orientation while in the opposite orientation it led to expression of silent genes (10).
However, the element causing most of the gross DNA rearrangements in the chromosome and in plasmids, is IS1. Jörg Reif and I presented evidence in 1975, that deletions are produced with a high frequency (11) such that the element is retained and can undergo subsequent rounds of deletions. Later, however, we showed that deletion formation can also be excisive, i.e. IS1 can be deleted concomitantly with adjecent genetic material (12). Besides deletions, IS1 also generates inversions as was shown by Cornelis and Saedler (13).
This may suffice as a reminder on the potential of IS elements in generating DNA rearrangements.

2. SCENARIO OF THE EVOLUTION OF A NEW GENE

It is generally believed that the evolution of a new gene requires the duplication of an existing gene. Subsequently mutations can occur in the one copy while retaining the original gene activity. However, there might be a problem if the gene duplicated is under particular control. Then the extra and mutated copy still might be under the same control. This might interfere with evolutionary processes.

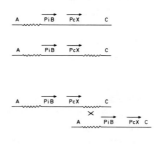

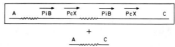

Legend to Fig. 1
A, B, X and C are genes, P indicates the promoter, i is inducible and c means constitutive expression, the arrows indicate direction of transcription. The wavy line represents the transposable element.

In Fig. 1 a transposon mediated duplication is diagrammed based on homologous or site specific recombination between transposed and hence displaced copies of the element. Hence a duplicated and a deleted chromosome arises.
Such a reaction seems to occur in IS1-mediated amplification of the r-determinant of certain R-factors (4).
In this scheme (Fig. 1) duplication is a consequence of a transposition event followed by cross over in out of register paired DNA molecules (chromosomes). It is assumed that gene B should evolve to give a new product. The expression of gene B is inducible but the neighbouring gene X is expressed constitutively. Therefore, the inducible expression signal of one copy of gene B has to be erased (Fig. 2), a reaction which we have seen many times in IS1-mediated gal deletions (11, 12). This results in a silent copy.
Selection for a newly evolved function occurs now either by accomulation of mutations in the silent gene followed by reactivation of this mutated gene or, more likely, reactivation of the silent gene occurs prior to the accumulation of mutations.
Reactivation can be achieved by excisive deletion (12) or by de-novo formation of a promoter using various mechanisms (14, 15). In the latter case there is continuous selection pressure, i.e. each mutation is probed for improvement of function by selection.
As can be seen from the bottom drawing in Fig. 2 erasure and reactivation result in a duplicated gene B where one copy is expressed constitutively. The reactions leading to this are all known to be mediated by IS-elements or other transposable elements.
However, mutations altering the protein structure were hitherto unknown to be catalysed ty transposable elements.
In the next paragraph I will describe experiments which although done with eukaryotic organisms, clearly show that transposable elements can cause alterations in protein structures.
It is suggested that similar reactions occur in prokaryotes. This, however has not extensively been tested experimentally due to the common belief

that the restoration of function of an IS or TN
induced mutation, in the codogenic segment of a
gene, invariably leads to a wildtype sequence.

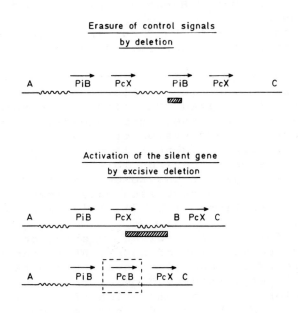

Legend to Fig. 2
The hatched boxed below the lines indicate the extend of deleted genetic material. Boxed in is the constitutively expressed duplicated gene.

3. PLANT TRANSPOSABLE ELEMENTS CAN CAUSE MUTATIONS
   ALTERING THE PROTEIN STRUCTURE

For approximately 4 years we have been studying
plant transposable elements at the Max-Planck-
Institut für Züchtungsforschung. My collaborators
H. Sommer, U. Bonas, K. Upadhyaya, E. Krebbers,
R. Piotrowiak and R. Hehl have cloned and analysed

the Tam1, Tam2 and Tam3 elements all integrated into the chalcone synthase gene of Antirrhinum majus. All integrated elements are flanked by small sequence duplications like bacterial transposons. In the case of Tam1, which is integrated into the promoter of the chs gene a 3 bp duplication is seen. In revertants of this line the 17 kb Tam1 element is excised but a small mutation is left behind (16).
Much more detailed information on the excision process of transposable elements has been obtained by my collaborators Zs. Schwarz-Sommer, A. Gierl and H. Cuypers who studied En(Spm)-induced revertants of wx m-8. This allele has a 2 kb DNA insert in an exon of the waxy gene of Zea mays (17).
Since wx m-8 responds to En(Spm) emitted signals this insert is envisioned as the receptive component of the En(Spm) transposable element system. Integration of this receptive component, termed Spm-I8 is accompanied by the formation of a 3 bp duplication at the target site.
In the absence of En in the genome Spm-I8 blocks the expression of the waxy gene, hence no particle bound UDPG starch transferase activity is detected, while in the presence of En Spm-I8 is excised. Excision of Spm-I8 during the development of the endosperm of the maize kernels leads to a variegated phenotype, i.e. sectors of cells that are mutant and other sectors of cells showing wildtype waxy activity (17). If an excision event has occured early enough in the plant development such that it occured in germ cells, then a stable full waxy$^+$ revertant is obtained. My colleagues have cloned and sequenced many somatic and a few germinal reversion events.
Besides precise excisions of the element leaving the duplication behind, wildtype sequences, small deletions and frameshift mutations were recovered, all these, however, were somatic events where correlation to waxy function is impossible.
Most remarkable, however, were the 2 germcell events studied thus far, which clearly have restored wildtype waxy function.
Fig. 3 shows that in these cases 3 extra bases compared to the wildtype sequence are left behind. This means that the UDPG-starch transferase now contains an extra amino acid, which apparently does not interfere with the function of the enzyme. In

one case a leucine and in the other case a serine is inserted. This event is not specific for germ-cells, since it also occurs in somatic tissue (fig 3).

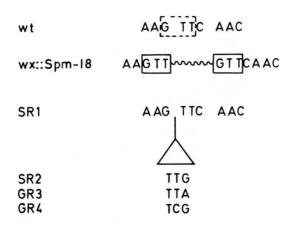

Legend to Fig. 3
Three codons of the waxy wildtype sequence are shown. The dashed box gives the position of integration for the Spm-I8 element. The duplications flanking the integrated element (wavy line) are boxed in. The sequence of 2 somatic and 2 germinal revertants are given. The additional triplets seen in three of the four revertants are indicated.

The sequence of all revertants obtained has lead to a model for transposition for at least some of the plant elements which differs from the well known replicative mode of bacterial transposition.

ROLE OF IS- AND COMPLEX TRANSPOSABLE ELEMENTS 111

EXCISION OF A PLANT TRANSPOSABLE ELEMENT

A. A transposase generates staggered nicks

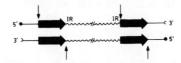

B. Hypothetical intermediates

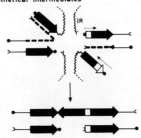

C. Possible excision products

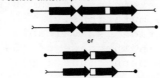

Legend to Fig. 4
A) Dots indicate the 5' end of the single stranded DNA, while half circles represent the 3' end.
Filled arrows are the duplications flanking the integrated elements, IR is inverted repeats, wavy line represents the element.
Vertical arrows indicate the position of the nicks.
B) Dotted arrows give the direction of the 5' exonucleolytic degradation, open parts of the filled arrows represent the extent of sequences removed.
Broken line indicates repair DNA synthesis.

4. EXCISION - INTEGRATION MODE OF TRANSPOSITION SEEN IN PLANT TRANSPOSABLE ELEMENTS

Clear genetic evidence compatible with an excision integration mode of transposition comes from the twin spot analysis in the modulator of pericarp (Mp) system of Zea mays. Molecularly this has not yet been analysed.
But the study of transposable element induced revertants reveals details of the excision integration process. The model devised is given in brief below. An extended version is in preparation. It is assumed that excision is initiated by staggered nicks at both ends of the integrated element (Fig. 4). The hypothetical intermediates drawn show how repair enzymes, polymerase and 5' exonuclease can operate on these substrates. In addition the Fig. 4 also features switching of the template during DNA synthesis, in a similar manner to that described for the formation of mini-insertions in IS2 (14, 18).
From the indermediates shown in Fig. 4 all revertant sequences can be deduced.
In any case the excision of an integrated plant element leads to mutations.
Not much information is available in bacterial systems, except in the case of IS4 were no sequence alterations have been found (19).

5. CONCLUDING REMARKS

Clearly IS- and Tn-element can cause all sorts of chromosomal rearrangements.
These have been reviewed on various occasions (20, 21).
In this article emphasis was given to their role in evolution of new genes. What seems to be missing or what has not yet been seen in the prokaryotic system is the capacity, as seen with plant transposable elements to serve as high frequency generators of mutations leading to altered protein structures, a process needed for the evolution of new functions.

REFERENCES

1. Saedler, H. (1977). The role of IS1 and IS2 in the evolution of the E.coli chromosome and of its plasmids. In "DNA Insertion Elements, Plasmids and Episomes", Cold Spring Harbor Laboratory p65
2. Saedler, H. and Ghosal, D. (1978). The role of IS-elements in E.coli. 28. Colloqium-Mosbach 1977 p. 41. Integration and Excision of DNA Molecules, Springer Verlag.
3. Saedler, H. et al. (1979). The role of gene rearrangement in evolution. Modern Trends in Human Leukemia III p. 507 Springer Verlag.
4. Saedler, H. and Starlinger, P. (1967). O^0 Mutations in the Galactose Operon in E. coli. I. Genetic Characterization. Molec. Gen. Genetics $\underline{100}$, 178.
5. Jordan, E. et al. (1968). O^0 and strong polar mutations in the gal Operon are insertions. Molec. Gen. Genetics $\underline{102}$, 353.
6. Hirsch, H.J. et al. (1972). Insertion mutations in the control region of the Galactose Operon of E. coli. II. Physical characterization of the mutations. Molec. Gen. Genetics $\underline{115}$, 266.
7. Fiandt, M. et al. (1972). Polar mutations in lac, gal and phage λ consist of a few DNA sequences inserted with either orientation. Mol. Gen. Genet. $\underline{119}$, 223
8. Saedler, H. and Heiß, B. (1973). Multiple copies of the insertion-DNA sequences IS1 and IS2 in the chromosome of E. coli K12. Molec. Gen. Genetics $\underline{122}$, 267.
9. Hu, S. et al. (1975). Electron microscope heteroduplex studies of sequence relations among bacterial plasmids: Identification and mapping of the insertion sequences IS1 and IS2 in F and R plasmids. J. Bacteriol. $\underline{122}$, 764
10. Saedler, H. et al. (1974). IS2, a genetic element for turn-off and turn-on of gene activity. Molec. Gen. Genetics $\underline{132}$, 265
11. Reif, H.J. and Saedler, H. (1975). IS1 is involved in deletion formation in the gal region of E. coli K12. Molec. Gen. Genetics $\underline{137}$, 17.
12. Sommer, H. et al. (1981). A new type of IS-mediated deletion. Mol. Gen. Genetics $\underline{184}$, 300

13. Cornelis, G. and Saedler, H. (1980). Deletion and an inversion induced by a resident IS1 of the lactose transposon Tn 951. Mol. Gen. Genetics 178, 367.
14. Ghosal, D. and Saedler, H. (1987). DNA sequence of the mini-insertion IS2-6 and its relation to the sequence of IS2. Nature 275, 611
15. Sommer, H. et al. (1979). IS2-43 and IS2-44: New alleles of the insertion sequence IS2 which have promoter activity. Mol. Gen. Genetics 175, 53.
16. Bonas, U. et al. (1984). The 17-kb Tam1 element of Antirrhinum majus induces a 3-bp duplication upon integration into the chalcone synthase gene. EMBO J. vol.3 no. 5, 1015.
17. Schwarz-Sommer, Z. et al. (1984). The Spm(En) transposable element controls the excision of a 2-kb DNA insert at the wx m-8 allele of Zea mays. EMBO J. vol.3, no.5, 1021.
18. Ghosal, D. et a. (1979). The DNA sequence of IS2-7 and the generation of mini-insertions by replication of IS2 sequences. 43. Cold Spring Harbor Symp. Quant. Biol.
19. Habermann, P. et al (1979). IS4 is found between eleven or twelve base pair duplications. Mol. Gen. Genet. 175, 369
20. Starlinger, P. and Saedler, H. (1976). IS-Elements in microorganisms. Current topics in microbiology and immunology vol. 75, 111 Springer Verlag
21. Calos, M.P. and Miller, J.H. (1980). Transposable elements. Cell 20, 579

5 S RIBOSOMAL RNA AS A TOOL FOR STUDYING EVOLUTION

R. De Wachter, E. Huysmans and A. Vandenberghe

Departement Biochemie
Universiteit Antwerpen (U.I.A.)
Antwerp, Belgium

INTRODUCTION

5 S ribosomal RNA is a nearly universally occurring component of the large ribosomal subunit. It is present in the ribosomes of eubacteria, archaebacteria and eukaryotes, as well as in the ribosomes of chloroplasts and plant mitochondria, but not in those of animal and fungal mitochondria. The chain length of 5 S RNA is usually close to 120 nucleotides, ranging from 112 to 130 nucleotides for more than 200 presently investigated structures. A single exception to this rule was found in the archaebacterium *Halococcus morrhuae*, whose ribosomes contain a structural equivalent of 5 S RNA with a chain length of 231 nucleotides.

5 S RNA was discovered in, and first isolated from, ribosomes of *Escherichia coli* in 1964 (1). Ever since its discovery, investigators have been intrigued by the nearly universal occurrence of this ribosome constituent and have tried to find a function for it. One of the most frequently cited proposals is that a more or less conserved sequence in one of the hairpin loops of 5 S RNA secondary structure, which happens to be the complement of the conserved GTψCG sequence in tRNA, would serve as a binding site for the latter molecule. However, this hypothesis has recently been disproven (2), and many other proposals for 5 S RNA function, reviewed in (3) and (4), remain experimentally unsubstantiated.

If the function of 5 S RNA remains elusive, its structure is nevertheless intensively studied. This is due in part to the circumstance that its isolation is relatively easy and that the RNA chain is short enough to be accessible to direct RNA sequence analysis, as outlined in the methods section. Hence 5 S RNA was the first ribosomal RNA for which complete primary structures could be determined. The first such structure was reported in 1967 (5) and at present, the nucleotide sequence is known for the 5 S rRNAs of more than 200 species and cell organelles. This rich collection of sequences has served two main purposes : the derivation of models for 5 S RNA secondary structure, and the study of evolution.

The question of 5 S RNA secondary structure will be shortly reviewed in this paper because it has some relevance to the use of 5 S RNA as an evolutionary marker. As a tool for studies in molecular evolution, 5 S RNA has some specific merits and disadvantages, when compared to other informational biopolymers. Its nearly universal occurrence is an undisputable advantage. Its small size makes it easily amenable to sequence analysis, but on the other hand means that it has a limited information content. A third aspect is its extremely conservative character. As discussed below, 5 S RNA evolves at an estimated rate of less than 2×10^{-10} mutations per nucleotide-year. A striking illustration of this conservativity is the finding that all hitherto investigated mammalian 5 S RNAs were found to be identical in structure.

The combination of universality, small size and low mutation rate make 5 S RNA a disappointing object for evolutionary studies on a small or intermediate time scale. On the other hand, it can serve a useful purpose for analyzing ancient divergence patterns among major taxa, and for reconstructing an outline of the evolutionary relationships between all presently living species and endosymbionts.

MATERIALS AND METHODS

Isolation and sequencing of 5 S RNA

The methods used for isolation and sequencing of 5 S RNA will only be described here in outline, with references to more detailed protocols published elsewhere.

Up to 1979, nearly all RNA sequencing work was carried out by the two-dimensional fingerprinting methods first introduced by Sanger *et al.* (6) and later perfected by Brownlee (7). In short, the method consists in digestion of the RNA with base-specific RNAses, separation of the resulting oligonucleotide sets, and sequencing of each oligonucleotide by exonuclease degradation. The succession of the oligonucleotides in the chain is than determined by partial digestion of the RNA with the same RNAses, fractionation of the digest by two-dimensional homochromatography (7) or gel electrophoresis (8), and analysis of the oligonucleotide composition of each partial degradation product. Besides being extremely tedious according to present-day standards, these methods are specific for, and require the availability of, uniformly ^{32}P-labeled RNA. As a result, only 5 S RNAs isolated from bacteria, vertebrate cell cultures, and plants were sequenced before 1980, because *in vivo* labeling by growth on [^{32}P] phosphate-containing media is possible in these cases.

The introduction of so-called gel-sequencing methods has changed this situation completely. After Maxam and Gilbert (9) published their well-known method for sequencing DNA-restriction fragments, methods were elaborated to sequence relatively small RNA molecules such as 5 S RNA according to the same principle. In short, the RNA is labeled *in vitro* with ^{32}P at one of the termini, degraded partially by a set of base-specific reactions, and the sets of degradation products are separated according to chain length on a thin polyacrylamide slab gel. The band patterns formed by the sets of radioactive degradation products, detected by autoradiography, allow to read the sequence directly.

Fig. 1 illustrates enzymatic methods for terminal *in vitro* labeling of 5 S RNA. The unesterified 3'-terminus can be labeled by reaction with [5'-^{32}P] cytidine bisphosphate in the presence of RNA ligase isolated from phage T4-infected *E. coli* (10) (Fig. 1a). The 5'-terminus can be labeled by reaction with [γ-^{32}P]-labeled ATP in the presence of polynucleotide kinase (11). For this purpose, however, the 5'-terminal phosphate residue has to be removed first by incubation with alkaline phosphatase (12) (Fig. 1b). Alternatively, an oligonucleotide such as (Ap)$_4$A can be ligated to the 5'-terminus, and the reaction product then 5'-terminally labeled in the presence of polynucleotide kinase (13) (Fig. 1c). Method (a) is the most straightforward, but the sequencing in the neighbourhood of the 5'-terminus often requires the preparation of 5'-end labeled 5 S RNA. In our experience, some 5 S RNA preparations are difficult to dephosphorylate (method b), a difficulty which can be avoided by applying method (c).

Partial base-specific digestion of the terminally labeled 5 S RNA can be achieved by a set of RNAses (14-17), each with a different specificity. Alternatively, the labeled RNA can be degraded by four different sets of chemical reactions (18), each more or less specific for one of the bases. The latter method is more easily reproducible and often serves to deduce the major part of the 5 S RNA sequence, but it can only be applied to 3'-labeled RNA. This is because the chemical degradation of an RNA chain at any one site yields a unique 3'-proximal fragment, but a set of chemically different 5'-proximal fragments. If the label is at the 5'-terminus, the resulting mixture of radioactive fragments

Symbols : pN .... N   5 S RNA chain with 5'-terminal phosphate and unesterified 3'-terminus
$\overset{*}{p}$ ^{32}P-labeled phosphate residue
$\overset{*}{p}Cp$ [5'-^{32}P] cytidine 5',3'-bisphosphate
$\overset{*}{p}ppA$ [γ-^{32}P] adenosine triphosphate

a) 3'-terminal labeling

$$pN \ldots N + \overset{*}{p}Cp \xrightarrow{\text{RNA ligase}} pN \ldots N\overset{*}{p}Cp$$

b) 5'-terminal labeling after dephosphorylation of 5 S RNA

$$pN \ldots N \xrightarrow{\text{alkaline phosphatase}} N \ldots N$$

$$\overset{*}{p}ppA + N \ldots N \xrightarrow{\text{polynucleotide kinase}} \overset{*}{p}N \ldots N + ppA$$

c) 5'-terminal labeling after ligation of oligo A to 5 S RNA

$$(Ap)_4A + pN \ldots N \xrightarrow{\text{RNA ligase}} (Ap)_4ApN \ldots N$$

$$\overset{*}{p}ppA + (Ap)_4ApN \ldots N \xrightarrow{\text{polynucleotide kinase}} \overset{*}{p}(Ap)_4ApN \ldots N$$

Fig. 1   In vitro terminal labeling of 5 S RNA with ^{32}P.

with slightly different electrophoretic mobilities gives rise to a diffuse band. Hence 5'-terminally labeled RNA can only be used for enzymatic degradation.

In case the chemical degradation method has yielded the entire sequence excepting the 5'-terminal nucleotide, it may suffice to identify the latter by complete hydrolysis with P$_1$ nuclease (19) of 5'-labeled RNA, if this can be obtained by method b (Fig. 1) :

$$\overset{*}{p}N \ldots N \xrightarrow{P_1 \text{ nuclease}} \overset{*}{p}N + xpN$$

The resulting radioactive nucleoside 5'-phosphate can be identified by thin layer chromatography (20). Alternatively, unlabeled 5 S RNA may be subjected to alkaline hydrolysis and the nucleoside bisphosphate resulting from the 5'-terminus identified by liquid chromatography (21) :

$$pN \ldots N \xrightarrow{OH^-} pNp + xNp + N$$

Although the methods available for sequencing of small RNAs are straightforward in principle, a number of problems often arise in practice. The first condition for success is to start with a homogeneous 5 S RNA. Many 5 S RNA preparations are heterogeneous in length, or in sequence, or both. After terminal labeling, the RNA should therefore carefully be fractionated by gel electrophoresis, which often will show the presence of more than one component. Gels with a polyacrylamide concentration of 6 to 8%, containing 7 M urea, pH 8.3 (18) usually give a good separation of components with identical sequence, but differing by the presence of one or more terminal bases. Components of identical chain length but different sequence can often be separated on 5% gels containing 4 M urea, pH 8.3 (22,23). The major part of the sequence of a purified 3'-labeled 5 S RNA can often be deduced from two or three sequencing gels of adequate polyacrylamide concentration run after chemical degradation (18). The gel concentration should be chosen at 6% to 8% for optimal separation of long degradation products (60-120 nucleotides), 12% for intermediate products (20-60 nucleotides) to 20% for the shortest chain lengths (1-20 nucleotides). The existence of a tight secondary structure in the examined 5 S RNA can result in the phenomenon of band compression (24), in which case sequencing gels can be run at elevated temperature (25) in an attempt to improve resolution.

The amount of 5 S RNA needed for a terminal labeling experiment is of the order of 10 μg. As a rule, about 100 μg of 5 S RNA should be sufficient for complete sequencing. In outline, the preparative procedure (13,26) is as follows. A suitable amount of animal or plant tissue (usually of the order of 100 g) or cell paste in the case of microorganisms (usually 2 to 20 g) is homogenized by an appropriate method. Cell or tissue debris are removed by low speed centrifugation, and ribosomes are precipitated from the supernatant by ultracentrifugation. Ribosomal RNA is isolated by phenol extraction of the ribosomes, followed by ethanol precipitation of the aqueous phase. An alternative method (22,27) which is shorter but yields a less pure ribosomal RNA preparation, consists in direct homogenisation of the tissue or cells in the presence of phenol. In both procedures the redissolved RNA is fractionated by electrophoresis on a preparative polyacrylamide slab gel and the 5 S RNA band eluted (26).

5 S RNA sequences available for comparative studies

The availability of the new sequencing methods has resulted in a sharp increase of the number of 5 S RNA sequences published since 1980. At the time of writing (July 1984) more than 200 sequences are available, 175 of which are compiled in the last annual review appearing in "Nucleic Acids Research" (28). It should be noted that the number of sequences reported is different from the number of species examined, because the same structure may be found in several related species, and conversely, some species contain several different 5 S RNAs due to polymorphism in their 5 S RNA genes. The latter phenomenon is detected with increasing frequency as the methods for fractionating 5 S RNAs of slightly different sequence are improved (23) and has probably gone unnoticed in many of the earlier studies.

The reliability of the reported data is a cause for concern. Sequencing errors are inevitably committed from time to time. Prior to 1980, sequencing of a 5 S RNA involved such an amount of work, that complete sequencing was often limited to a single representative of a set of related species, other structures being derived from observed differences in the oligonucleotide catalogs. In this way however, any mistake committed in the ordering of oligonucleotides by partial digestion is repeated in all examined species. Several errors due to erroneous oligonucleotide ordering in earlier studies have come to light when some structures were reexamined by gel sequencing methods. A notorious example is a *Chlorella* 5 S RNA sequence that formed the basis of a fashionable eukaryote-specific 5 S RNA secondary structure model (29) until it was proven to be completely wrong (30). Errors have also been committed on the basis of gel-sequencing, and are probably due to experimental difficulties such as heterogeneity or band compression mentioned above. Sequence corrections have been published for 19 of the approximately 200 structures hitherto reported, although only a fraction of them has been reexamined. The problem of accuracy of reported results is probably a general one in nucleic acid sequence analysis. Actually in the case of 5 S RNA errors show up rather frequently because of the relative conservativity of primary structure and chain length and the strong conservativity of secondary structure. Several sequencing errors have been discovered and corrected (13,23, 31,32) on the basis of distortions brought about in a sequence alignment and superimposed base pairing scheme (33, 28) and additional corrections (unpublished) are forthcoming.

Construction of phenograms

The 5 S RNA phenogram appearing in the Results section was constructed from an alignment of 234 5 S RNA sequences. The alignment of 175 of these sequences has been published (28), and part of it is illustrated in the Results section.

Additional sequences used in the computation were taken mainly from papers in the 1984 issues of "Nucleic Acids Research" and unpublished work from our laboratory. The sequence alignment comprises 148 positions. The computer program that constructs the phenogram first calculates a dissimilarity matrix. The dissimilarity D_{AB} between any pair of sequences, A and B, is calculated according to the expression

$$D_{AB} = -\frac{3}{4} \ln\left[1 - \frac{4}{3}\left(\frac{S}{I+S}\right)\right]\frac{I+S}{N} + \frac{G}{N}$$

where N, S, I, and G are defined as follows :
  I the number of alignment positions where sequences A and B contain an identical nucleotide.
  S the number of positions where A is substituted with respect to B.
  G the number of positions where A contains a gap and B a nucleotide or vice versa.
  N the number of positions where at least one of the two sequences contains a nucleotide (N = I + S + G).

The first term of the above equation comprises a correction for multiple mutations per site (34). As a result, the dissimilarity between two extremely divergent sequences can be larger than 1 (or 100 percent) since it is possible that they have diverged by more than one mutation per site on the average. Starting from the dissimilarity matrix, sequences are then clustered by the weighted pairwise grouping method (35).

RESULTS

Secondary structure of 5 S RNA

A brief account of our studies on 5 S RNA secondary structure will be given here because these have allowed us to construct a 5 S RNA sequence alignment (33,28) that we estimate superior to a previous proposal (36). Since the construction of a well-founded sequence alignment is a prerequisite for the deduction of a meaningful evolutionary tree, secondary structure considerations have some bearing on the use of 5 S RNA as an evolutionary marker. Moreover the secondary structure, although universal in outline, shows limited taxon-specific variations (37,38,39). These variations, although not yet fully exploited, may be useful in future evolutionary studies. The secondary structure models discussed below have been deduced by comparative examination of sequences, an approach which in our view has been much more fruitful than experimental probing of 5 S RNA secondary structure. The latter experiments, although very numerous, will not be considered here.

Secondary structures for 5 S RNA were proposed as soon as the first sequences were reported, but contrary to the situation with tRNA, where the cloverleaf model gained acceptance very soon, complete confusion prevailed for several years. Eight years after the first 5 S RNA was sequenced, a review (3) listed 15 primary structures and 20 proposals for secondary structure models. The dust settled somewhat in 1975, when Fox and Woese (40) published a secondary structure model assuming the existence of 4 helices in bacterial 5 S RNAs. The structure of eukaryotic 5 S RNAs remained in discussion until 1980, when a consensus was reached that the folding is similar to that of bacteria, but involves the existence of a fifth helix (30,41). In 1982 however, several research groups independently arrived at the conclusion (33,38,42) that 5 S RNA secondary structure is in fact universal and contains 5 helices in all species. It is worth mentioning that this conclusion had been anticipated in one of the early papers (43) on 5 S RNA secondary structure.

The existence of the five areas of sequence complementarity in 5 S RNAs isolated from eukaryotes, eubacteria, chloroplasts and archaebacteria is illustrated in the alignment of Fig. 2, which contains a 5 S RNA sequence from each of these groups.

In retrospect, a number of causes can be cited to account for the fact that the derivation of a detailed and universal secondary structure model for 5 S RNA has been so laborious, although the molecule is not much larger than tRNA.

(1) Contrary to the situation in tRNA, whose cloverleaf structure contains only hairpin loops and multibranched loops, 5 S RNA also contains interior- and bulge loops. This increases the complexity of the structure.
(2) The probable occurrence of odd base pairs (or non-standard base pairs : all pairs other than G·C, A·U and G·U) in certain molecules and helices (13,23,26,31,32,33,39) violates certain rules (44) often used for the prediction of RNA secondary structure.
(3) A number of sequencing errors (discussed above) have delayed the insight that eubacterial and eukaryotic 5 S RNAs show homologous sets of sequence complementarities.
(4) In fact there is not a single 5 S RNA secondary structure, but the molecule can alternate between a set of slightly different structures.

The latter assertion requires some comment. It has been shown (33) that one of the hairpin structures in the universal secondary structure model can always be drawn in two shapes, one usually carrying a bulge, the other containing an internal loop. Hence it was suggested that these alterna-

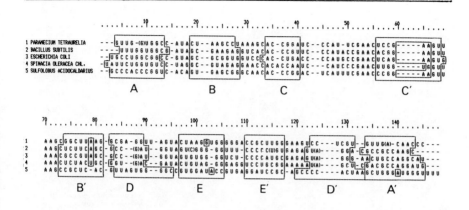

Fig. 2 Excerpt from the 5 S rRNA sequence alignment. Sequences from a eukaryote (1), two eubacteria (2,3), a chloroplast (4) and an archaebacterium (5) are aligned to maximize primary and secondary structure homology. Complementary areas A-A', B-B', etc. are boxed and correspond with the helical areas labeled A, B, etc. on the secondary structure models in Fig. 3. Positions remaining empty in all 5 sequences listed here are required to accomodate certain sequences in the complete alignment (28).

tive shapes, possible in all known 5 S RNAs, reflect the existence of a dynamic equilibrium in the secondary structure, with the molecule switching periodically from one shape to the other. This observation was recently extended (39) to the second hairpin existing in 5 S RNA secondary structure.

The universal 5 S RNA secondary structure model, and the existence of equilibrium structures, is illustrated in Fig. 3 for a eubacterial, a chloroplast, a eukaryotic, and an archaebacterial 5 S RNA. One of the equilibria, consisting in a switch of hairpin C between shapes C1 and C2, is completely universal. Another equilibrium involving alternative shapes of hairpin E exists only in eukaryotic, and possibly in certain archaebacterial 5 S RNAs. Moreover, the sets of structures among which hairpin E can alternate in eukaryotic 5 S RNAs seems to be taxon-specific. A more thorough discussion of these secondary structure equilibria, and a comparison with previously proposed dynamic models for 5 S RNA, has been published elsewhere (39). If this view of 5 S RNA structure is correct, then possibly the function of 5 S RNA can be likened to that of a molecular spring, which by flip-flopping

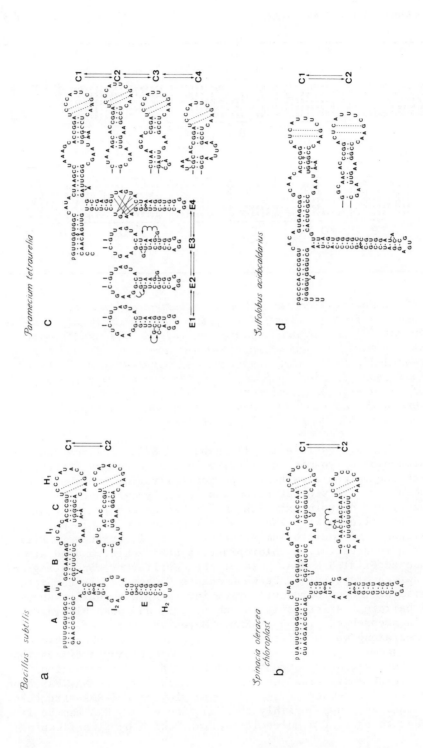

Bacillus subtilis

Spinacia oleracea chloroplast

Paramecium tetraurelia

Sulfolobus acidocaldarius

between alternative shapes keeps other parts of the surrounding molecular machinery (the ribosome) temporarily in alternative states.

Evolution of 5 S rRNA

5 S RNA has been used for phylogenetic reconstruction since 1973, when a tree based on an alignment of five sequences was published (45). Table 1 gives a list of 19 papers on the reconstruction of evolutionary trees, and mentions the number of species included and the employed computation method. Fig. 4 shows a simplified version of our most recent phylogenetic tree constructed by WPGMA clustering, as described in the methods section, from an alignment of 234 5 S RNA sequences. Of these, 162 are from eukaryotic species, 57 from eubacteria, 6 from archaebacteria and 9 from organelles. For a more detailed discussion of the eukaryotic branches of the tree, the reader is referred to previous papers (60,23,32). The prokaryotic branches are shown in more detail in Figs. 5 to 8.

No attempt is made in the following discussion to compare our results to previous 5 S RNA trees published by other research groups (see Table 1), mainly because most of these have been constructed from much smaller sets of sequences. The prokaryotic part of the tree will rather be compared with the classification found in the 8th edition of Bergey's manual (62) and the one based on T_1 ribonuclease cataloguing of 16 S rRNA (63). A schematic comparison of the three classifications is shown in Table 2 for the 57 eubacterial species whose 5 S RNA structure was included in the elaboration of our tree (Fig. 4 to 8). As can be seen from Fig. 4, the prokaryotic domain is divided in two parts corresponding to the eubacteria and archaebacteria (64). The archaebacteria are clustered closer to the eukaryotes than to the eubacteria. The lowest branch of the eubacteria is formed by the

---

Fig. 3 Secondary structure models for 5 S RNA.
Examples are given for 5 S RNA from a eubacterium (a), a chloroplast (b), a eukaryote (c), and an archaebacterium (d). Universal structural parts, labeled only in Fig. 3a, are helices A to E, loops M (multibranched), I_1 and I_2 (interior) and H_1 and H_2 (hairpin). Bases thought to form non-standard base pairs are connected by a losenge. Equilibrium shapes of the secondary structure are labeled C_1, C_2 etc. in area I_1-C, and E_1, E_2 etc. in area I_2-E. Additional base pairing opportunities are indicated by dotted lines in loop H_1 and by continuous lines in loop I_2. Two-pointed arrows indicate bulges that can migrate along a helix (39).

TABLE 1

Previous evolutionary trees constructed from 5S RNA sequences. Column 3 gives some information of the method of tree construction : WPGMA, UPGMA (35) and the method of Fitch and Margoliash (F&M) (61) are average linkage clustering methods. Other trees are constructed by some kind of minimization procedure to select the most parsimonious tree out of all possible tree topologies (pars.).

Authors	Number and range[a] of sequences	Method of tree construction	Ref.
Sankoff et al..(1973)	5	pars.	45
Hori (1975)	17	WPGMA	46
Hori (1976)	19	WPGMA	47
Denis and Wegnez (1978)	12 (V)	pars.	48
Schwartz and Dayhoff (1978)	15	pars.	49
Hori and Osawa (1979)	54	WPGMA	36
Osawa and Hori (1979)	61	WPGMA	50
Hori and Osawa (1980)	71	WPGMA	51
Hinnebusch et al. (1981)	11	pars., F&M	52
Küntzel et al. (1981)	47	F&M	53
Fox et al. (1982)	14 (A)	UPGMA	54
Sankoff et al. (1982)	19 (B)	pars.	55
Küntzel (1982)	53	F&M	56
Hori et al. (1982)	95	WPGMA	57
Kumazaki et al. (1983)	15 (P)	UPGMA	58
Küntzel et al. (1983)	71	F&M	59
Huysmans et al. (1983)	82 (E)	WPGMA	60
Chen et al. (1984)	218[b]	WPGMA	23
Vandenberghe et al. (1984)	213[b]	WPGMA	32

([a]) Trees limited to certain taxa are indicated by one of the following abbreviations : B, prokaryotes; A, archaebacteria; E, eukaryotes; P, protozoa; V, vertebrates.
([b]) Only the outline of the tree is shown in the paper.

cyanobacteria. After the cyanobacteria are separated from the core of the eubacteria, the latter bifurcate into two main branches. One of them contains only gram negative bacteria. The other contains the gram positive bacteria as well as some gram negative bacteria. Beside the branches mentioned above, there are two additional branches appended at the bottom of the eubacterial lineage. The lowest is formed by the chloro-

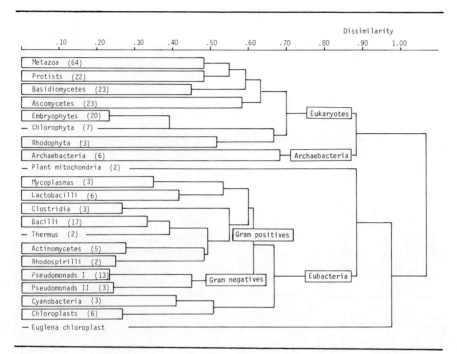

Fig. 4 *Outline of the evolutionary tree derived by WPGMA clustering of 234 5 S rRNA sequences.*
Only major taxa are displayed. The number of sequences in each taxon is mentioned in brackets. The right end of the box enclosing a taxon is situated at the first divergence point within the taxon. The Chlorophyta contain two branches diverging at dissimilarity 0.53, one of which also leads to the embryophytes. The prokaryote branches are elaborated further in figs. 5-8.

plast 5 S rRNA of *Euglena gracilis* and will be discussed below. The other contains plant mitochondrial sequences. This position is in good agreement with a bacterial origin of these organelles, as postulated by the endosymbiotic theory (65). It is impossible to assign the mitochondrial sequences to any particular branch of the eubacteria. Consequently the descent of mitochondria from bacteria related to *Paracoccus* (66) is not confirmed.

If we compare this general tree topology with the classification of Bergey's manual (Table 2), we find the two schemes to be fundamentally different. In the manual, the archaebacteria are not recognized as a unit. The different archaebacterial lineages are scattered throughout the eubacteria. The cyanobacteria are excluded from the eubacteria,

TABLE 2

Comparison of eubacterial classifications

The species mentioned in the first column are those appearing in the trees of fig. 5-8. In the case of Mycobacteria, Thermi, Bacilli, Lactobacilli, Streptococci and Mycoplasmas, only the genus is mentioned.

Eubacterial 5 S rRNA sequences	Classification according to Bergey's manual (62)	Classification based on 5 S rRNA			Classification based on 16 S rRNA (63)		
		Mitochondria			Mitochondria		
Chloroplasts		Chloroplasts		Cyanobacteria	Chloroplasts		Cyanobacteria
Synechococcus lividus	not included	Cyanobacteria			Cyanobacteria		
Anacystis nidulans							
Prochloron sp.							
Proteus vulgaris	fac. anaerobic gram negatives	Pseudomonads I		Pseudomonads	Purple sulphur and relatives		Purple bacteria
Escherichia coli							
Photobacterium phosphoreum							
Beneckea harveyi							
Azotobacter vinelandii	aerobic gram negatives						
Pseudomonas fluorescens							
Pseudomonas aeruginosa							
Pseudomonas cepacia		Pseudomonads II			Purple non sulphur II		
Alcaligenes faecalis							
Aquaspirillum serpens	spiral and curved bacteria						
Rhodospirillum rubrum	Rhodospirillales	Rhodospirilli			Purple non sulphur I		
Paracoccus denitrificans	aerobic gram negatives			Actinomycetes			
Mycobacterium	Actinomycetes	Actinomycetes			Actinomycetes		Actinomycetes
Streptomyces griseus							
Micrococcus lysodeikticus	non endosporous gram positives						
Thermus	aerobic gram negatives	Thermus		Clostridia	not included in the study		
Bacillus	endosporous gram positives	Bacilli			Bacilli		Clostridia
Clostridium oceanicum		Clostridia			Clostridia I		
Clostridium pasteurianum							
Clostridium butyricum							
Clostridium bifermentans	non endosporous gram positives	Lactobacilli			Clostridia II		
Lactobacillus					Lactobacilli		
Streptococcus					Streptococci		
Mycoplasma	Mycoplasmatales				Mycoplasmas		
Spiroplasma sp.							

although it was fully recognized that, beside their pigmentation, they are typical eubacteria (67). On the contrary, when the classification derived from the 5 S RNA tree is compared with that based on 16 S RNA cataloguing (Table 2), a much better correspondence is seen.

The detailed structure of the cyanobacterial branch of the 5 S RNA tree is shown in Fig. 5. Along with the cyanobacteria, this lineage contains the chloroplast sequences of higher plants. We included the plastidial sequence of *Euglena gracilis* in this subtree, although, as mentioned above, it clusters way down the eubacterial branch. We believe the remoteness of this sequence may be caused by a high evolutionary rate, a tendency displayed by all plastidial sequences. The close relationship of the chloroplast sequences and the cyanobacteria provide further evidence for a cyanobacterial descent of these eukaryotic organelles. Another interesting species included among the cyanobacteria is *Prochloron* sp. These prokaryotic organisms were once thought to be the intermediate form between prokaryotes and eukaryotes by a cyanophycean-chlorophycean link. They were therefore classified in a new division, the Prochlorophyta (68). This view is clearly not supported by the 5 S rRNA analysis. The picture described above is comparable with that based on

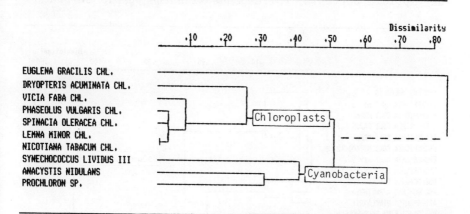

Fig. 5 *Subtree of cyanobacterial and chloroplast 5 S rRNA sequences.*
The dashed line connecting the chloroplast sequence of Euglena gracilis to the cyanobacterial and higher plant chloroplast sequences indicates that the former sequence does not belong to the cyanobacterial and chloroplast 5 S rRNA cluster. This is further explained in the text.

16 S RNA oligonucleotide catalogs.

The subtree in Fig. 6 gives the structure of the branch comprising most gram negative bacteria. Comparison (Table 2) of the obtained classification with that of Bergey's Manual, shows a number of differences. Firstly, the genus *Pseudomonas* is less homogeneous than previously thought. On the basis of 5 S RNA structure it is split into two widely separated groups which seem to correspond with two of the four sections recognized in Bergey's Manual. Future sequencing work on this genus may uncover whether the correspondence extends to the other sections. Secondly, in the manual, *Aquaspirillum* is classified as a spiral-shaped bacterium in a separate group, whereas 5 S RNA analysis reveals its true affinity to the pseudomonads. This points to the unreliability of morphological characters to define major taxa of eubacteria, as already observed by others (69). The relatedness of *Azotobacter* to the pseudomonads is confirmed. Some genera which in Bergey's Manual are classified as having uncertain affiliation could be assigned to their closest relatives. This is the case of *Alcaligenes* and *Beneckea*. The classification obtained on the basis of 5 S RNA structure can be correlated with that based on 16 S RNA catalogs as follows (Table 2). The group named pseudomonads I in the 5 S RNA classification, which contains the fluorescent pseudomonads, the enteric and the vibrioid bacteria, corresponds with the purple sulphur

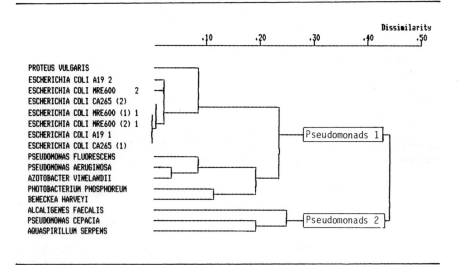

Fig. 6 Subtree of the 5 S rRNA sequences from gram negative bacteria.

bacteria and relatives in the 16 S RNA classification. The group named pseudomonads II corresponds with the purple non sulphur bacteria II as defined on the basis of 16 S RNA. This comparison can only be superficial since the photosynthetic and chemolithotrophic bacteria are poorly represented in the 5 S RNA sequence collection.

The subtree containing the gram positive bacteria is drawn in Fig. 7a. There are some striking differences between this subtree and the classification of the gram positives as discerned in Bergey's Manual or with 16 S rRNA catologs (Table 2). Fig. 7a shows that the lactic acid bacteria and mycoplasmas cluster as the lowest branch in the subtree. This branch may be excluded from its true position by the deviating structure of the mycoplasmal 5 S rRNA. These molecules are characterized by deletions in the area of helix E in the secondary structure model, which makes it very difficult to align sequences unequivocally in this area. We employed two procedures to examine this problem. If the mycoplasmal sequences are omitted completely in the tree construction, the lactic acid bacteria cluster close to the bacilli. This enlarges the distance between bacilli and actinomycetes, resulting in the latter being clustered with the clostridia l. The same shift in tree topology is observed if the mycoplasmal sequences are conserved, but the positions in the area of helix E are excised from the alignment. The mycoplasmas then remain clustered to the lactobacilli in the altered topology described above, which is illustrated in Fig. 7b.

A second point of difference with the 16 S RNA classification is the dispersion of the clostridia over the different branches of the gram-positive bacteria. There are two possible explanations for this finding. Considering the evolutionary success of the clostridial phenotype, it is possible that it originated more than once during evolution. This would mean that the topology of Fig. 7 reflects true phylogeny. Alternatively, the present-day clostridia could be the descendants of a large parental group of bacteria from which the other gram positive bacteria, with the possible exeption of the actinomycetes, evolved. This implies that bacilli evolved from clostridia ancestral to *C. oceanicum* and *C. tyrobutyricum*, and the lactic acid bacteria evolved from ancestors of *C. bifermentans*.

A third fact requiring comment is the occurrence of the gram negative genera *Rhodospirillum*, *Paracoccus* and *Thermus* in the gram positive branch. This is not easily explainable considering the homogeneity of the gram positive bacteria. Actually, the 5 S RNAs of gram negative bacteria, including those found among the gram positives, are easily distinguishable from the latter on the basis of a slightly differ-

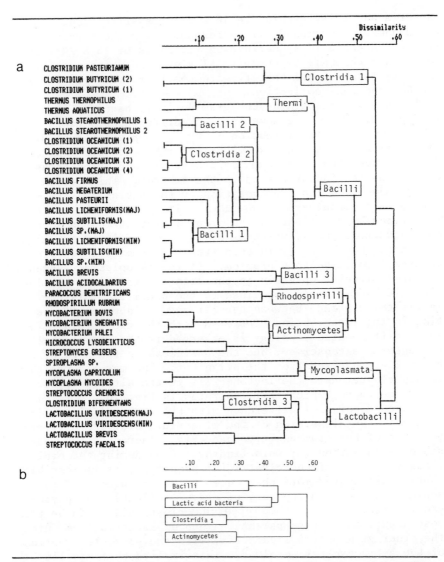

Fig. 7 Subtree of the 5 S rRNA sequences from gram positive bacteria.
The 5 S rRNA sequences of Clostridium tyrobutyricum are omitted from the tree since they are identical to those of C. oceanicum. The tree in (a) also comprises some gram negative species. The simplified tree in (b) shows rearrangements occuring if the mycoplasmal sequences are omitted or if the positions forming the E helix in the secondary structure of 5 S rRNA are excised from the alignment as explained in the text.

ent secondary structure. The occurrence of the rhodospirilli as one of the lower branches of the gram-positive subtree may imply that the gram-positives as a whole originated from a gram-negative branch also ancestral to the rhodospirilli.

The archaebacterial subtree is shown in Fig. 8. It is similar to the one obtained on the basis of 16 S RNA catalogs except for the branching point of the halobacteria, which in the latter classification are closest to the Methanobacteriales. In the 5 S RNA tree the halobacteria diverge from the Methanomicrobiales.

DISCUSSION

Reliability of 5 S RNA as a phylogenetic marker

How reliable is the picture of bacterial evolution provided by the 5 S RNA tree ? One of the most important conditions that must be satisfied, is that the average rate of mutation should be approximately the same in the different bacterial lineages. In order to evaluate this question, we calculated the average dissimilarity, as defined in the methods section, between 5 S RNAs of each of the eubacterial groups mentioned in Fig. 5 to 7 and a number of eukaryotic reference groups mentioned in Fig. 4. The results are listed in Table 3. For each dissimilarity value, the normalized deviation from the mean is also listed in Table 3. The last column lists this deviation, averaged over seven eukaryotic reference groups. The results point to a relatively constant mutation rate for all eubacterial groups except the chloroplasts which have a higher rate.

Whereas the rate of evolution seems to be relatively uniform within the eubacterial primary kingdom, an appreciable difference was found with the eukaryotic evolutionary rate. An estimate of the absolute mutation rate was made in both

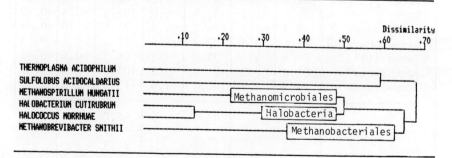

Fig. 8 *Subtree of the archaebacterial 5 S rRNA sequences.*

TABLE 3

Compared dissimilarity of eubacterial groups with respect to eukaryotic reference groups.

D_{xy} Average dissimilarity between a eubacterial group and a eukaryotic reference group.

m_D Mean of the average dissimilarities, D_{xy}, of the eubacterial groups relative to one eukaryotic reference group.

s_D Standard deviation on m_D.

d_x Normalised deviation of the average dissimilarity D_{xy} from its mean m_D:

$$d_x = (D_{xy} - m_D)/s_D$$

$\overline{d_x}$ Mean of the normalised deviations, d_x, averaged over all reference groups.

Similar calculations to those shown for metazoa, Ascomycetes and Rhodophyta were made for 4 other eukaryotic groups: protists, Basidiomycetes, Chlorophyta and embryophytes. The mean normalized deviation, $\overline{d_x}$, is averaged over all seven reference groups.

eubacterial groups	Eukaryotic reference groups						$\overline{d_x}$
	Metazoa		Ascomycetes		Rhodophyta		
	D_{xy}	d_x	D_{xy}	d_x	D_{xy}	d_x	
Clostridia 1	83.8	1.2	90.3	0.7	123.7	3.8	1.30
Clostridia 2	72.7	-0.4	80.1	-0.4	87.7	-1.0	-0.36
Clostridia 3	70.6	-0.6	87.6	0.4	101.2	0.8	0.01
Bacilli 1	73.5	-0.2	83.5	0.0	101.2	0.8	-0.07
Bacilli 2	64.8	-1.5	68.1	-1.7	81.1	-1.8	-1.17
Bacilli 3	69.6	-0.8	74.4	-1.0	91.5	-0.5	-0.47
Lactobacilli	69.7	-0.8	74.9	-1.0	90.5	-0.6	-0.73
Mycoplasmas	83.1	1.1	89.4	0.6	95.5	0.1	0.54
Actinomycetes	82.8	1.1	95.6	1.3	100.9	0.8	0.59
Thermi	66.6	-1.2	83.7	0.0	91.4	-0.5	-0.77
Pseudomonads I	69.3	-0.9	76.2	-0.8	81.6	-1.8	-0.94
Pseudomonads II	79.5	0.6	78.1	-0.6	97.7	0.4	0.73
Rhodospirilli	70.5	-0.7	95.0	1.3	88.3	-0.9	-0.61
Cyanobacteria	84.3	1.3	91.1	0.8	105.2	1.4	1.30
Chloroplasts	99.5	3.4	100.4	1.9	113.6	2.5	2.27
m_D	75.2		83.6		95.0		
s_D	7.12		9.06		7.52		

cases on the basis of geological data. In the case of bacteria, the only point of reference chosen was the occurrence of stromatolite beds, starting 3.1×10^9 years BP. and predominantly formed by cyanobacteria (70). We assumed this age to correspond to the time of divergence between cyanobacteria and other eubacteria and calculated a mean difference between the two groups, corrected for multiple and back mutation, of 0.60 ± 0.09 substitutions per nucleotide. This points to an evolutionary rate of 1.0×10^{-10} mutations per nucleotide-year. Similar calculations were made for eukaryotic 5 S RNAs, based on several paleontological points of reference concerning the metazoa and plants. The resulting average evolutionary rate is 1.7×10^{-10} mutations per nucleotide-year, which is twice the estimate for the eubacterial rate. This means that the picture of the divergence of the three primary kingdoms given by the general 5 S RNA tree (Fig. 4) may be quite distorted.

Due to the limited chain length of 5 S RNA there is a rather large spread on the average mutational distance measured between any pair of clusters. The relative standard error on such mutational distances often amounts to 10% for the lower branches. A solution to this problem could be the use of larger ribosomal RNAs such as 16 S - 18 S or 23 S - 28 S RNA. However, since the effort involved in sequencing of these molecules is much larger, a collection of structures as diverse as that of 5 S RNA will not be available in the near future. A short cut to this problem consists in the oligonucleotide catalog analysis of 16 S RNA (63,69) but in this case the full information content of the molecule is not exploited. The validity of the similarity coefficient used in such studies for measuring divergence among ancient branches has been criticized (50).

Evolution of prokaryotes

Based primarily on molecular data, eukaryotes, archaebacteria and eubacteria were placed as three separate lineages beside each other (64). However, geological data situate the emergence of eukaryotes at least 1×10^9 years after that of prokaryotic organisms (71). The archaebacteria are thought to be as old as or older than the eubacteria (72). Considering the higher evolutionary rate of the eukaryotes, it is very well imaginable that they originated somewhere within the prokaryote lineages. The 5 S RNA data support the eubacterial origin of plant mitochondria and chloroplasts with both organelles evolving from gram-negative lineages. The gram-negative bacteria are the oldest eubacteria represented by a 5 S rRNA sequence. This stands in strong constrast to the classical opinion about prokaryote evolution, which situates the oldest bacteria among the gram-positives. The clostridia have been designated as the most ancient bacteria on the basis of their obligate anaerobic and fermentative metabolism (73). Alternatively, the mycoplasmas have been postulated to be the oldest. Two lines of evidence substantiated this hypothesis. These were the lack of cell walls (74) and the small genome size (75) of the mycoplasmas. If the gram-positive bacteria are relieved from their ancestral position, the fermentative character of the metabolism of clostridia and lactic acid bacteria may be considered to result from a secondary loss of their respiratory chains. It seems that the whole branch of gram-positive bacteria evolves towards specialisation as saprophytes and parasites. The mycoplasmas which are derived from lactic acid bacteria, have gone furthest in this direction (76). They eliminated all superfluous genetic information, a parsimony which can be seen even in their 5 S RNAs, which due to the reduction in size of helix E have the shortest chain lengths among all known species.

ACKNOWLEDGMENTS

Our research was supported in part by grants from the Nationaal Fonds voor Wetenschappelijk Onderzoek.

REFERENCES

1. Rosset, R., Monier, R. and Julien, J. (1964). Les ribosomes d'*Escherichia coli*. I. Mise en evidence d'un RNA ribosomique de faible poids moléculaire, *Bull. Soc. Chim. Biol.* 46, 87-109.
2. Pace, B., Matthews, E.A., Johnson, K.D., Cantor, C.R. and Pace, N.R. (1982). Conserved 5 S rRNA complement to transfer RNA is not required for protein synthesis, *Proc. Natl. Acad. Sci. USA* 79, 36-40.
3. Erdmann, V.A. (1976). Structure and function of 5 S and 5.8 S RNA, *Prog. Nucl. Acid Res. Mol. Biol.* 18, 45-90.
4. McDougall, J. and Nazar, R.N. (1983). Tertiary structure of the eukaryotic ribosomal RNA. Accessibility of phosphodiester bonds to ethylnitrosourea modification, *J. Biol. Chem.* 258, 5256-5259.
5. Brownlee, G.G., Sanger, F. and Barrell, B.G. (1967). Nucleotide sequence of 5 S ribosomal RNA from *Escherichia coli*, *Nature* 215, 735.
6. Sanger, F., Brownlee, G.G. and Barrell, B.G. (1965). A two-dimensional fractionation procedure for radioactive nucleotides, *J. Mol. Biol.* 13, 373-398.

7. Brownlee, G.G. (1972). Determination of sequences in RNA. *In* "Laboratory Techniques in Biochemistry and Molecular biology" (Eds T. Work and E. Work), Vol. 3, part 1, p. 90. Elsevier, Amsterdam.
8. De Wachter, R. and Fiers, W. (1972). Preparative two-dimensional polyacrylamide gel electrophoresis of ^{32}P labeled RNA, *Anal. Biochem.* 49, 184-197.
9. Maxam, M.A. and Gilbert, W. (1977). A new method for sequencing DNA, *Proc. Natl. Acad. Sci. USA* 74, 560-564.
10. England, T.E. and Uhlenbeck, O.C. (1978). 3'-Terminal labeling of RNA with T4 RNA ligase, *Nature* 275, 560-561.
11. Richardson, C.C. (1965). Phosphorylation of nucleic acid by an enzyme from T4 bacteriophage-infected *Escherichia coli*, *Proc. Natl. Acad. Sci. USA* 54, 158-165.
12. Ursi, D., Vandenberghe, A. and De Wachter, R. (1982). The sequence of the 5.8 S ribosomal RNA of the crustacean *Artemia salina*. With a proposal for a general secondary structure model for 5.8 S ribosomal RNA, *Nucl. Acids Res.* 10, 3517-3530.
13. Dams, E., Vandenberghe, A. and De Wachter, R. (1983). Sequences of the 5 S rRNAs of *Azotobacter vinelandii*, *Pseudomonas aeroginosa* and *Pseudomonas fluorescens* with some notes on 5 S RNA secondary structure, *Nucl. Acids Res.* 11, 1245-1252.
14. Donis-Keller, H., Maxam, A.M. and Gilbert, W. (1977). Mapping adenines, guanines and pyrimidines in RNA, *Nucl. Acids Res.* 4, 2527-2538.
15. Simoncsits, A., Brownlee, G.G., Brown, R.S., Rubin, J.R. and Guilley, H. (1977). New rapid gel sequencing method for RNA, *Nature* 269, 833-836.
16. Donis-Keller, H. (1979). Site specific enzymatic cleavage of RNA, *Nucl. Acids Res.* 7, 179-192.
17. Boguski, M.S., Hieter, P.A. and Levy, C.C. (1980). Identification of a cytidine-specific ribonuclease from chicken liver, *J. Biol. Chem.* 255, 2160-2163.
18. Peattie, D.A. (1979). Direct chemical method for sequencing RNA, *Proc. Natl. Acad. Sci. USA* 76, 1760-1764.
19. Silberklang, M., Gillum, A.M. and RajBhandary, U.L. (1979). Use of in vitro ^{32}P labeling in the sequence analysis of nonradioactive tRNAs, *Methods Enzymol.* 59, 58-109.
20. Volckaert, G. and Fiers, W. (1977). A micro method for base analysis of ^{32}P-labeled oligoribonucleotides, *Anal. Biochem.* 83, 222-227.
21. Vandenberghe, A. and De Wachter, R. (1982). High pressure liquid chromatography determination of the 5'-terminal residue of small RNA molecules, *J. Liq. Chromatogr.* 5, 2079-2084.

22. Bartnik, E., Strugala, K. and Stepien, P.P. (1981). Cloning and analysis of recombinant plasmids containing genes for *Aspergillus nidulans* 5 S rRNA, *Current Genetics* 4, 173-176.
23. Chen, M.-W., Anné, J., Volckaert, G., Huysmans, E., Vandenberghe, A. and De Wachter, R. (1984). The nucleotide sequences of the 5 S rRNA of seven molds and a yeast and their use in studying ascomycete phylogeny, *Nucl. Acids Res.* 12, 4881-4892.
24. Kramer, F.R. and Mills, D.R. (1978). RNA sequencing with radioactive chain-terminating ribonucleotides, *Proc. Natl. Acad. Sci. USA* 75, 5334-5338.
25. Nazar, R.N. and Wildeman, A.G. (1981). Altered features in the secondary structure of *Vicia faba* 5.8 S rRNA, *Nucl. Acids Res.* 9, 5345-5358.
26. Fang, B.-L., De Baere, R., Vandenberghe, A. and De Wachter, R. (1982). Sequences of three molluscan 5 S ribosomal RNAs confirm the validity of a dynamic secondary structure model, *Nucl. Acids Res.* 10, 4679-4685.
27. Kumazaki, T., Hori, H. and Osawa, S. (1982). The nucleotide sequence of 5 S ribosomal RNA from sea anemone *Anthopleura japonica*, *FEBS Lett.* 146, 307-310.
28. Erdmann, V.A., Wolters, J., Huysmans, E., Vandenberghe, A. and De Wachter, R. (1984). Collection of published 5 S and 5.8 S ribosomal RNA sequences, *Nucl. Acids Res.* 12, r133-r166.
29. Vigne, R. and Jordan, B.R. (1977). Partial enzyme digestion studies on *Escherichia coli, Pseudomonas, Chlorella, Drosophila*, HeLa and yeast-5 S RNAs support a general class of 5 S RNA models, *J. Mol. Evol.* 10, 77-86.
30. Luehrsen, K.R. and Fox, G.E. (1981). Secondary structure of eukaryotic cytoplasmic 5 S ribosomal RNA, *Proc. Natl. Acad. Sci. USA* 78, 2150-2154.
31. Dams, E., Londei, P., Cammarano, P., Vandenberghe, A. and De Wachter, R. (1983). Sequences of the 5 S rRNAs of the thermo-acidophilic archaebacterium *Sulfolobus solfataricus (Caldariella acidophila)* and the thermophilic eubacteria *Bacillus acidocaldarius* and *Thermus aquaticus*, *Nucl. Acids Res.* 11, 4667-4676.
32. Vandenberghe, A., Chen, M.-W., Dams, E., De Baere, R., De Roeck, E., Huysmans, E. and De Wachter, R. (1984). The corrected nucleotide sequences of 5 S RNAs from six angiosperms, *FEBS Lett.* 171, 17-23.
33. De Wachter, R., Chen, M.-W. and Vandenberghe, A. (1982). Conservation of secondary structure in 5 S ribosomal RNA: a uniform model for eukaryotic, eubacterial, archaebacterial and organelle sequences is energetically favourable, *Biochimie* 64, 311-329.

34. Kimura, M. and Ohta, T. (1972). On the stochastic model for estimation of mutational distance between homologous proteins, *J. Mol. Evol.* 2, 87-90.
35. Sneath, P.H.A. and Sokal, R.R. (1973). "Numerical Taxonomy". Freeman, San Fransisco.
36. Hori, H. and Osawa, S. (1979). Evolutionary change in 5 S rRNA secondary structure and a phylogenetic tree of 54 5 S rRNA species, *Proc. Natl. Acad. Sci. USA* 76, 381-385.
37. Dams, E., Vandenberghe, A. and De Wachter, R. (1982). Nucleotide sequences of three poriferan 5 S ribosomal RNAs, *Nucl. Acids Res.* 10, 5297-5302.
38. Delihas, N. and Andersen, J. (1982). Generalized structures of the 5 S ribosomal RNAs, *Nucl. Acids Res.* 10, 7323-7344.
39. De Wachter, R., Chen, M.-W. and Vandenberghe, A. (1984). Equilibria in 5 S ribosomal RNA secondary structure, *Eur. J. Biochem.* 143, 175-182.
40. Fox, G.E. and Woese, C.R. (1975). 5 S RNA secondary structure, *Nature* 256, 505-507.
41. Hori, H., Osawa, S. and Iwabuchi, M. (1980). The nucleotide sequence of 5 S rRNA from a cellular slime mold *Dictyostelium discoideum*, *Nucl. Acids Res.* 8, 5535-5539.
42. Böhm, S., Fabian, H. and Welfle, H. (1982). Universal structural features of prokaryotic and eukaryotic ribosomal 5 S RNA derived from comparative analysis of their sequences, *Acta Biol. Med. Germ.* 41, 1-16.
43. Nishikawa, K. and Takemura, S. (1974). Structure and function of 5 S ribosomal ribonucleic acid from *Torulopsis utilis*, *J. Biochem.* 76, 935-947.
44. Tinoco, I., Borer, P.N., Dengler, B., Levine, M.D., Uhlenbeck, O.C., Crothers, D.M. and Gralla, J. (1973). Improved estimation of secondary structure in ribonucleic acids, *Nature New. Biol.* 246, 40-41.
45. Sankoff, D., Morel, C. and Cedergren, R.J. (1973). Evolution of 5 S rRNA and the non-randomness of base replacement, *Nature* 245, 232-234.
46. Hori, H. (1975). Evolution of 5 S rRNA. *J. Mol. Evol.* 7, 75-86.
47. Hori, H. (1976). Molecular evolution of 5 S rRNA, *Molec. Gen. Genet.* 145, 119-123.
48. Denis, H. and Wegnez, M. (1978). Evolution of the 5 S rRNA genes in vertebrates, *J. Mol. Evol.* 12, 11-15.
49. Schwartz, R.M. and Dayhoff, M.O. (1978). Origins of prokaryotes, eukaryotes, mitochondria and chloroplasts, *Science* 199, 395-403.
50. Osawa, S. and Hori, H. (1979). Molecular evolution of ribosomal components. *In* "Ribosomes, structure, function and

genetics" (Eds Chambliss, G., Craven, G.R., Davies, J., Kahan, L., Nomura, M.) pp. 333-355. University Park Press, Baltimore.
51. Hori, H., Osawa, S. (1980). Recent studies on the evolution of 5 S rRNA. *In* "Genetics and evolution of RNA polymerase, tRNA and ribosomes" (Eds Osawa,S., Ozeki,H., Uchida,H. and Yuri,T.) pp. 539-551. University Press,Tokyo.
52. Hinnebusch, A.G., Klotz, L.C., Blanken, R.L. and Loeblich, A.R. (1981). An evaluation of the phylogenetic position of the dinoflagellate *crypthecodinium cohnii* based on 5 S rRNA characterisation, *J. Mol. Evol.* 17, 334-347.
53. Küntzel, H., Heidrich, M. and Piechulla, B. (1981). Phylogenetic tree derived from bacterial, cytosol and organelle 5 S rRNA sequences, *Nucl. Acids Res.* 9, 1451-1461.
54. Fox, G.E., Luehrsen, K.R. and Woese, C.R. (1982). Archaebacterial 5 S ribosomal RNA, *Zbl. Bakt. Hyg.* I. Abt. Orig. C3, 330-345.
55. Sankoff, D., Cedergren, R.J. and Mackay, W. (1982). A strategy for sequence phylogeny research, *Nucl. Acids Res.* 10, 421-431.
56. Küntzel, H. (1982). Phylogenetic trees derived from mitochondrial, nuclear, eubacterial and archaebacterial rRNA sequences : Implications on the origin of eukaryotes, *Zbl. Bakt. Hyg.* I. Abt. Orig. C3, 31-39.
57. Hori, H., Itoh, T. and Osawa, S. (1982). The Phylogenetic structure of the metabacteria, *Zbl. Bakt. Hyg.* I. Abt. Orig. C3, 18-30.
58. Kumazaki, T., Hori, H. and Osawa, S. (1983). Phylogeny of protozoa deduced from 5 S rRNA sequences, *J. Mol. Evol.* 19, 411-419.
59. Küntzel, H., Piechulla, B. and Hahn, U. (1983). Consensus structure and evolution of 5 S rRNA, *Nucl. Acids Res.* 11, 893-900.
60. Huysmans, E., Dams, E., Vandenberghe, A. and De Wachter, R. (1983). The nucleotide sequences of the 5 S rRNAs of four mushrooms and their use in studying the phylogenetic position of basidiomycetes among the eukaryotes, *Nucl. Acids Res.* 11, 2871-2880.
61. Fitch, W.M. and Margoliash, E. (1967). Construction of phylogenetic trees, *Science* 155, 279-284.
62. Buchanan, R.E. and Gibbons, N.E. (Eds) (1974) "Bergey's manual of determinative bacteriology". Williams and Wilkins, Baltimore.
63. Stackebrandt, E. and Woese, C.R. (1981). The evolution of prokaryotes. *In* "Molecular and cellular aspects of microbial evolution". (Eds Carlile, Collins and Moseley) pp. 1-31. University Press, Cambridge.
64. Woese, C.R. and Fox, G.E. (1977). Phylogenetic structure

of the prokaryotic domain : the primary kingdoms, *Proc. Natl. Acad. Sci. USA* 74, 5088-5090.
65. Sagan, A. (1967). On the origin of mitosing cells. *J. Theor. Biol.* 14, 225-274.
66. John, P. and Whatley, F.R. (1975). *Paracoccus denitrificans* and the evolutionary origin of the mitochondrium, *Nature* 254, 495-498.
67. Stanier, R.Y. and Cohen-Bazire, G. (1977). Phototrophic bacteria : The cyanobacteria, *Annu. Rev. Microbiol.* 31, 225-274.
68. Lewin, R.A. (1976). Prochlorophyta as a proposed new division of algae, *Nature* 261, 697-698.
69. Fox, G.E., Stackebrandt, E., Hespell, R.B., Gibson, J., Maniloff, J., Dyer, T.A., Wolfe, R.S., Balch, W.E., Tanner, R.S., Magrum, L.S., Zablen, L.B., Blakemore, R., Gupta, R., Bonen, L., Lewis, B.J., Stahl, D.A., Luehrsen, K.R., Chen, K.N. and Woese, C.R. (1980). The phylogeny of prokaryotes, *Science* 209, 457-463.
70. Cloud, P. (1976). Beginnings of biospheric evolution and their biogeochemical consequences, *Paleobiology* 2, 351-387.
71. Schopf, J.W. (1976). How old are the eukaryotes ?, *Science* 193, 47-49.
72. Matheson, A.T. and Yaguchi, M. (1982). The evolution of the archaebacterial ribosome, *Zbl. Bakt. Hyg. I. Abt. Orig.* C3, 192-199.
73. Broda, E. (1975). "The evolution of the bioenergetic processes" University Press, Oxford.
74. Edward, D.G. (1960). Biology of pleuropneumonia like organisms, *Ann. N.Y. Acad. Sci.* 79, 308-311.
75. Wallace, D.C. and Morowitz, H.J. (1973). Genome size and evolution, *Chromosoma* 40, 121-126.
76. Maniloff, J. (1983). Evolution of wall-less prokaryotes, *Annu. Rev. Microbiol.* 37, 477-499.

EVOLUTION OF LIGHT ENERGY CONVERSION

K. Krishna Rao, Richard Cammack and David O. Hall

Department of Plant Sciences
King's College
London, England

1. INTRODUCTION

The development in our biosphere of organisms with the capacity to harvest sunlight and convert the light-energy to chemical energy marked a major step in biological evolution. In recent years there has been a tremendous upsurge in research aimed at understanding the structure, function and evolution of the light-harvesting, energy converting and electron transport apparatus of photosynthetic organisms. Concurrent with the acquisition of basic knowledge on photosynthesis derived from these studies are the investigations by photochemists and photobiologists on artificial photocatalytic systems which could mimic the natural process carried out by bacteria and plants. It is almost impossible in this short review to give due credit to all original publications in this rapidly expanding field. Instead we have used as reference information contained in the Proceedings of the Sixth International Congress on Photosynthesis held in Brussels, 1983 (1), in the Proceedings of the Symposium on the Oxygen evolving system of Photosynthesis held in Japan 1983 (2) and in the excellent book on Photosynthesis edited by Govindjee (3).

2. CLASSIFICATION OF PHOTOSYNTHETIC ORGANISMS

The overall reaction of photosynthesis is the same in all photosynthesisers, an oxidation-reduction sequence initiated by light reactions, in which carbon dioxide (low energy) is reduced to carbohydrates (high energy) at the expense of hydrogen donors. It can be represented by the general equation proposed by Van Niel (4):

$$2H_2A + CO_2 \longrightarrow (CH_2O) + 2A + H_2O$$

143

TABLE 1

Characteristics of photosynthetic organisms (6)

Group	Electron Donors	Photosynthetic Pigments	Growth conditions and other properties
Green photosynthetic bacteria (Chlorobiaceae)	H_2S $Na_2S_2O_3$ H_2	Bchl a, Bchl c, Bchl d or e carotenoids	Light, autotrophic. Strict anaerobes, non motile (except *Chloroflexus*) e.g.: *Chlorobium limicola*
Purple sulphur bacteria (Chromateaceae)	H_2S $Na_2S_2O_3$ H_2 organic substrates	Bchl a or b carotenoids	Light, autotrophic or dark, heterotrophic, strict anaerobes, some are motile, e.g. *Chromatium*
Purple non-sulphur bacteria (Rhodospirillaceae)	H_2 organic substrates	Bchl a or b carotenoids	Light, autotrophic or heterotrophic. Dark, aerobic respirer, e.g.: *Rhodospirillum rubrum*
Cyanobacteria (Blue-green algae)	H_2O	Chlorophyll a Phycobilins	Light, autotrophic, some species will grow heterotrophically in dark, e.g.: *Anacystis nidulans*
Eukaryotic algae	H_2O	Chlorophyll a, chlorophyll b or chlorophyll c, Phycobilins (red algae) carotenoids	Light, autotrophic, some will grow heterotrophically in dark, e.g.: *Chlorella*
Higher plants	H_2O	Chlorophyll a chlorophyll b carotenoids	Light, autotrophic

Depending on the nature of H_2A, photosynthetic organisms are divided into two broad groups.

(1) The photosynthetic bacteria, which are obligate anaerobes when grown in the light using H_2A a substrate more reducing than water, e.g., sulphide, thiosulphate, organic molecules such as acetate, malate, succinate, lactate, etc., or even H_2 gas. There are three major classes of photosynthetic bacteria (5): (a, the green or brown bacteria (represented by the family Chlorobiaceae); (b, the purple sulphur bacteria (Chromateaceae) and (c, the purple non-sulphur bacteria (Rhodospirillaceae). The Chloroflexaceae occupy an intermediate position between the Chlorobiaceae and the Rhodospirillaceae.

(2) Cyanobacteria, algae and plants which can grow aerobically in light and in which H_2A can be water with the result that oxygen is evolved as a byproduct. Two light reactions are involved in oxygenic photosynthesis. The major characteristics of photosynthetic organisms are given in Table 1.

3. COMPARATIVE PHOTOCHEMISTRY OF BACTERIA AND PLANTS

Understanding and elucidating the basic similarities between bacterial- and plant-type photosynthesis has been a major aim in photosynthesis research ever since Van Niel first proposed his hypothesis. Strong similarities exist, as we will show later, in the phenomena of light absorption, primary photochemistry, electron transport, energy coupling and the composition of protein components in the two groups. It has been recognized for some time that there is a striking functional homology between the electron acceptors of purple photosynthetic bacteria and of PSII in plants (7). In both cases two specially bound quinones acting in series function to transfer reducing equivalents in pairs from the reaction centre to the bulk quinone pool, i.e., the one electron event of photochemistry is converted to a two-electron process.

1st flash $\quad Q_A Q_B \xrightarrow{h\nu} Q_A^- Q_B \longrightarrow Q_A Q_B^-$

2nd flash $\quad Q_A Q_B^- \xrightarrow{h\nu} Q_A^- Q_B^- \xrightarrow{2H^+} Q_A Q_B H_2 \longrightarrow Q_A Q_B + [2H]$

Also there are many similarities between photosynthetic electron transport in green bacteria, for example *Chlorobium limicola*, and PSI of plants.

The primary event in all photosynthetic processes is a

light induced electron transfer from a donor species D (also called P) to an acceptor A. The donor is a special type of BChl pair in photosynthetic bacteria and a special type of Chl (or may be a Chl pair) in plants; in both cases the donor pigment is conjugated to protein and is located in a reaction centre (RC) complex. The energy stored in the charge separation (D^+A^-) caused by the photon absorption is used by the organism for electron transport reactions ultimately to provide energy (ATP, NADH, NADPH). In this paper we will briefly describe the characteristics of the light harvesting (LH) and RC components of bacteria and plants and then attempt to compare and contrast the two systems to find an evolutionary link between them.

4. BACTERIAL PHOTOSYNTHESIS

Purple photosynthetic bacteria

The isolation of the photosynthetic apparatus of purple bacteria is well documented (8,9). Membrane preparations, called chromatophores, from *Rhodopseudomonas spheroides* consist of 64% protein, 25% phospholipids and 4.6% BChl. Each chromatophore contains approximately 42 RC complexes, 500 LH complexes, 1000 carotenoids and 1000 ubiquinone molecules.

Reaction Centres In recent years by the selective use of detergents (Triton X-100, SDS, LDAO, CTAB, etc.,) it has been possible to isolate pure reaction centres from chromatophores and study their spectral and structural properties. Although there are some minor disagreements in the reported composition of the RC, mainly due to the differences in the nature of the detergents used to separate the RC from the chromatophore membranes, the preparations were all free from antenna pigments and other external components. Typical compositions of some RC preparations are given in Table 2.

The most fully characterized RC is that from *Rh. sphaeroides* (9). The RC unit has a M_r of approx. 80 kD and is built up of three polypeptides of M_r 21 kD, 24 kD and 32 kD designated as the L (light), M (medium) and H (heavy) subunits respectively. Each RC unit contains 4 BChl, 2BPh, two UQ (one easily removable and the other tightly bound) and one Fe^{2+}. The H subunit alone can be dissociated from the RC by treatment with chaotropic agents without causing any change in the spectral or photochemical

TABLE 2

A Comparison of the Composition of Some Isolated Photochemical Reaction Centers

Species	Rh. sphaeroides	Rh. viridis	Chloroflexus aurantiacus
Pigment content per P870 or P960 molecule	4 BChl a 2 BPh a 2 UQ_{10} 1 carotenoid	4 BChl b 2 BPh b 1 UQ_{10} 1 MQ_7 1 carotenoid	3 BChl a 3 BPh a No carotenoid
<u>Subunit composition</u> <u>Number per</u> P870 or P960	3 polypeptides (L, M and H)	4 (L, M, H and two c type cytochrome(s))	2 (L and M?)
Apparent size of subunits from SDS-PAGE (from amino acid analyses)	H - 32 kD M - 24 kD L - 21 kD	H - 35 kD M - 28 kD L - 24 kD cytochrome - 38 kD	M - 30 kD L - 28 kD

Adapted from ref (10)

properties of the residual RC. This suggests that all the pigments are associated with the L and M subunits of the RC complex. Partial amino acid sequences of the L and M subunits have been determined; they are very non polar and show homologies in many domains. When the sequence of the M subunit was examined for potential membrane-spanning regions, five hydrophobic segments which could span the membrane in a helix were identifiable (11). The possible site of primary quinone binding was probed by photoaffinity labelling of RC proteins with an azido analogue of UQ. The results suggest that the primary quinone binding site is located either on the M subunit or very close to it (within 0.5 nm). Although the H subunit is not active in the photochemistry of RC, i.e., for the charge separation between D^+ and Q_A^-, reconstitution studies have shown that the H subunit is essential for effective electron transfer from the primary to secondary qinone ($Q_A \longrightarrow Q_B$).
  The topology of the RC in the chromatophore membrane has been studied by the techniques of spectroscopy (12), antibody binding, selective proteolysis and protein

labelling (8,11). The data suggest that a major portion of the H subunit is exposed on the interior of the chromatophore along with a minor segment of the M subunit.

Light-harvesting complex The antenna pigment (LH) complex has been identified only by its absorption and circular dichroism spectra (8,10). In SDS-PAGE of *Rh. spheroides* LH complex two polypeptides of M_r 9 kD and 12 kD have been identified - there is a possibility that the complex contains three polypeptides. The smallest functional unit of the B800 - B850 pigment-protein complex is estimated to be 3 BChl and 1 carotenoid.

Photochemistry and electron transport Parson and coworkers (13,14) have studied the reaction sequences following light absorption in the RC of purple bacteria. Two of the four BChl in the RC form a special dimer (P) that has an intense absorption band in the near infra red spectral region. When P is raised to an excited singlet state (P*) it releases an electron within a few picoseconds, which is captured by an acceptor complex (I) presumably consisting of another BChl, and one of the BPh; P is left in the form of a radical cation P^+. The electron from the reduced acceptor (I^-) is subsequently transferred to the quinone Q_A in about 200 picoseconds and from Q_A^- to Q_B in about 100 microseconds. The electron transfer reactions are very efficient with almost 100% quantum yield.

In all purple bacteria (and also in green bacteria) the immediate electron donor to P is a c-type cytochrome (Fig.1). In *Chromatium vinosum*, *Rh. viridis*, *Rh. palustris*, and *Thiocapsa pfennigi* the RC is closely bound to two hydrophobic cytochromes c with different red ox potentials (E_m = ca. 0.35v and 0.0v). In *Rhodospirillum rubrum*, *Rh. sphaeroides* and *Rh. capsulata* the immediate donor is a soluble Cyt c; these bacteria do not seem to have the membrane bound hydrophobic Cyt c (7).

The reducing power available from the photo reactions of purple bacteria (E_m of Q_A/Q_A^- = ca. 0.17v) is not sufficient for the direct reduction of either CO_2 (E_m of CO_2/CH_2O = -0.45v) or even of NAD (E_m NAD/NADH$^+$ ≅ -0.32v). Hence the purple bacteria conserve energy by the synthesis of ATP through a cyclic electron transport pathway. NAD reduction is then accomplished by an ATP-consuming reverse electron transport.

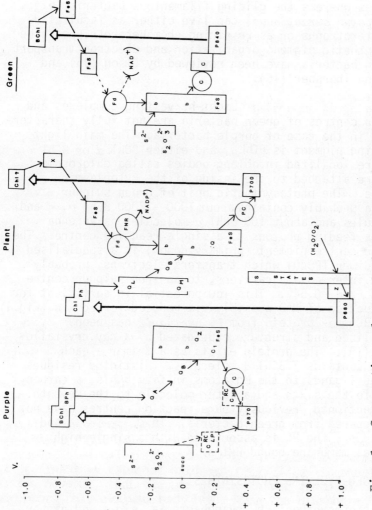

Fig. 1. Comparative scheme of electron transport in photosynthetic bacteria and plants. Adapted from (7). For explanation of symbols see List of Abbreviations.

Green photosynthetic bacteria

The green bacteria include two groups, viz., green sulphur bacteria (Chlorobiaceae) and gliding filamentous bacteria (Chloroflexaceae). The green sulphur bacteria (e.g., *Chlorobium limicola, Prosthecochloris aestuarii*) are strict anaerobes whereas the gliding filamentous bacteria (e.g., *Chloroflexus aurantiacus*) can live either as facultative photoheterotrophs or as respiring chemoheterotrophs. The photosynthetic pigment organization and electron transport in green bacteria have been reviewed by Olson (15) and Olson and Thornber (16).

Photosynthetic apparatus Light-harvesting complexes and reaction centres of green bacteria are not fully characterized as in the case of purple bacteria. The main light harvesting pigment is BChl c, and either BChl d or BChl e - these are localized in oblong bodies called chlorosomes which are attached to the inside of the cytoplasmic membrane. The photosynthetic unit of green sulphur bacteria probably contains about 1000 - 2000 BChl c, d and e molecules and about 100 BChl a molecules as antenna pigments feeding photons to a single reaction centre. The reaction centre pigment (primary donor) is a specialised BChl a dimer (P840) which transfers electrons, probably through intermediate carriers, to an iron-sulphur centre with an E_m = -0.54v. This low potential is sufficient for the reduction of NAD possibly via reduced ferredoxin (Fig.1)

The BChl a- protein from *P. aestuarii* has been crystallized and structure elucidated by X-ray crystallography (15). The protein exists as a trimer - each subunit contains 7 BChl a molecules. Histidine residues serve as ligands to the Mg atoms in five BChls, a carbonyl oxygen to the sixth and a water molecule to the seventh.

As mentioned previously pure reaction centres have not been prepared from green bacteria. In *C. limicola* and *P. aestuarii* the RC is associated with a single high potential membrane bound Cyt c.

5. OXYGEN-EVOLVING PHOTOSYNTHESIS

During photosynthesis by cyanobacteria, algae and higher plants, water is oxidized to molecular oxygen and the water-derived electrons and protons are used for the photoreduction of NADP to NADPH. Photosynthetic bacteria cannot use water as substrate; plants acquired this remarkable ability during the course of evolution by the addition

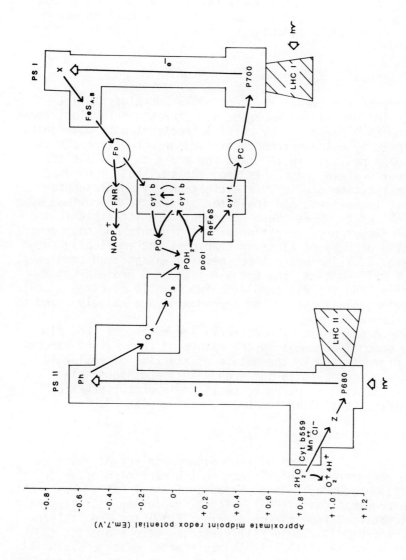

Fig. 2. Pathway of photosynthetic electron transport in chloroplasts. Adapted from (38). For explanation see List of Abbreviations.

of a second photosystem. The plant photosynthetic apparatus
thus developed an enzyme system, a water oxidase, which
could bind water and release oxygen during the photochemistry of PSII. As proposed by Hill and Bendall (17) the
concerted action of two photosystems provides sufficient
energy for the uphill transport of electrons from water
(E_m = 0.8v) to ferredoxin (E_m -0.42v), the electron donor
to NADP.

Photosynthetic apparatus

Photosynthetic events in green algae and higher plants
occur in specialised membrane structures, thylakoids, found
in the chloroplasts. By careful treatment with detergents
followed by gel electrophoresis and/or chromatography it
has been possible in the last few years to separate the
pigment-protein components from the thylakoid membranes.
Reconstitution studies, investigations by spectrophoto-
metric techniques, and inhibitor studies have provided some
insight as to where these components are localized in the
membranes, the organization of the pigments and components
and the pathway of electron transport (schematically shown
in Fig. 2). There are three membrane-associated complexes,
viz., PSII complex, the Cyt *b*-Reiske Fe-S protein-Cyt *f*
complex, and the PSI complex. Two soluble electron transfer
proteins, plastocyanin and ferredoxin, are loosely bound to
the membrane.

The prokaryotic cyanobacteria and eukaryotic red algae
show some differences in the nature of their light-harvesting pigments and in the nature of stacking of thylakoid
membranes compared to the green algal and higher plant
chloroplasts; in other respects their photosynthetic
machinery is similar to that of higher plants.

Mechanism of water oxidation

Photosystem II Much of the research in recent years has
been concentrated in elucidating the mechanism of water
oxidation coupled to the photoreactions of PSII. Reasonably
pure preparations of PSII, with O_2 evolution activity, have
been obtained from cyanobacteria (18,19) and from chloroplasts (20-22). These PSII particles were subjected to
polyacrylamide gel electrophoresis after detergent-solubili-
zation of the membrane polypeptides. It is now accepted
that almost all the chlorophyll molecules (both antenna and
reaction centre) are bound to polypeptides. The antenna
chlorophyll molecules are conjugated to a 43 kD polypeptide;

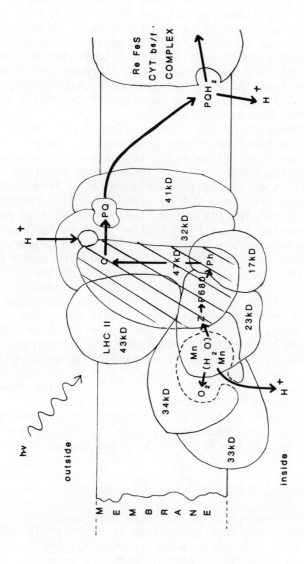

Fig. 3. Proposed model for the structural organization of the main components of PSII. Adapted from (23). For explanations see text.

this pigment-protein complex is designated LHC II (Figs.2,3). PSII reaction centre chlorophyll is built into a polypeptide M_r ca. 47 kD. This primary donor of PSII, P680, is now believed to be a special type of Chl a molecule. (By analogy with BChl a dimers, the primary donors of photosynthetic bacteria, it was postulated that P680 of PSII and P700 of PSI were also specialised Chl a dimers. However, recent data from low temperature ESR studies of the spin polarized triplet signals arising from illuminated PSI and PSII preparations (24) and a comparison of redox potential and ENDOR experiments on P680 and synthetic chlorophyll analogues (25) indicate that the reaction centre Chl a in both photosystems may be monomeric).

When light absorbed by the antenna pigments of LHC II is channelled to the reaction centre chlorophyll, rapid electron transfer occurs from P680* to the primary quinone Q_A via a pheophytin molecule. The transferred electron crosses the full span of the thylakoid membrane from inside to outside within less than 200 picoseconds (23). From fluorescence and triplet state ESR studies the site of the intermediate carrier Ph has been located near to P680 and close to the inner surface of the membrane. The primary quinone Q_A as well as the secondary quinone Q_B are bound to the 32 kD polypeptide - this polypeptide is also the binding site for some common herbicides. The midpoint redox potential of the Ph/Ph$^-$ couple has been determined by optical and ESR spectrophotometry to be -614 mV. From this E_m value and from the value of the energy of an absorbed photon at 680 nm the E_m of the P680$^+$/P680 couple can be calculated to be +1.1 eV which is sufficient to oxidise water (26,27).

The immediate electron donor to P680$^+$, Z, is also found in the 47 kD RC protein. Although Z has not been separated from the RC protein it is thought to be a bound quinone molecule. The reduction of P680$^+$ by Z is very rapid and complex. Oxidized Z, (Z$^+$), promotes the oxidation of water.

The water-oxidizing enzyme

The structure and mechanism of action of the water-oxidizing enzyme, sometimes symbolised as Y, is still not fully understood. Treatments which selectively block the O_2 evolution capacity of chloroplasts, for example, washing with alkaline Tris or high salt are accompanied by loss of manganese from the thylakoid membrane. Reconstitution of such treated chloroplasts with added Mn partially restores O_2 evolution capacity. The functioning of the enzyme complex was thought to require 4 Mn atoms but very recent studies have shown that only two Mn atoms are indispensable; the other two are probably essential to maintain the conformation of the enzyme and can be replaced by divalent metals such as calcium (28).

Several models for the catalytic centre of the enzyme have been proposed. All these models assume that each of the two essential manganese atoms is bound by at least bidentate ligation to an apoprotein which determines the redox potential and kinetics of electron and proton transfer by the holoenzyme (29,23). The protein matrix is thought to be important for the geometrical array of two manganese atoms in order to allow O-O bond formation at the potential level of H_2O_2. It has been known for some time from flash photolysis experiments that the water-oxidizing enzyme extracts electrons from water in steps and stores them in different oxidation states S_0, S_1, S_2, S_3 and S_4 prior to the evolution of a molecule of oxygen.

Reconstitution studies using Mn-depleted PSII particles and algal mutants grown in Mn-deficient medium strongly suggest that the catalytic site with the bidentate Mn is embedded in a 34 kD polypeptide which is bound to the inner surface of the thylakoid membrane. Three other polypeptides of Mr 17 kD, 24 kD and 33 kD are associated with the 34 kD polypeptide and are involved, albeit indirectly, with O_2 evolution. In addition a cyt b_{559} which switches between a high-potential and a low-potential form (ΔE_m ca. 300 mV) has also been implicated in O_2 evolution (30).

Nugent and Atkinson (31) have measured the functional size of PSII from particles of *Phormidium laminosum* (a thermophilic cyanobacterium) using the technique of radiation inactivation and obtained a value of 125 kD. The minimum functional unit may be composed of the 47 kD RC polypeptide, the 32 kD quinone- and herbicide-binding protein, and the 34 kD Mn-binding protein.

The cytochrome b complex The second membrane-bound complex (Figs. 1 and 2) that has been solubilised from the thylakoid membranes is the cytochrome *b* complex which contains cyt *b*, cyt *f* and the Reiske iron-sulphur protein. Hauska et al. have reviewed (32) the preparation, characterization and function of this complex. It has two major functions: (a) the transport of PSII-derived electrons from reduced plastoquinone pool to PSI via plastocyanin and (b) transduction of protons (for ATP synthesis) by the operation of a Q cycle in which the plastoquinone pool and the cyt b participate.

The PSI complex The PSI complex has been fairly well characterized (9,33). The components of this complex are the light-harvesting Chl-protein, reaction centre Chl-protein, the primary (?) acceptor X and two membrane-bound iron-sulphur proteins which finally transfer electrons to soluble ferredoxin. The PSI RC complex of Swiss chard contained six different polypeptides with apparent Mr of 70, 25, 20, 18, 16, and 8 kD. The 70 kD polypeptide may function as the P700-RC with one P700 per two 70 kD polypeptides. Amino acid composition of 70 kD polypeptide shows low content of polar residues and is similar to the LHM oligomer of photosynthetic bacteria. The PSI RC isolated from the thermophilic cyanobacterium *Mastigocladus laminosus* was found to resemble the RC of green algae and higher plants in molecular mass, P-700 content, and polypeptide patterns by SDS-PAGE (34). Recent ESR studies point to the existence of some intermediate carrier(s) between P700 and X - one of these possibly may be a Chl a molecule (9).

Organization of the two photosystems in the chloroplast membranes

Until a few years ago it was assumed that in the thylakoid membranes the two photosystems were arranged side by side for the efficient transfer of absorbed light energy and for the transport of electrons. However, recent studies (35,36) on the inside-out vesicles prepared from chloroplasts by mechanical fragmentation and phase partition techniques indicate lateral heterogeneity in the location of the two photosystems. These studies show that most of the PSII is located in the appressed partition membranes of the grana stacks whereas PSI is restricted to the exposed membranes of the end granal and stromal lamellae. There is a certain proportion of PSII in the stroma lamellae with a composition, probably slightly different from that in the grana stacks. Distribution of absorbed light quanta between the two photosystems is controlled by the phosphorylation and dephosphorylation of the PSII light harvesting chlorophyll-protein complex - this complex under certain circumstances exists as a separate unit which can change its relative distribution between appressed and non-appressed membranes (37,38).

6. PLANT AND BACTERIAL PHOTOSYNTHETIC ELECTRON TRANSPORT - A COMPARISON

The electron transport chains of plant and two bacterial photosystems are shown in Fig. 1. A close examination reveals many common features between the three types of photosystems. Photosynthetic bacteria, algae, and plants regulate the amount of light-harvesting components synthesized in response to environmental light intensity conditions. All LH Chls are conjugated to proteins. The main distinction among light-harvesting properties of photosynthetic cells lies in the patterns of pigment localization. In the green-sulfur bacteria the LH function is located in non-membranous particulate structures, chlorosomes, embedded in the cell membrane and whose numbers and structures vary with the light intensity. In the purple bacteria the LH and RC complexes are together within a membrane system. In cyanobacteria and red algae the water-soluble phycobilins are found in the phycobilisomes localized on the outer surface of the photosynthetic membranes (39). The phycobilisomes contain about 300 to 800 tetrapyrrole chromophores which absorb light over much of the visible spectrum. The energy absorbed by the phycobilisomes are transferred to the RC with an efficiency approaching 100%. The LH complexes of PSI and PSII of higher plants have already been described.

The RC of bacteria and plants are membrane-bound proteins containing Chl or BChl. The primary photoreaction in all cases is a light-induced electron transfer from a donor species P to an acceptor A. The initial donor is a specialized Chl a molecule in plants and a BChl a dimer in bacteria; the primary acceptor appears to be a Ph or BPh-Fe-quinone complex. The relatively low energies of excited Chl a and BChl a at the RC helps to minimise the wasteful decay of the excited states via fluorescence with the result that the quantum yields of electron transfer reactions are high at the RC complex. Wasteful back reactions are also prevented by conducting the electron transfer from P* through a series of donor-acceptor species stabilised for increasingly longer lifetimes and separated by increasingly longer distances (9,14,40).

There are highly conserved patterns of sequences of amino acids which are common to the L and M subunits of the RC proteins of *Rh. capsulata* and the 32 kD herbicide binding protein of PSII isolated from spinach and tobacco leaves (41). The homology between these polypeptides suggest a common ancestor for these bacterial and plant proteins.

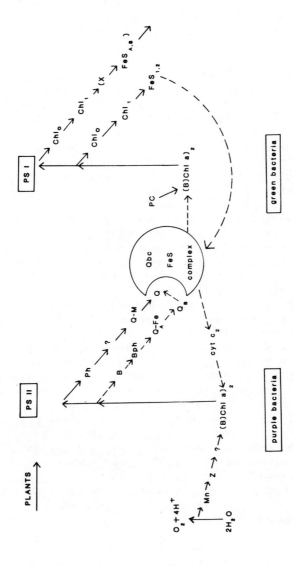

Fig. 4. Photosynthetic electron transport: similarities between the plant and bacterial systems. Adapted from (42). For explanations see List of Abbreviations.

In plant as well as bacterial photosystems a quinone-cyt b-cyt c (cyt f) assembly functions as a gate between a two electron carrier and one electron carriers via a Q-cycle in which protons are transported across the membrane. The final product of photosynthetic electron transport in the green-sulphur bacteria is reduced pyridine nucleotide (NADH) as also is NADPH in the PSI of plants. In this respect, and in the content and organization of electron transfer intermediates the green bacterial photosystem resembles plant PSI. The purple bacterial photosynthesis, on the other hand, is more analogous to PSII of plants (Figs. 1,4).

7. EVOLUTION OF PHOTOSYNTHETIC SYSTEMS

In the last twenty years extensive data has accumulated relating to the earliest Precambrian fossil record from morphological, electron microscopic, and carbon and sulphur isotopic studies of phytoplankton, cyanobacterial stromalites, etc., found in Precambrian deposits in various parts of the world. From these data Schopf (43) has proposed a "best guess scenario" for the events in biological evolution in which (a) the origin of life occurred earlier than 3.5×10^9 years ago, (b) the development of anaerobic chemoheterotrophy, anaerobic chemoautotrophy and anaerobic photoautotrophy probably also occurred earlier than 3.5×10^9 years ago, and (c) the origin of aerobic photoautotrophy occurred about 3×10^9 years ago. Thus within a span of one billion years after the formation of the Earth, abiogenic and biogenic (enzymatic) pathways for the synthesis and assembly of prophyrins, chlorophylls, ferredoxins, cytochromes and membrane structures were developed. Based on studies of sequences of ferredoxins, cytochromes, superoxide dismutases, nucleic acids, etc., from various organisms, Dayhoff and Schwarz (44) and Matsubara and Hase (45) have constructed evolutionary trees which in many respects follow evolutionary trees drawn from taxonomical considerations. We have proposed a simplified scheme (Fig. 5) for the origin of life and evolution based on sequences, comparative biochemistry and the geological record (46).

Evolution of energy-transducing systems

The biochemical evolution of energy-transducing systems is generally considered to have followed the sequence: fermentation ⟶ photosynthesis ⟶ respiration (47,48). Anaerobic fermentations probably were the earliest energy supplying processes and some of the catalysts and products

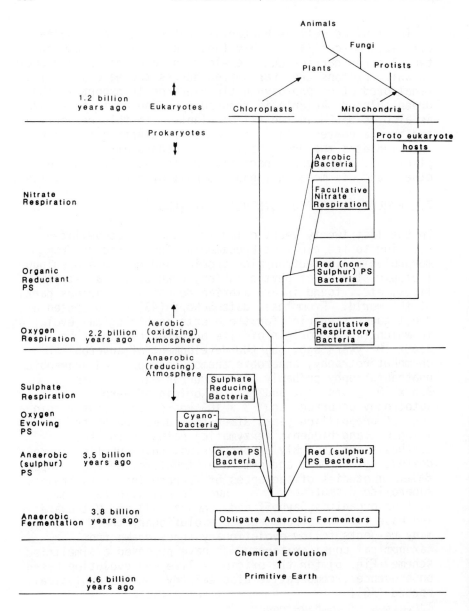

Fig. 5. Simplified evolutionary tree drawn from protein sequences and other data (46).

of fermentations such as ferredoxins, ATP and NADH are constituents of the energy metabolism of every other type of organism. From a comparison of the available sequences of ferredoxins from fermentative and photosynthetic bacteria we have proposed that photosynthetic bacteria would have evolved from heterotrophic fermenters such as the clostridia (49). Gest has advanced a hypothesis to explain the mechanism of transition from heterotrophs to phototrophs based on studies of the redox balances of fermentations in photosynthetic and non-photosynthetic bacteria (50). He uses the term "electrophosphorylation" to designate any process in which electron flow in membranes is the driving force for phosphorylation of ADP. Evolution in anaerobes followed the sequence: (a) fermentations in which the red-ox is internally balanced (e.g., pyruvate, lactate), (b) cytoplasmic fermentations in which an accessory electron acceptor is needed for redox balance (e.g. fumarate), and (c) fermentations, of the fumarate type, using a membranous electron flow. The establishment of a membrane-associated electron flow resulted in energy conservation through an electrophosphorylation module with Fe-S proteins, quinones and cyt b participating as catalysts. According to Gest the earliest photophosphorylation system could have evolved by the fusion of a membrane-bound pigment (Mg-porphyrin) complex into a fermentative anaerobe operating an electrophosphorylation module. The development of a 'photosynthetic unit' in which a photoactivatable pigment complex, embedded vectorially in membranes, could replace organic molecules as an electron source for energy conversion was a significant event in the course of evolution - one that linked biological systems to the inexhaustible energy supply from the sun.

Once the photosynthetic pigment was incorporated into the cell membrane the next step would have been to devise a control mechanism to coordinate the energy flow from the extremely fast photochemical reactions to the rather slower biochemical reactions of the cell. The photosynthesizers solved this problem, as described earlier, by stabilizing the light-induced separation of charges via a cyclic electron flow through a series of intermediates of increasing life time. The addition of a c-type cytochrome to the electron flow cycle completed the evolution of the prototype of the purple sulphur bacteria. The unity of the bioenergetic processes is evident when one compares the respiratory and photosynthetic electron transport chains (Fig. 6). The Q-b-c complex plays a pivotal role in all electron transport chains, probably in conjunction with an Fe-S protein. Electrons are fed into the complex *via* a quinone pool and leave *via* a c-type cytochrome in respiratory and photosynthetic bacteria, and *via* plastocyanin (or a cytochrome in a few instances) in cyanobacteria and plants (7,42). Amino acid sequences of cyt b of complex III from five different mitochondrial sources and the chloroplast cyt b_6 from spinach show a high degree of homology (51). The hydropathy plots for the bovine, mouse and *Anacystis nidulans* cyt b sequences resemble those of human and yeast cytochromes b.

Gest favours the idea that photosynthetic and respiratory phosphorylation systems developed independently from a common ancestor. Broda (47) however has advocated the view that all aerobic respiratory electron transport chains originated through modifications of the electron transport chains of photosynthetic bacteria.

Consideration of the morphological and compositional similarities of the present-day prokaryotic photosynthetic and cyanobacteria coupled with the available fossil records of these types of microorganisms has led Olson (52) to propose that "the major steps in the evolution of photosynthetic systems at the molecular level were probably completed by the end of the middle Precambrian era". One line of descent led to the development of anaerobic photosynthesizers whereas the other line of descent had led to the evolution of aerobic O_2-evolving photosynthesizers, *viz.*, cyanobacteria. In this context it is relevant to mention that there is an extant group of prokaryotes, typified by *Oscillatoria limnetica*, which can switch its energy metabolism from one of O_2-evolving aerobic photosynthesis (with water as electron donor) to one of anaerobic photosynthesis (sulphide as electron source) depending on the availability of substrate in the environment (53). The question as to whether BChl originated from Chl a or *vice versa* has still not been settled (52,54).

Olson (52) also has proposed an interesting hypothesis for the evolution of PSII in cyanobacteria. He postulates that the protoalgae would have derived its energy by the oxidation of a set of weak, nitrogenous, electron donors such as N_2H_4, NH_2OH, NO and NO_2^-, before the development of the water-oxidizing enzyme system. Probably the appearance of a Mn complex replacing Fe in cyt c would have facilitated the catalytic oxidation of NO to NO_2^-.

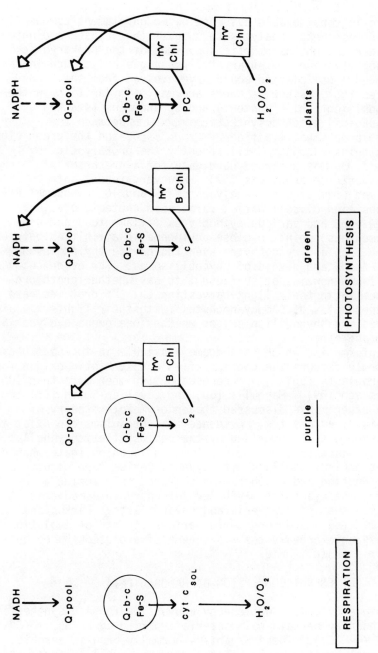

Fig. 6. A comparison of the respiratory and photosynthetic electron transport chains. Adapted from (42).

Origin of chloroplasts

There is considerable evidence based on pigment composition and RNA sequence analysis to suggest that close homologies between the photosynthetic apparatus of the prokaryotic cyanophytes and the eukaryotic red algae exist such that the red algal plastid would have evolved from the cyanophytes (55). Likewise there are many similarities in the composition (Chl b content) and arrangement (stacked thylakoids) of photosynthetic components between the prochlorophytes, typified by *Prochloron*, and the green algal chloroplasts (56). Until recently the prokaryotic *Prochloron* was a strong candidate for a precursor of eukaryotic chloroplasts (57). However, recent data derived from sequence analyses of 5S and 16S ribosomal RNA, composition of cell walls, carotenoid content, etc., of *Prochloron* and various cyanophytes (filamentous and nonfilamentous) point to close phylogenetic affinities between *Prochloron* and cyanophytes even though the thylakoids of *Prochloron* are devoid of phycobilisomes. The evidence available now is not sufficient to say whether prochlorophytes (with their light-harvesting Chl a/b proteins) arose independently of the cyanophytes (with their light-harvesting phycobilisomes) or whether one group evolved from the other.

Information which has become available in the past decade strongly favours the concept of an endosymbiotic origin for chloroplasts (55,58) as proposed by Schimper more than 100 years ago (59). Margulis (60) and Cavalier-Smith (55) and many others have discussed the endosymbiotic theory of origin of eukaryotes. Evidence that organisms can exist as endosymbionts is provided by the photosynthetic cyanelles, e.g., *Cyanophora paradoxa* (61) which is an obligate photoautotrophic flagellate with cyanobacterium-type structure and function and which can be cultured only inside a host cell. The cyanelles may be evolutionary intermediates between the cyanobacteria and chloroplasts. Larkum and Barrett (62) have considered various aspects of evolution of oxygenic prokaryotes - a scheme of evolution based on this is shown in Fig. 7.

8. LIGHT ENERGY CONVERSION BY HALOBACTERIA

A unique type of light energy conversion for ATP synthesis is shown by certain halobacteria typified by *Halobacterium halobium* (63). These organisms normally grow in aerobic, extremely saline environments. However, when cultured at

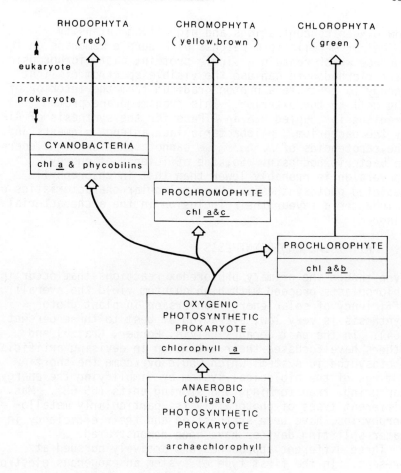

Fig. 7. A simplified scheme for the evolution of oxygenic photosynthetic prokaryotes and of eukaryotic algae. Redrawn from (62).

low oxygen concentrations and high light intensity
H. halobium cells form patches of a purple membrane on their
surface which contain a single protein, bacteriorhodopsin.
Bacteriorhodopsin can use the visible spectrum of light
energy to generate a proton gradient from the interior of
the cell to the exterior. This transmembrane proton
gradient is coupled *via* an ATPase for the synthesis of ATP
by the bacterium. Halobacteria lack antenna pigments and
the carotenoids of *H. halobium* cannot transfer light energy
to bacteriorhodopsin. So, the efficiency of energy
conversion is probably lower than that in chlorophyll-
mediated photosynthesis. Many molecular characteristics of
halobacteria favour their inclusion in the archebacterial
kingdom.

9. ARTIFICIAL PHOTOSYNTHESIS

Even though the primary photoredox reactions that occur in
chloroplasts proceed with high quantum yield the overall
efficiency of solar energy conversion in plant photosynthesis is very low, averaging at best to three percent
(64). In the past decades Calvin, Porter, Gratzel and
others have focussed their attention in devising artificial
photosynthetic systems which would overcome the short-
comings of the biological system by simplifying the energy
capturing, transforming, and storing units (65-68). Many
different types of photocatalysts, particularly metallo-
porphyrins, have been synthesized and their efficiency in
water-splitting devices have been demonstrated.

Three different approaches are actively pursued at
present. In the first type of system an exogenous electron
mediator is photoreduced *via* thylakoid membranes or cyanobacterial cells and then the reduced mediator is coupled to
a catalyst for the production of hydrogen from water or for
the reduction of NADP. The second approach is to use
synthetic molecular assemblies such as micelles and lipid
vesicles as photoreaction media. These assemblies
simulate the membranous environment of the biological
systems and allow photosensitized electron transport to
hydrophobic and amphiphilic molecules incorporated in them.
Finally, light-induced water-splitting systems based on
semiconductor colloidal particles and ultrafine redox
catalysts are studied for the simultaneous generation of
oxygen and hydrogen. This type of system probably mimicks
best the conditions on earth before the advent of photosynthetic cells. Although the efficiency and stability of
these artificial systems, so far reported, are low, research

activity in this field is expanding rapidly and the outlook seems promising. After all, we should remember that it took half a billion years for nature to evolve a stable light-energy conversion system.

Acknowledgements

We thank Mrs. V. Lilleberg for typing this MS and Miss P. Taylor for drawing and Mr. H. Edge for photographing the figures.

References

1. Sybesma, C. Ed. (1984) "Advances in Photosynthesis Research", Vols. I to IV. Martinus Nijhoff/Dr. W. Junk Publishers, The Hague.
2. Inoue, Y., Murata, N., Crofts, A.R., Renger, G., Govindjee and Satoh, K. Eds. (1983). "The Oxygen Evolving System of Photosynthesis". Academic Press, Tokyo.
3. Govindjee, Ed. (1982) "Photosynthesis", Vol. 1. Academic Press, New York.
4. Van Niel, C.B. (1941). The bacterial Photosynthesisers and their importance for the general problem of Photosynthesis. *In* "Advances in Enzymology" (Eds. E.F. Nord and C.H. Werkman), Vol. 1. pp. 263-328. Interscience Publishers, New York.
5. Pfennig, N. (1978). General physiology and ecology of photosynthetic bacteria. *In* "The Photosynthetic Bacteria" (Eds. R.K. Clayton and W.R. Sistrom), pp. 3-18. Plenum Press, New York.
6. Rao, K.K., Hall, D.O. and Cammack, R. (1981). The Photosynthetic apparatus. *In* "Biochemical Evolution" (Ed. H. Gutfreund). pp. 150-202. Cambridge University Press, Cambridge.
7. Wraight, C.A. (1982). Current Attitudes in Photosynthesis research. pp. 19-61. *In* Ref. 3.
8. Kaplan, S. and Arntzen, C.J. (1982). Photosynthetic membrane structure and function. pp. 65-151. *ibid*.
9. Okamura, M.Y., Feher, G. and Nelson, N. (1982). Reaction centres. pp. 195-272. *ibid*.
10. Thornber, J.P., Cogdell, R.J, Seftor, R.E.B, Pierson, B.K. and Tobin, E.M. (1984). Comparative biochemistry of chlorophyll-protein complexes. pp. 25-32, Vol. 2, Ref. 1.
11. Feher, G. and Okamura, M.Y. (1984). Structure and function of the reaction centre from *Rhodopseudomonas sphaeroides*. pp.155-164. Vol. 2, Ref.1.

12. Breton, J. and Vermeglio, A. (1982). Orientation of photosynthetic pigments *in vivo*. pp. 153-194. Ref. 3.
13. Parson, W.W., Holten, D, Kirmaier, C, Scherz, A. and Woodbury, N. (1984). Primary reactions in bacterial reaction centers. pp. 187-194, Ref. 1. Vol. 1.
14. Parson, W.W. and Ke, B. (1982). Primary photochemical reactions. pp. 331-385, Ref. 3.
15. Olson, J.M. (1980). Chlorophyll organization in green photosynthetic bacteria. *Biochim. Biophys. Acta.* 589, 30-45.
16. Olson, J.M. and Thornber, J.P. (1979). Photosynthetic reaction centers. *In* "Membrane Proteins in Energy Transduction" (Ed. R.A. Capaldi). pp. 279-340. Marcel Dekker Inc., New York.
17. Hill, R. and Bendall, F. (1960). Function of the two cytochrome components in chloroplasts: a working hypothesis. *Nature* 186, 136-137.
18. Stewart, A.C. and Bendall, D.S. (1979). Preparation of an active, oxygen evolving photosystem 2 particle from a blue-green alga. *FEBS Lett.* 107, 308-312.
19. Ford, R.C., Evans, M.C.W. and Atkinson, Y.E. (1984). Kinetics of $P680^+$ reduction in oxygen-evolving photosystem 2 preparations from lettuce and *Phormidium laminosum*, pp, 147-150, Vol, 1, Ref. 1.
20. Vernon, L.P. and Shaw, E.R. (1971). Subchloroplast fragments: Triton X-100 method. *In* "Methods in Enzymology". Vol. 23 (Ed. A. San Pietro). pp. 277-289. Academic Press, New York.
21. Murata, N., Miyao, M, Omata, T. and Kuwabara, T. (1984). The oxygen evolution system of photosystem particles from spinach chloroplasts. pp. 329-332. Vol. 1, Ref. 1.
22. Berthold, D.A., Babcock, G.T. and Yocum, C.F. (1981). A highly resolved, oxygen evolving photosystem 2 preparation from spinach thylakoid membranes: ESR and electron transport properties. *FEBS Lett.* 134, 231-234.
23. Junge, W. and Föster, U. (1984). Final Report of Contract ES-D-033-D, on Solar Energy R & D in the European Community. To be published by E.E.C.
24. Rutherford, A.W. and Mullet, J.E. (1981). Reaction centre triplet states in Photosystem I and Photosystem II. *Biochim. Biophys. Acta* 635, 225-235. Rutherford, A.W., Paterson, D.R. and Mullet, J.E. (1981). A light-induced spin-polarized triplet detected by EPR in Photosystem II reaction centers. *Biochim. Biophys. Acta.* 635, 205-214.
25. Davis, M.S., Forman, A. and Fajer, A. (1979). Ligated chlorophyll cation radicals: their function in Photosystem II of plant photosynthesis, *Proc. Natl.*

Acad. Sci. USA. 76, 4170-4174.
26. Klimov, V.V. (1984). Charge separation in Photosystem II reaction centers: the role of pheophytin and manganese. pp. 131-138, Vol. 1, Ref. 1.
27. Govindjee (1984). Photosystem II: the oxygen evolving system of photosynthesis. pp. 227-238, Vol. 1, Ref. 1.
28. Yamamoto, Y. and Nishimura, M. (1984). Structure of the O_2-evolution enzyme complex in a highly active O_2-evolving Photosystem II preparation. pp. 333-336. Vol. 1, Ref. 1.
29. Renger, G., Eckert, H.J. and Weiss, W. (1983). Studies on the mechanism of photosynthetic oxygen formation. pp. 73-80, Ref. 2.
30. Butler, W.L. and Matsuda, H. (1983). Possible role of cytochrome b_{559} in Photosystem II. pp. 113-117, Ref.2.
31. Nugent, J.H.A. and Atkinson, Y.E. (1984). Estimation of the functional size of Photosystem II. *FEBS Lett.* 170, 89-93.
32. Hauska, G., Hurt, E, Gabellini, N. and Lockau, W. (1983). Comparative aspects of quinol-cytochrome c/plastocyanin oxidoreductases, *Biochim. Biophys. Acta* 726, 97-133.
33. Takahashi, Y. and Katoh, S. (1982). Functional subunit structure of Photosystem I reaction centre in *Synechococcus sp. Arch. Biochem..Biophys.*. 219, 219-227.
34. Nechushtai, R. and Nelson, N. (1984). Photosystem I reaction center from cyanobacterium, green alga and higher plants. pp. 85-93. Vol. 2, Ref. 1.
35. Andersson, B. and Anderson, J.M. (1980). Lateral heterogeneity in the distribution of chlorophyll protein complexes of the thylakoid membranes of spinach chloroplasts. *Biochim. Biophys. Acta.* 503, 462-472.
36. Barber, J. (1984). Lateral heterogeneity of proteins and lipids in the thylakoid membrane and implications for electron transport. pp. 91-98, Vol. 3, Ref. 1.
37. Allen, J.F., Bennett, J. Steinback, K.E. and Arntzen, C.J. (1981). Chloroplast protein phosphorylation couples plastoquinone redox state to distribution of excitation energy between photosystems. *Nature* 291, 21-25.
38. Staehelin, L.A. and Arntzen, C.J. (1983). Regulation of chloroplast membrane function: protein phosphorylation changes the spatial organization of membrane components. *J. Cell Biol.* 97, 1327-1337.

39. Glazer, A.N. (1984). Phycobilisome. A macromolecular complex optimized for light energy transfer. *Biochim. Biophys. Acta* 768, 29-51.
40. Shuvalov, V.A. (1984). Main features of the primary charge separation in photosynthetic reaction centers. pp. 93-100, Vol. 1, Ref. 1.
41. Hearst, J.E. and Sauer, K. (1984). Protein sequence homologies between portions of the L and M subunits of reaction centers of *Rhodopseudomonas capsulata* and the 32 kD herbicide-binding polypeptide of chloroplast thylakoid membranes and a proposed relation to quinone binding sites. pp. 355-359. Vol. 3, Ref. 1.
42. A.J. Hoff (1984). Electron transport chains of plant and bacterial photosystems. pp. 89-91. Vol. 1, Ref. 1.
43. Schopf, J.W. (1983). The early Palaeontological record. Abstract B1-1, "Book of Abstracts", 7th International Conference on the Origins of Life. Mainz.
44. Schwartz, R.M. and Dayhoff, M.O. (1981). Chloroplast origins: inferences from protein and nucleic acid sequences. *Ann. N.Y. Acad. Sci.* 361, 260-272.
45. Matsubara, H. and Hase, T. (1983). Phylogenetic consideration of ferredoxin sequences in plants, particularly algae. *In* "Proteins and Nucleic acids in Plant Systematics". (Eds. U. Jensen and D.E.Fairbrothers) pp. 168-181. Springer-Verlag, Berlin.
46. Hall, D.O. (1979). Solar energy use through biology - past, present and future. *Solar Energy* 2, 307-328.
47. Broda, E. (1975). The Evolution of the Bioenergetic Processes. Pergamon Press, Oxford.
48. Cammack, R., Rao, K.K. and Hall, D.O. (1981). Metalloproteins in the evolution of photosynthesis. *Biosystems* 14, 57-80.
49. Hall, D.O., Rao, K.K. and Cammack, R. (1975). The iron-sulphur proteins: structure, function and evolution of a ubiquitous group of proteins. *Science Progress* 62, 285-317.
50. Gest, H. (1980). The evolution of biological energy transducing systems. *FEMS Microbiology Letters*. 7, 73-77.
51. Widger, W.R., Cramer, W.A, Herrman, R.G. and Trebst, A. (1984). Sequence homology and structural similarity between cytochrome b of mitochondrial complex III and the chloroplast b_6-f complex: position of the cytochrome b hemes in the membrane. *Proc. Natl. Acad. Sci. USA* 81, 674-678.

52. Olson, J.M. (1978). Precambrian evolution of photosynthetic and respiratory organisms. *In* "Evolutionary Biology". (Eds. M.K. Hecht, W.C. Steere and B. Wallace). Vol. 11. pp. 1-37. Plenum Pub. Corp., New York.
53. Padan, E. (1979). Facultative anoxygenic photosynthesis in cyanobacteria. *Ann. Rev. Plant Physiol.* 30, 27-40.
54. Rebeiz, C.A. and Lascelles, J. (1982). Biosynthesis of pigments in plants and bacteria. pp. 699-780. Ref. 3.
55. Cavalier-Smith, T. (1981). Origin and early evolution of eukaryotes. *In* "Molecular and Cellular Aspects of Microbial Evolution". (Eds. M.J. Carlile, J.F. Collins and B.E.B. Moseley). pp. 33-84. Cambridge University Press, Cambridge.
56. Giddins, T.H., Withers, N.W. and Staehelin, L.A. (1980). Supramolecular structure of stacked and unstacked regions of the photosynthetic membranes of *Prochloron* sp., a prokaryote. *Proc. Natl. Acad. Sci, USA.* 77, 352-356.
57. Lewin, R.A. (1984). *Prochloron* - a status report, *Phycologia* 23, 203-208.
    Seewaldt, E. and Stackebrandt, E. (1982). Partial sequence of 16S ribosomal RNA and the phylogeny of *Prochloron, Nature* 295, 618-620.
58. Whatley, J.M. and Whatley, F.R. (1984). Evolutionary aspects of the eukaryotic cell and its organelles. *In* "Encyclopaedia of Plant Physiology, Vol. 17, Cellular Interactions". (Eds. H.F. Linskens and J. Heslop-Harrison). pp. 18-58. Springer-Verlag, Berlin.
59. Schimper, A.F.W. (1883). Uber die entwickelung der chlorophyllkorner und farbkorper. *Botanische Zeitung.* 41, 105-114.
60. Margulis, L. (1981). Symbiosis in Cell Evolution. W.H. Freeman and Co., San Francisco.
61. Bothe, H. and Floener, L. (1978). Physiological characterization of *Cyanophora paradoxa*, a flagellate containing cyanelles in endosymbiosis. *Z. Naturforsch.* 33c, 981-987.
62. Larkum, A.W.D. and Barrett, J. (1983). Light-harvesting processes in algae. *In* "Advances in Botanical Research" (Ed. H.W. Woolhouse). Vol. 10, pp. 1-189. Academic Press, London.
63. Stoeckenius, W., Lozier, R.H. and Bogomolni, R.A. (1979). Bacteriorhodopsin and the purple membrane of halobacteria. *Biochim. Biophys.Acta* 505, 215-78.

64. Bolton, J.R. and Hall, D.O. (1979). Photochemical conversion and storage of solar energy. *Ann. Rev. Energy* 4, 353-401.
65. Calvin, M. (1983). Artificial photosynthesis: quantum capture and energy storage. *Photochem. Photobiol.* 37, 349-360..
66. Gratzel, M. (1982). Artificial photosynthesis, energy- and light-driven electron transfer in organized molecular assemblies and colloidal semiconductors. *Biochim. Biophys. Acta* 683, 221-244.
67. Porter, G. (1983). Efficiency limitations in photochemical conversion. *In* "Photochemical Conversions". (Ed. A.M. Braun). pp. 169-233, Presses Polytechniques Romandes, Lausanne.
    Krasnovsky, A.A. (1983). Pathways of biological conversion of solar energy. pp. 49-72, *ibid*.
    Rao, K.K. and Hall, D.O. (1983). Photobiological production of fuels and chemicals. pp. 1-48, *ibid*.
68. Magliozzo, R.S. and Krasna, A.I. (1983). Hydrogen and oxygen photoproduction by titanate powders. *Photochem. Photobiol.* 38, 15-21.
    Mercer-Smith, J.A. and Mauzerall, D.C. (1984). Photochemistry of porphyrins: a model for the origin of photosynthesis. *Photochem. Photobiol.* 39, 397-405.
    Gratzel, M. ed. (1983) "Energy Resources through Photochemistry and Catalysis". Academic Press, New York.

LIST OF ABBREVIATIONS

BChl, bacteriochlorophyll; BPh, bacteriopheophytin;
Chl, chlorophyll; ESR, electron spin resonance;
Fd, ferredoxin; FNR, ferredoxin-NADP reductase;
LHC, light-harvesting complex; P, primary donor chlorophyll;
PC, plastocyanin; Ph, pheophytin; Q, quinone;
RC, reaction centre; Re-FeS, Rieske iron-sulphur protein;
SDS-PAGE, sodium dodecyl sulphate-polyacrylamide gel
electrophoresis; UQ, ubiquinone.

Explanation of electron transport schemes given in Figs 1, 2 and 4

Fig. 1. The ordinate indicates measured or estimated midpoint redox potentials. Symbols in rectangular boxes represent components in membrane protein complexes, symbols in circles denote soluble or peripheral membrane components. CHP, high potential and CLP, low potential cytochromes; Q_H and Q_L are high and low potential forms of the primary quinone acceptor. For other symbols see above.

Fig. 2. The ordinate indicates approximate midpoint potentials of the red ox carriers. Components in boxes are combined into integral membrane protein complexes; components in circles are water-soluble proteins. For explanation of other symbols see above.

Fig. 4. MW, the water oxidizing enzyme complex; Z, the electron donor to P680, Q_A and Q_B quinone acceptors; Chl_0 and Chl_1 chlorophyll-containing pigments that probably function as the earliest and secondary acceptors respectively in PSI of chloroplasts and in green sulphur bacterial system. For explanations of other symbols see above.

THE EVOLUTION OF BACTERIAL RESPIRATION

C. W. Jones

Department of Biochemistry
University of Leicester
Leicester, England

INTRODUCTION

It is now generally accepted that the principal function of the respiratory chain in chemotrophic bacteria is to catalyse a series of membrane-associated redox reactions via which reducing equivalents (electrons, hydride ions or hydrogen atoms) are transferred in a thermodynamically spontaneous manner from a reductant to an appropriate oxidant, concomitant with the ejection of H^+ across the respiratory (cytoplasmic) membrane at one or more sites. The free energy released by respiration reflects the difference in redox potential between the donor and acceptor couples, and is conserved in the form of a trans- or intra-membrane electrochemical potential difference of H^+, the so-called protonmotive force ($\Delta\bar{\mu}_{H^+}$ or Δp), which is variably composed of a chemical potential difference (ΔpH) and a membrane potential ($\Delta\psi$)(1-4). There is increasing evidence that different mechanisms of respiration-linked proton translocation may be employed by different energy coupling sites or by the same site in different organisms. These include the direct redox arm, loop and cycle mechanisms (in which the translocated species is the electron and net proton translocation results from associated protolytic reactions; the stoichiometry of H^+ ejection to electron transfer, the $\rightarrow H^+/2e^-$ quotient, is therefore fixed and predictable) and the indirect, conformational redox pump mechanism (in which the translocated species is the proton, and the $\rightarrow H^+/2e^-$ quotient cannot be predicted).

The respiratory chains of chemotrophic bacteria can between them utilise a very wide range of donors and acceptors, both organic and inorganic. Two main types of respiration have been recognised, viz. anaerobic respiration (in which relatively weak oxidants such as fumarate and various oxyanions of nitrogen and sulphur are employed as terminal acceptors) and aerobic respiration (in which the terminal acceptor is the relatively strong oxidant, molecular oxygen).

The protonmotive force generated by respiration is subsequently used to drive various energy-dependent membrane functions, quantitatively the most important of which is ATP synthesis (respiratory chain phosphorylation or oxidative phosphorylation). This process is dependent on a membrane-bound ATP phosphohydrolase complex which is composed of a transmembrane H^+ channel (BF_0) and a headpiece which binds adenine nucleotides and inorganic phosphate (BF_1). During respiration the $BF_0.BF_1$ complex acts predominantly as an ATP synthase, i.e. it uses the free energy of the protonmotive force to disequilibrate the $ATP \rightleftharpoons ADP + Pi$ reaction in favour of ATP concomitant with the inward movement of H^+. Respiratory chain phosphorylation is thus a spatially-oriented, membrane-bound process in which energy transfer between the redox and ATP phosphohydrolase systems is effected by a circulating proton current.

In contrast, during fermentation ATP synthesis is catalysed by soluble cytoplasmic enzymes in a series of scalar redox or clastic reactions involving the sequential, stoichiometric formation of covalent intermediates (substrate level phosphorylation). Under these conditions, the membrane-bound ATP phosphohydrolase complex acts as an ATPase rather than as an ATP synthase, i.e. it generates a protonmotive force by hydrolysing some of the ATP made by substrate-level phosphorylation (5).

The protonmotive force generated by respiration or ATP hydrolysis is also responsible for driving the active transport of various solutes, most importantly the uptake of essential nutrients and the ejection of unwanted cations (e.g. Na^+)(6). Since active transport is fully reversible, the passive efflux of solutes from the cell (e.g. some fermentation products) can be coupled to the generation of a protonmotive force (7). Other membrane functions for which the protonmotive force provides a source of energy include reversed electron transfer, motility and pyrophosphate synthesis, although the importance of these processes varies widely between different species of bacteria and different modes of growth.

Because of the relatively small amount of work that has been carried out on the respiratory systems of archaebacteria, this paper will concentrate on the evolution of respiration in the eubacteria.

THE FIRST CELLS AND THE ADVENT OF FERMENTATION

The earth was formed approximately 4.5×10^9 years ago, at which time it had a hot and highly reducing, anaerobic atmosphere that was rich in methane, carbon dioxide, ammonia, hydrogen, hydrogen sulphide, formaldehyde and water. As the earth cooled, it was constantly exposed to electrical storms, ultraviolet radiation and volcanic activity; the atmosphere as a result probably became somewhat less reducing and more complex organic molecules began to accumulate. These included various hydrocarbons, sugars and amino acids, some of which later polymerised to form polysaccharides, proteins, nucleic acids and lipids. It was from this myriad of simple and complex molecules that the first primitive cells arose approximately 4.0×10^9 years ago (8,9).

The ubiquitous occurrence of sugars and their various derivatives in current day organisms, together with the remarkably high stability of the glucose molecule and the widespread ability of cells to use it as a source of carbon and energy, makes it extremely likely that the earliest cells were able to conserve energy by catalysing the anaerobic fermentation of glucose (10,11). These earliest fermentations were probably of the homolactic type in which glucose was converted to two molecules of lactate via an ancestral version of the Embden-Meyerhof glycolytic pathway. This relatively simple metabolic sequence allows the net synthesis of two molecules of ATP via substrate level phosphorylation concomitant with the reduction of two molecules of NAD^+ to NADH (E_0 -320mV), which are subsequently reoxidised during the terminal reduction of pyruvate to lactate. This process is thus not only energetically inefficient but also metabolically inflexible since all of the glucose carbon is of necessity used to ensure the required oxidation-reduction balance (Fig. 1a). Homolactic fermentation is today restricted to a relatively small group of bacteria which include *Streptococcus* and some subgenera of *Lactobacillus*, but several lines of evidence indicate that these organisms are not very closely related to the original homolactic fermenters.

It is obvious that at this stage of cellular evolution any organism that could develop an enhanced efficiency of energy and/or carbon conservation would grow significantly faster than its rivals and hence establish a competitive advantage. The abundance of iron and sulphide in the early biosphere is compatible with the view that some of these

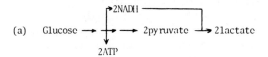

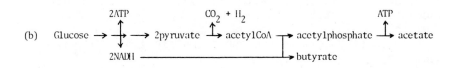

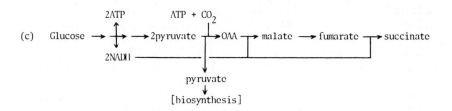

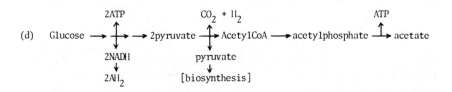

Fig. 1 The possible evolution of fermentation reactions (a) internally-balanced, single-end product, (b) internally-balanced, multiple end-products, (c) and (d) accessory oxidant-dependent (externally-balanced). In (d), A is an accessory oxidant which is reduced to AH_2.

primitive organisms were able to enhance their efficiency of energy conservation using simple iron-sulphur proteins. By analogy with their descendants in extant clostridia, these early iron-sulphur proteins probably contained two covalently-bound [4Fe-4S] centres (E_m < -400mV), each of which carried one electron (13). The appearance of these low redox potential centres in pyruvate dehydrogenase, ferredoxin and hydrogenase thus allowed the catabolic role of pyruvate to be extended from that of a simple oxidant to that of a precursor of acetyl CoA, the additional reducing power released by this oxidative decarboxylation reaction being released into the atmosphere as H_2. Such organisms were able to enhance their ATP yield by converting some of their acetyl CoA to acetyl phosphate and then using the latter to drive an additional substrate level phosphorylation reaction with the concomitant formation of acetate. Overall oxidation reduction balance was achieved by using the remainder of the acetyl CoA (or pyruvate) to oxidise the NADH produced earlier in the fermentation pathway and so produce a mixture of end-products, composed principally of C_2 to C_4 acids and alcohols. Many fermentations of this type are catalysed by extant organisms, and are particularly well exemplified by the mixed-acid fermentations of the *Enterobacteriacea* and by the butyrate or butanol-acetone fermentations of various species of *Clostridium* (Fig. 1b).

H^+ EJECTION VIA END-PRODUCT EFFLUX AND ATP HYDROLYSIS

Since these primitive cells predominantly assimilated their nitrogen in the form of ammonium ions and obtained their energy as the result of various fermentation reactions, both of which lead to intracellular acidification, they were faced with the major problem of maintaining their internal pH at a value close to neutral which would thus allow their cytoplasmic enzymes to function. This problem appears to have been solved by the development of two processes that were capable of ejecting H^+, viz. end-product efflux and ATP hydrolysis.

The first of these was a membrane-bound, solute transport system which catalysed the efflux of an acidic fermentation product down its own concentration gradient in co-transport with H^+ (e.g. anion$^-$.H^+ symport or possibly anion$^-$.$2H^+$ symport); such a system has recently been identified in several organisms including the homolactic fermenter *Streptococcus cremoris*. Depending on the exact solute.H^+ stoichiometry this process would not only alleviate the acidification of the cytoplasm, but also lead to the generation of a useful ΔpH or $\Delta \bar{\mu}_{H^+}$ (7). The efficacy of this system is limited, however, since it only operates effectively under conditions where $[\text{anion}^-]_{in} > [\text{anion}^-]_{out}$; obviously, as the concentration of the fermentation end-product in the surrounding medium builds up, the capacity for further efflux diminishes.

This probably provided the selection pressure for the evolution of the second method of pH regulation in these primitive cells, viz. H^+ ejection by a membrane-bound ATP phosphohydrolase ($BF_0.BF_1$) complex at the expense of some of the ATP generated by substrate level phosphorylation. It has been suggested (14) that the BF_0 and BF_1 components may have had quite separate origins, the former as a membrane-bound protein capable of facilitating passive (energy-independent) H^+ leakage from the cell, and the latter as a soluble pyrophosphatase which perhaps served to hydrolyse the pyrophosphate released during the activation of amino acids. The subsequent condensation of these two components into a single complex, and the transition of BF_1 from a pyrophosphatase to an ATPase for thermodynamic reasons, thus allowed the advent of active (energy-dependent) H^+ ejection and hence, in combination with a separate K^+ uptake system, the generation of a substantial ΔpH and the maintenance of a near-neutral cytoplasm. Since there is abundant evidence that the ATP phosphohydrolase complex has been extremely well conserved during its subsequent evolution, it is likely that the early enzyme was rather similar to its modern day counterpart in clostridia, i.e. BF_0 was composed of multiple copies of a single hydrophobic proteolipid and BF_1 was made up of three or four different hydrophilic proteins according to the exact growth conditions (one of these, in current terminology, was a β subunit that was probably responsible for binding ATP). In this respect it is interesting to note that the ATP phosphohydrolase complex of *Streptococcus* is more closely akin to those of respiratory organisms than clostridia, an observation that lends further credence to the view that this organism may be a degenerate aerobe.

ACCESSORY OXIDANT-DEPENDENT FERMENTATIONS

In all of the fermentation reactions outlined above, oxidation-reduction balance is achieved by various internal reactions (e.g. the reduction of an endogenously-formed oxidant or the release of carbon dioxide and H_2). In contrast, the further evolution of fermentation pathways to produce an enhanced efficiency of carbon conservation or an increased metabolic flexibility, almost certainly occurred via the development of additional, externally-balanced oxidation-reduction reactions through the use of accessory oxidants.

The most important fermentations of this type, at least during the early stages of evolution, were probably those which involved the condensation respectively of carbon dioxide and acetate (the latter in the form of acetyl CoA) with C_3 and C_2 intermediates of metabolism to yield relatively oxidised C_4 compounds that were capable, either immediately or after further transformations, of reoxidising NADH. The use of carbon dioxide as an accessory oxidant was a particularly important evolutionary event since it was almost certainly responsible for the appearance of haemoproteins and for the development of the tricarboxylic acid cycle. Both of these events were dependent on the condensation of carbon dioxide with pyruvate to form oxaloacetate, and the subsequent reduction of the latter to form malate, and hence fumarate, which in turn could act as a terminal oxidant (Fig. 1c). Since the resultant succinate readily gives rise to succinylCoA, the biosynthesis of various haemoproteins including cytochromes, sirohaem and eventually catalase and peroxidase became possible. Although the earliest cytochromes were almost certainly restricted to those of the b-type (E_m < 50mV), their appearance significantly extended the range of redox carriers available to the cell and was fully in tune with the slightly less-reduced nature of the biosphere by this time.

Although these carbon dioxide- and acetate-dependent fermentations actually consumed ATP during their initial fixation or activation reactions, they nevertheless caused a significant increase in metabolic efficiency since they allowed valuable C_3 intermediates such as pyruvate (or, more usually, phosphoenol pyruvate in extant organisms) to be spared for biosynthetic purposes. They also allowed certain organisms to use substrates such as lactate or glycerol which they were otherwise incapable of fermenting. Several examples of accessory oxidant-dependent fermentation have been recognised in modern day anaerobes and photoanaerobes, including those which use not only carbon dioxide, acetate or fumarate, but also various oxyanions of sulphur and nitrogen, dimethyl sulphoxide (DMSO) or trimethylamine-N-oxide (TMAO) as terminal oxidants (Fig. 2d). It should be noted, however, that the use of the latter group of acceptors only evolved much later when the biosphere had become significantly more oxidising.

THE ADVENT OF ANAEROBIC RESPIRATION

The very widespread ability of extant anaerobes and facultative anaerobes to catalyse anaerobic respiration using fumarate as a terminal electron acceptor, and the obvious involvement of fumarate in carbon dioxide-dependent fermentations of the type outlined above, suggest that fumarate reduction may have provided the earliest evolutionary link between fermentation and anaerobic respiration (Fig. 2)(10,11,13). It is likely that fumarate reductase was originally a relatively simple flavoprotein or iron-sulphur flavoprotein, which was located entirely in the cytoplasm and which reduced exogenously-supplied fumarate to succinate at the expense of NADH, i.e. it exhibited both NADH-oxidising and fumarate-reducing properties. This enzyme then evolved into a more complex NADH-fumarate reductase system which contained separate NADH dehydrogenase (Fe-S, FAD) and fumarate reductase (Fe-S, FAD) enzymes, perhaps in response to kinetic pressures; by analogy with extant systems, the iron sulphur centres were probably of the higher redox potential [2Fe-S] and single [4Fe-4S] types. The subsequent acquisition of cytochrome $\underline{b}$, menaquinone (E_m -70mV) or in some cases demethyl-menaquinone, further increased the complexity of the system and led to it becoming associated with the cytoplasmic membrane to form a recognisable respiratory chain endowed with the future potential for accepting reducing equivalents from other dehydrogenases. The newly-formed respiratory chain at first probably only exhibited scalar properties, but later (perhaps much later) acquired the necessary topographical organisation for it to catalyse vectorial H^+ movement. Fumarate respiration at the expense of NADH thus served a dual function, viz. the reoxidation of excess NADH to ensure overall redox balance and the ejection of H^+ to supplement existing methods of pH regulation.

The latter function was probably further satisfied by the subsequent evolution of analogous systems to the NADH-fumarate reductase system that oxidised low redox potential fermentation products, viz. hydrogen ($E_0^\prime$ -420mV) and formate ($E_0^\prime$ -432mV). It is likely that the hydrogenase and formate dehydrogenase enzymes that were responsible for the initial oxidation of these substrates descended from soluble counterparts that catalysed hydrogen evolution and carbon dioxide reduction in ancestors of modern-day clostridia. They were subsequently incorporated into the cytoplasmic membrane, probably with their substrate-binding sites close to the outer surface of the membrane, and were used as alternatives to NADH dehydrogenase for channelling reducing equivalents to menaquinone and hence to the fumarate reductase on the opposite side of the membrane. In extant organisms capable of fumarate respiration, both enzymes contain one or more low redox potential iron-sulphur centres;

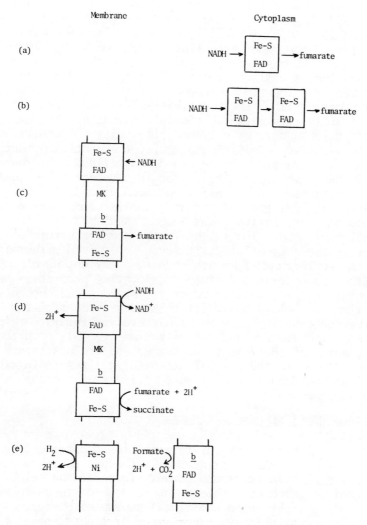

Fig. 2 *The possible evolution of fumarate respiration.
(a) a soluble NADH-fumarate reductase which catalyses
accessory oxidant-dependent fermentation (see Fig. 1d), (b)
the appearance of separate NADH dehydrogenase and fumarate
reductase enzymes, (c) the incorporation of menaquinone (MK)
and, in some cases, cytochrome b to form a membrane-associated respiratory chain, (d) the appearance of vectorial
proton translocation, possibly via a redox loop mechanism,
and (e) the incorporation of hydrogenase and formate
dehydrogenase into the membrane to allow proton translocation to occur via a redox arm mechanism.*

hydrogenase also contains nickel, whereas formate dehydrogenase additionally contains molybdenum (and possibly also selenium in some cases) and is closely associated with a specialised, low redox potential cytochrome $\underline{b}$ (E_m -200mV).

It is not unreasonable to suspect that these early respiratory chains utilised relatively simple direct methods of proton translocation. Indeed, all three fumarate reductase systems outlined above exhibit $\rightarrow H^+/2e^-$ quotients of approximately 2 in extant bacteria, indicating that proton translocation occurs at only one site, viz. the equivalent of site 1 in each case. Furthermore, the spatial orientation of the fumarate reductase and the various dehydrogenases in the membrane indicates the involvement of a simple protonmotive redox arm (half-loop) mechanism during the oxidation of hydrogen and formate, and is commensurate with the use of protonmotive redox loop mechanism during the oxidation of NADH. The slightly more complex nature of the latter system suggests that it may have evolved rather later, i.e. the scalar NADH-fumarate reductase may not have acquired its vecorial, proton-translocating properties until the other two protonmotive redox arm systems had become established. Some support for this idea comes from the observation that in the more highly evolved fumarate respirers (e.g. facultative anaerobes such as *E.coli*), both the hydrogenase and the formate dehydrogenase are positioned in the membrane with their substrate-binding sites facing the cytoplasm such that proton translocation cannot occur via a redox loop mechanism.

LIGHT-DEPENDENT ELECTRON TRANSFER

The diminishing levels of fermentation substrates, and the paucity of terminal oxidants, in the biosphere at this stage of evolution (8,9) undoubtedly favoured the selection of organisms which were able to spare some of these resources by using light as an additional source of energy. A plausible scenario for the appearance of light-dependent electron transfer (10,11,13,15) is that it was predicated on the pre-existence of organisms that were capable both of synthesising porphyrin and of catalysing a primitive form of anaerobic respiration, most probably fumarate respiration.

The replacement of iron by magnesium in the centre of the porphyrin molecule, together with relatively small changes in the structure of the ring system, thus allowed the formation of a primitive bacteriochlorophyll that was capable of undergoing a photochemical charge separation (BChl $\rightarrow$ BChl$^+$ + e$^-$). In order to prevent the released electron from simply recombining with BChl$^+$ via a rapid and entirely wasteful back reaction, the initially soluble bacteriochlorophyll-protein complex became embedded in the respiratory membrane in such a way that the electron was transferred to an appropriate primary acceptor located on one side of the membrane, and the resultant $BChl^+$ was reduced by an appropriate primary donor on the other side. These acceptor /donor requirements were most probably met respectively by a very low redox potential iron-sulphur protein (probably derived from a 2[4Fe-4S] ferredoxin; E_m < -400mV) and by either cytochrome b or a specialised high redox potential [2Fe-2S] protein (also called the Rieske protein after its discoverer in mitochondria; E_m 280mV). The condensation of much of the anaerobic respiratory chain with bacteriochlorophyll thus resulted in the formation of a proton-translocating, cyclic electron transfer system ($\rightarrow H^+/2e^-$ = 2?). This clearly provided an enormous evolutionary advantage, since it allowed the primitive phototroph to combat the internal acidification that resulted from fermentation in a manner that was independent both of ATP and exogenous substrates.

This advantage was further strengthened by the subsequent appearance of the first c-type cytochrome (probably related to *Chlorobium* c_{555}; E_m 250mV) which presumably improved the kinetics of cyclic electron transfer by replacing cytochrome b or the iron-sulphur protein as the primary donor. Haem c probably evolved from haem b via small changes in the nature of the substituents at positions C_2 and C_4 on the porphyrin ring (-CH = CH_2 replaced by $-CH_2.CH_3$), but unlike haem b was very firmly attached to the apoprotein via covalent thioether bridges to cysteinyl residues. This was a particularly important development, since the topographical organisation of these primitive cyclic electron transfer systems demanded that the nascent cytochrome c be inserted in the cytoplasmic membrane close to the outer surface whence, in the absence of covalent bonding, some of the haem c might be irretrievably lost to the external medium (16).

The ability to synthesise c-type cytochrome also presaged two further events. The first of these was the appearance of light-dependent, non-cyclic electron transfer to NAD^+ at the expense of the hydrogen sulphide that was present in large amounts in the biosphere. The second was the formation of a discrete, multicomponent cytochrome bc_1 complex (composed of tightly membrane-bound cytochrome c_1, two cytochromes b and the high redox potential iron sulphur protein) which, in association with the quinone (presumably menaquinone or a close relative, chlorobium quinone) allowed the development of the protonmotive quinone cycle (17). The latter was a particularly important inovation in terms of energy conservation since it effectively doubled the efficiency of proton translocation hitherto obtainable at a single energy coupling site via protonmotive redox arm or loop mechanisms ($\rightarrow H^+/2e^- = 4$ cf.2). These redox systems were probably fairly closely related to those present in extant green photosynthetic bacteria (*Chlorobiaceae*, *Chloroflexaceae*).

The subsequent evolution of these early photosynthetic systems into redox chains that are more closely related to those present in current day purple sulphur bacteria (*Chromatiaceae*) and purple non-sulphur bacteria (*Rhodospirillaceae*) involved several major changes. These included the development of bacteriochlorophylls that were capable of utilising other regions of the electromagnetic spectrum, the replacement of low redox potential iron-sulphur proteins with bacteriopheophytins, the appearance of quinone-iron complexes as secondary acceptors and the replacement of menaquinone with ubiquinone ($E_m \simeq 0mV$). In addition, cytochrome $\underline{c}_{555}$ (MW 9.5K) was replaced in the *Rhodospirillaceae* by cytochrome $\underline{c}_2$ (MW $\geqslant$ 9.5K); thus, in the current terminology, a small (S) class I cytochrome $\underline{c}$ was replaced by one that was either short (S), medium (M) or long (L) according to the exact genus and/or species of the organism. Primary and tertiary structure analyses of these different $\underline{c}$-type cytochromes have shown that they have well-conserved amino acid sequences and very similar folding patterns, indicating that they have probably evolved from a common, short ancestor by very slight changes and/or sequential chain insertions of up to a total of approximately 40 amino acids (18, 19).

In the *Rhodospirillaceae*, non-cyclic electron transfer occurs principally at the expense of relatively weak, organic reductants (e.g. succinate). However, since the redox potential of the quinone-iron complex is too low to allow the direct, light-dependent reduction of NAD^+ by succinate, this reaction is effected by reversed electron transfer via succinate dehydrogenase, ubiquinone/cytochrome b and NADH dehydrogenase at the expense of the protonmotive force generated by cyclic electron transfer. There is extremely good evidence, both from conventional analytical studies and from more recent gene cloning and sequencing work, that succinate dehydrogenase is very closely related to fumarate reductase, and that the former (in these photosynthetic bacteria) almost certainly evolved from the latter (in the fumarate respirers) via small changes in kinetic properties (20).

Cyclic electron transfer in the blue-green bacteria (*Cyanobacteriaceae*), which more closely resembles that present in the green bacteria than in the purple bacteria, probably evolved via relatively minor modifications to bacteriochlorophyll, ubiquinone, cytochrome $\underline{b}$ and cytochrome $\underline{c}_1$ (to form chlorophyll, plastoquinone, cytochrome $\underline{b}_6$ and cytochrome $\underline{f}$), together with the appearance probably of pheophytin as the primary acceptor and of a small copper protein, plastocyanin (E_m 370mV), as the primary donor. The small amount of structural information that is currently available on cytochrome $\underline{c}_{554}$ from blue-green bacteria indicates that it is a short (S*) class I cytochrome $\underline{c}$ with a similar folding pattern, but a very different amino acid sequence, to cytochromes $\underline{c}_{555}$ and $\underline{c}_2$(S). Very importantly, non-cyclic electron transfer to $NADP^+$ in the blue-green bacteria requires the participation of an additional, very high redox potential chlorophyll which catalyses the light dependent transfer of electrons from water to plastoquinone (photosystem II) with the concomitant release of molecular oxygen. The appearance of oxygenic electron transfer in cyanobacterial ancestors was of signal importance to the evolution of bacterial respiration since it heralded the eventual onset of aerobiosis.

PROTONMOTIVE FORCE-DEPENDENT ENERGY CONSERVATION

The advent of anaerobic respiration and photosynthetic electron transfer, and hence the ability of cells actively to extrude H^+ without recourse to ATP hydrolysis, was clearly a particularly important step in the evolution of bacterial energy conservation. Furthermore, since the protonmotive force generated by these redox reactions is at least as high as that produced by ATP hydrolysis, the inward movement of H^+ with concomitant ATP synthesis became thermodynamically feasible. The selection pressure for such a change to become established was particularly strong since it led to a significantly higher yield of ATP than that produced by substrate-level phosphorylation alone. The role of the ATP phosphohydrolase complex thus changed from that of a protonmotive force-generating ATPase (high ATPase activity, regulated by various metabolites; low ATP synthase activity, inhibited by ATP) to that of a protonmotive force-utilising ATP synthase (high ATP synthase activity, low ATPase activity)(14).

Analyses of $BF_0 \cdot BF_1$ complexes from extant facultative anaerobes capable of catalysing anaerobic respiration (e.g. *Escherichia coli*, *Salmonella typhimurium*) indicate that this change in function was probably the result of several structural alterations. Thus BF_0 acquired multiple copies of two additional hydrophobic proteins (subunits a and b in current terminology), BF_1 evolved into a five subunit enzyme, and the overall complex attained the approximate subunit stoichiometry $\alpha_3\beta_3\gamma\delta\epsilon a_2 b_2 c_{9-12}$ which appears to be characteristic of modern day respiratory and photosynthetic bacteria. Unfortunately, no structural studies have yet been carried out on the ATP phosphohydrolase complexes of obligate anaerobes such as *Clostridium formicoaceticum* or *Vibrio succinogenes*, both of which catalyse fumarate respiration and may be regarded as intermediates between the obligately fermentative clostridia and the facultative anaerobes. In this context, however, it is interesting to note that the $BF_0.BF_1$ complexes of the very few obligate aerobes that have so far been examined in any detail (e.g. *Micrococcus lysodeikticus*, *Mycobacterium phlei*) are very similar to those of the facultative anaerobes, emphasising yet again the highly conserved nature of this enzyme complex.

SULPHATE RESPIRATION

The ability of primitive photosynthetic bacteria to oxidise reduced sulphur compounds via non-cyclic electron transfer was a particularly important milestone in the evolution of bacterial respiration since it heralded the eventual accumulation of large quantities of sulphate (9), and hence the appearance of sulphate respiration (dissimilatory sulphate reduction). The available evidence indicates that bacteria capable of sulphate respiration are descended from fermentative bacteria, probably via organisms that had the ability to catalyse fumarate respiration. Indeed, amongst the extant sulphate respirers (e.g. *Desulfovibrio*, *Desulfotomaculum*), some species have retained their ability to ferment pyruvate or reduce fumarate, and most are capable of using lactate or hydrogen as respiratory chain substrates.

The term sulphate respiration is in a sense a misnomer, since the terminal acceptor is not sulphate itself. The latter is too weak an oxidant to be useful in terms of respiratory chain energy conservation (E_0' -480mV) and must therefore be converted at the expense of ATP to a stronger oxidant, adenosine phosphosulphate (APS; E_0' -60mV). The hydrogen-APS reductase system contains a hydrogenase (Fe-S) that is loosely attached to the periplasmic surface of the membrane, menaquinone and perhaps also cytochrome b, and an APS-reductase (Fe-S, FAD) that is on the cytoplasmic side. It thus resembles the hydrogen-fumarate reductase system both in its spatial organisation and in its ability to translocate protons via a redox loop mechanism ($\rightarrow H^+/2e^-$ =2); the redox carrier composition of the two systems is also broadly similar except that during sulphate respiration the hydrogenase to menaquinone step is often mediated by a low redox potential, four-haem cytochrome c_3 (E_m -300mV), and the menaquinol to APS reductase step involves a specialised, high redox potential ferredoxin. The sulphite generated during the reduction of APS is subsequently reduced to sulphide via an analogous respiratory chain to that described above except that APS reductase is replaced by (bi)sulphite reductase (Fe-S, sirohaem). A plethora of minor redox carriers have also been identified in these organisms including flavodoxins, rubredoxins, and iron-sulphur proteins that contain molybdenum (21).

In addition to cytochrome c_3, the sulphate respirers also contain two other c-type cytochromes, viz. cytochrome c_7 (three haem, low E_m) and cytochrome c_{553} (monohaem, E_m +50mV); all these have low molecular weights (<13K) and are thus classified as short (S) or medium (M) type I cytochromes c. In terms of their size, their as yet incompletely analysed primary sequences and their partly-predicted folding patterns, they obviously show some relationship to cytochrome c_{555} in the green photosynthetic bacterium, *Chlorobium*. It is not difficult, therefore, to envisage that all of these c-type cytochromes evolved from haem b and a common ancestral apoprotein, and that the propensity to low E_m values and multiple haems in the sulphate respirers, but not in *Chlorobium*, reflects subtle differences in the structure and/or axial-binding of the haem as dictated by the quite different functions of the c-type cytochromes in these two groups of organisms.

Interestingly, the recently discovered genus *Desulfuromonas* catalyses anaerobic respiration to sulphur (dissimilatory sulphur reduction) but not to sulphate; the small amount of information currently available on the respiratory chains of these organisms indicates, not unexpectedly, that they are closely related to those of the sulphate respirers.

AEROBIC RESPIRATION

Bacterial respiratory chains capable of using oxygen as a terminal electron acceptor can be grossly divided into three major types on the basis of their redox carrier composition and the nature of the carbon/energy source for growth. These are (i) the respiratory chains of chemoheterotrophic bacteria which contain cytochrome oxidases d (plus a_1) and/or o, but lack a high redox potential cytochrome c (Type I chains), (ii) the respiratory chains of chemoheterotrophic bacteria which contain cytochrome oxidases aa_3 and/or o plus a high redox potential cytochrome c (Type II chains), and (iii) the respiratory chains of chemolithotrophic bacteria (Type III chains)(3,22-24).

In contrast to the essentially monophyletic evolution of anaerobic respiration and light-dependent electron transfer, there is evidence that the subsequent evolution of aerobic respiration was probably multiphyletic, i.e. the three major types of respiratory chains evolved in parallel from several different ancestors which were present in the biosphere when ancestors of modern-day cyanobacteria started to release oxygen.

(i) Type I aerobic respiratory chains

There is increasing evidence that Type I aerobic respiratory chains evolved from the primitive respiratory chains of early fumarate-respiring bacteria. A plausible scenario for the conversion of the latter to aerobic respiration (Fig. 3) is that, with the advent of aerobic conditions, the cytochrome b associated with fumarate reductase underwent some form of gene duplication and that one of the products acquired a reasonably high affinity for molecular oxygen which it reduced to water. It is likely that the emergence of this autoxidisable b-type cytochrome (now called cytochrome oxidase o; E_m usually < 270mV) was similarly followed by the appearance of cytochrome oxidases d (E_m 280mV) and a_1. All three of these cytochrome oxidases are widely distributed in extant bacteria, d always being accompanied by a_1 and o, whereas the latter often occurs alone. Cytochrome oxidase d contains both haem b (protohaem) and the closely-related haem d (chlorin), the latter being responsible for binding oxygen in a very high affinity reaction. It may therefore have evolved in order to allow organisms to take maximal advantage of low concentrations of oxygen for respiratory chain energy conservation and/or to act as an oxygen scavenger and so protect oxygen-labile enzymes from inactivation. In the latter context, it is interesting to note that cytochrome oxidase d is the major terminal oxidase in the few known species of aerobic nitrogen fixers, and has also been detected in a few species of obligately anaerobic, sulphate-reducing bacteria. There is conflicting evidence concerning the status of cytochrome a_1 (haem a) as an oxidase in chemoheterotrophic bacteria. Oxidase function has been established in *Acetobacter pasteurianum*, but remains doubtful in *E.coli* and *K.aerogenes*.

The further evolution of this type of respiratory chain involved the appearance of succinate dehydrogenase via gene duplication of fumarate reductase and, in some organisms, a change in the type of quinone present. The former event allowed facultative anaerobes to catalyse the aerobic oxidation of succinate as well as anaerobic respiration to fumar-

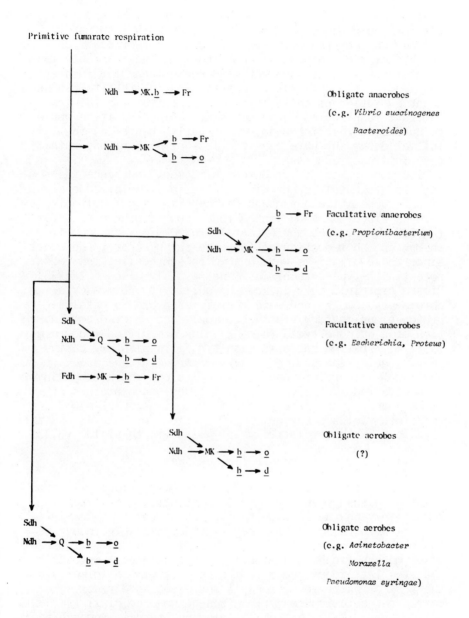

Fig. 3 *The possible evolution of type I aerobic respiratory chains. Abbreviations: Ndh, NADH dehydrogenase; Fr. fumarate reductase; Fdh, formate dehydrogenase; Sdh, succinate dehydrogenase*

ate. In extant Gram positive bacteria (e.g. *Propionibacterium*) menaquinone is the sole quinone in both respiratory chains, whereas the Gram negative *Enterobacteriaceae* contain the higher redox potential ubiquinone in their aerobic respiratory chain, but have retained menaquinone for fumarate respiration (25,26). It is easy to imagine that the subsequent evolution of obligate aerobes occurred via the loss of fumarate respiration; however, since this transition appears to have been restricted to Gram negative bacteria, the initial event may have been the loss of menaquinone rather than fumarate reductase.

The appearance of cytochrome oxidases, and hence the capacity to catalyse aerobic respiration, conferred a significant competitive advantage on the primitive facultative anaerobes since the NADH-linked reduction of oxygen to water (E_0' 820mV) releases approximately three times as much free energy as the reduction of fumarate to succinate. However, since these organisms were unable to synthesise high redox potential c-type cytochromes (and possibly also a [2Fe-2S] Rieske protein), they lacked the ability to form a cytochrome bc_1 complex and were therefore unable to catalyse proton translocation via the commonly-encountered combination of a quinone cycle (site 2) plus a modified redox arm (site 3). Extant bacteria with type I respiratory chains (e.g. *E.coli*, *Acinetobacter calcoaceticus*, *Klebsiella pneumoniae*) therefore use a less-efficient redox loop or primitive redox pump mechanism to translocate protons at sites 1 and 2, and lack site 3. The oxidation of NADH therefore yields a maximum $\rightarrow H^+/O$ quotient of only 4, and the organisms exhibit relatively low molar growth yields (3,22).

(ii) Type II aerobic respiratory chains

A comparison of the redox carrier compositions, electron transfer pathways and cytochrome c structures of type II respiratory chains in facultative phototrophs (*Rhodospirillaceae*) with those in Gram negative aerobes (e.g. many species of *Pseudomonas*, together with *Paracoccus* and various other methanotrophs and methylotrophs) strongly suggests that the latter descended from ancestors of the former via the loss of bacteriochlorophyll and associated redox carriers (bacteriopheophytin, iron-quinone complex) (10,15, 18). It is likely that obligately phototrophic ancestors of extant purple non-sulphur bacteria acquired the ability to use molecular oxygen as a terminal acceptor for respiration by developing o-type cytochrome oxidases in a manner similar to that described for the evolution of Type I aerobic respiratory chains. Interestingly, however, many of the type

II chains contain both a low redox potential ($E_m \simeq 250mV$) and a high redox potential ($E_m \simeq 400mV$) cytochrome oxidase $\underline{o}$, the former accepting electrons from cytochrome $\underline{b}$ and the latter from cytochrome $\underline{c}_2$. The further evolution of these respiratory chains was characterised by the appearance of cytochrome oxidase $\underline{aa}_3$ which, in extant bacteria, contains two molecules of haem $\underline{a}$ ($E_m \simeq 220$ and $375mV$) and two atoms of copper. This event did not constitute a particularly large evolutionary jump since haem $\underline{a}$ has a similar structure to, and is synthesised from, haem $\underline{b}$. However, as cytochrome oxidase $\underline{aa}_3$ is now a very widely-established terminal oxidase in extant bacteria, it presumably offered significant kinetic and/or energetic advantages over cytochrome oxidase $\underline{o}$. The advantages in terms of proton translocation of using a high redox potential $\underline{c} \rightarrow \underline{aa}_3$ pathway compared with a low redox potential $\underline{b} \rightarrow \underline{o}$ pathway are obvious, and will be discussed in more detail later. Less obvious is the fact that cytochrome oxidase $\underline{aa}_3$ reacts extremely rapidly with the cytochrome $\underline{bc}_1$ complex (via cytochrome $\underline{c}_2$ or $\underline{c}$) and also, at least under conditions of oxygen sufficiency, with molecular oxygen.

The subsequent loss of the photosynthetic apparatus from the redox systems of selected facultative phototrophs thus produced the respiratory chains which are characteristic of many extant Gram negative aerobes, in particular oxidase test-positive species of *Pseudomonas*, and *Pc.denitrificans*. Interestingly, following growth under the appropriate conditions, some of these organisms can oxidise methanol via a pyrroloquinoline quinone-linked methanol dehydrogenase which donates electrons to a specialised $\underline{c}$-type cytochrome. This property thus adds further credence to the view that *Pc. denitrificans*, the facultatively methylotrophic pseudomonads (e.g. *Ps.AM1*, *Ps.extorquens*) and the obligate methylotrophs (e.g. *Methylophilus methylotrophus*) are closely related to the purple non-sulphur bacteria, since several of the latter are also capable of oxidising methanol (e.g. *Rps.acidophila*).

Structural analyses of cytochrome $\underline{c}$ from various species of bacteria strongly supports the broad evolutionary pathway outlined above in the sense that the short (S) and long (L) cytochromes $\underline{c}_2$ of various facultative phototrophs can be correlated with the (S) cytochromes $\underline{c}_{551}$ or $\underline{c}_H$ of *Pseudomonas*, *A.vinelandii* and *M.methylotrophus*, and with the (L) cytochrome $\underline{c}_{550}$ of *Pc.denitrificans* respectively.

The evolutionary pathways of Type II respiratory chains in Gram positive bacteria (e.g. *Micrococcus*, *Staphylococcus*, *Bacillus*, coryneforms and the *Actinomycetales*) are very much less clear since the electron transfer pathways and cytochrome $\underline{c}$ structures of these organisms have been the subject

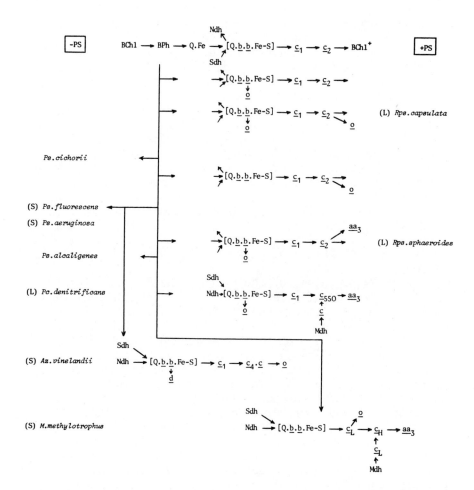

Fig. 4 The possible evolution of type II aerobic respiratory chains in Gram negative bacteria. Abbreviations: S, short; M, medium; L, long; Mdh, methanol dehydrogenase; BChl, bacteriochlorophyll; BPh, bacteriopheophytin; others as in Fig. 3. S, M and L refer to the size of the loosely membrane bound, $\underline{c}$-type cytochromes ($\underline{c}_2$, $\underline{c}_{550}$, $\underline{c}_H$ etc.).

of considerably less study. Very strikingly, however, all of these respiratory chains contain menaquinone, rather than ubiquinone, as their sole isoprenoid quinone component.

In view of this it seemed possible that these type II respiratory chains of Gram positive bacteria might have evolved from ancestors of primitive, menaquinone-containing type I respiratory chains similar to those present in extant species of *Propionibacterium*. In terms simply of redox carrier composition this idea has some merits since the transition could have occurred quite plausibly through the acquisition of c-type cytochromes and cytochrome oxidase aa_3 (possibly via the haem a-containing cytochrome oxidase a_1), and the loss of fumarate reductase and cytochrome oxidase d. However, all of these type II chains so far examined behave in a manner compatible with the presence of a cytochrome bc_1 complex (and probably also with a protonmotive quinone cycle) neither of which are present in any of the modern relatives of possible type I ancestors. This evolutionary pathway therefore seems unlikely.

A more plausible scenario, therefore, is that these type II respiratory chains of Gram positive bacteria evolved, like those of Gram negative bacteria, from facultative phototrophs via the loss of photosynthetic capacity; in this case from ancestors either of extant purple bacteria (some of which contain small amounts of menaquinone in addition to ubiquinone) or green bacteria (in which menaquinone is the major quinone component in association with minor amounts of chlorobium quinone)(26). Unfortunately our current knowledge both of the respiratory systems of facultatively phototrophic green bacteria (*Chloroflexaceae*), and of the cytochrome c structures of these organisms and the Gram positive aerobes, is so scant as to preclude a more detailed discussion.

Type II respiratory chains from both Gram negative and Gram positive chemoheterotrophs exhibit energy conservation properties commensurate with the presence of three energy coupling sites, i.e. the oxidation of NADH yields an $\rightarrow H^+/O$ quotient of approximately 6 in most organisms. This is achieved via site 1 (NADH-quinone reductase; $\rightarrow H^+/2e^- = 2$) plus a combination of sites 2 and 3 ($\rightarrow H^+/2e^- = 4$); site 2 is probably a protonmotive quinone cycle (quinol-cytochrome c reductase; $\rightarrow H^+/2e^- = 4$, $\rightarrow charge/2e^- = -2$), whereas site 3 is a modified redox arm which lacks the external protolytic reaction (cytochrome c linked to cytochrome oxidase o or aa_3; $\rightarrow H^+/2e^- = 0$, $\rightarrow charge/2e^- = -2$). During the oxidation of the site 3 substrate, methanol, this protolytic reaction is catalysed by the externally-located methanol dehydrogenase such that energy conservation is characterised by net proton ejection ($\rightarrow H^+/O = 2$, $\rightarrow$charge/O = -2).

However, $\rightarrow H^+/O$ quotients significantly greater or less than 6 for the oxidation of NADH have been observed for several type II respiratory chains. Low values are usually caused by the presence of an active cytochrome $\underline{b} \rightarrow \underline{o}$ branch ($\underline{b} \rightarrow \underline{d}$ in the case of the honorary type II system of $Az.$ $vinelandii$), which thus avoids site 3. Higher values are attributed either to the presence at site 1 of a proton-pumping NADH dehydrogenase ($\rightarrow H^+/2e^- > 2$, e.g. $Pc.denitrificans$) and/or to the presence at site 3 of a proton pumping cytochrome oxidase $\underline{aa_3}$ ($\rightarrow H^+/2e^- = 2$, $\rightarrow$charge/$2e^- = -4$; e.g. $Pc.denitrificans$, $Bacillus\ stearothermophilus$) which acts both as a modified redox arm and also as an indirect proton pump (27).

(iii) Type III aerobic respiratory chains

The type III respiratory chains present in the Gram negative, chemolithotrophic bacteria oxidise, according to species, a wide range of inorganic reductants including hydrogen, various compounds of nitrogen and sulphur, and ferrous iron. Since all of these donors were available in the biosphere at the onset of oxygenic photosynthesis, the ability to use them as donors for aerobic respiration was clearly of benefit to the newly-evolving aerobes (indeed, the use of certain inorganic reductants for aerobic respiration quite possibly preceded the use of organic donors).

This benefit was extremely variable, however, since the E_0 values of these substrates cover an extremely wide range (e.g. $2H^+/H_2$ - 420mV cf. Fe^{3+}/Fe^{2+} 780mV), and hence their oxidation yields a highly variable amount of energy. Furthermore, since these organisms can all grow autotrophically, and thus require copious amounts of NADH for carbon dioxide fixation, energy-dependent reversed electron transfer to NAD^+ is a major respiratory chain function in all of the chemolithotrophs except some of the hydrogen bacteria. However, irrespective of whether forward or reversed electron transfer predominates, these type II respiratory chains closely resemble the type II chains of Gram negative chemotrophs as modified by the addition of various oxido-reductases capable of catalysing the oxidation of inorganic donors.

It is extremely likely, therefore, that the Gram negative thiobacilli are descended from sulphur-oxidising purple bacteria by the loss of the photosynthetic apparatus, the acquisition of cytochrome oxidases $\underline{aa_3}$ and $\underline{o}$, and some modification of the sulphide- and thiosulphate-oxidising enzyme systems to include the capacity to oxidise sulphur

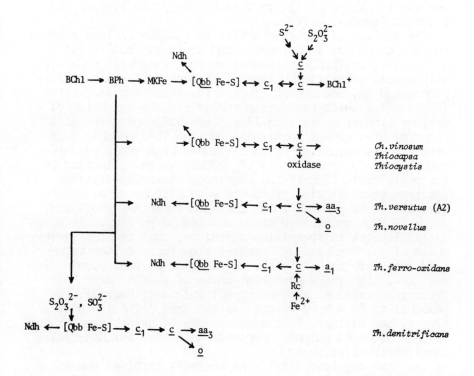

Fig. 5. The possible evolution of respiratory chains in the genus Thiobacillus. Abbreviations: Rc. rusticyanin; others as in Fig. 3.

and sulphite (Fig. 5). This view is supported by recent reports that some species of purple sulphur bacteria catalyse the aerobic oxidation of sulphide and thiosulphate (e.g. *Chromatium vinosum, Thiocapsa roseopersicina, Thiocystis violacea*), as also can a few species of purple non-sulphur bacteria. The oxidation of reduced sulphur compounds by most of the thiobacilli occurs predominantly via cytochrome c-linked enzymes, thus allowing energy conservation only at site 3. However, in *Thiobacillus denitrificans* electrons are transferred to the respiratory chain at the flavin or quinone level, which thus not only gives rise to significantly enhanced energy conservation via the additional use of site 2, but also allows this organism, rarely amongst the thiobacilli, to obtain energy via nitrate respiration (see below)(28).

The ability of the thiobacilli to oxidise various reduced sulphur compounds to sulphate, and hence to acidify the surrounding environment, endows many of them with moderately acidophilic properties. Indeed, acidophily is particularly pronounced in, and indeed crucial to, those few species of *Thiobacillus* which can obtain energy via the oxidation of ferrous iron (e.g. *Thiobacillus ferro-oxidans*). The acid external environment maintains the iron in a soluble form prior to its oxidation by an extremely high redox potential copper protein, rusticyanin (E_m 680mV) which is located on the periplasmic side of the membrane. Inwardly-directed electron transfer via cytochrome $\underline{c}$ to cytochrome $\underline{a}_1$ subsequently allows protons to be consumed from the cytoplasm, which is thus maintained at an acceptable pH (29). It seems fairly logical to speculate, therefore, that the respiratory chain of *Th.ferro-oxidans* evolved from those of non iron-oxidising thiobacilli by the acquisition of rusticyanin plus modified, high redox potential forms of cytochrome $\underline{c}$ and cytochrome oxidase $\underline{a}_1$. One cannot rule out, however, the possibility of a more direct descent from primitive phototrophs, now extinct, that were capable of using both ferrous iron and reduced sulphur compounds as donors for non-cyclic electron transfer.

The Gram negative nitrifying bacteria catalyse the oxidation of ammonia to nitrite (e.g. *Nitrosomonas, Nitrosococcus*) and the subsequent oxidation of nitrite to nitrate (e.g. *Nitrobacter, Nitrococcus*). The origin of the nitrifiers is much more obscure than that of the thiobacilli since no extant photosynthetic bacteria can use ammonia as a reductant for non-cyclic electron transfer in a manner analogous to the use of various reduced sulphur compounds by the purple bacteria. Nevertheless, as has been pointed out previously (8,9), the anaerobic oxidation of ammonia to dinitrogen could have served as a perfectly good source of reducing power for primitive photosynthetic bacteria. Indeed, it might be envisaged that, following the appearance of molecular oxygen in the biosphere, the enzyme responsible for this reaction was no longer synthesised (as is the case with the oxygen-sensitive nitrogenase of many facultative anaerobes) and became replaced by an alternative system which was actually dependent on molecular oxygen. Thus, in *Nitrosomonas*, ammonia is converted to hydroxylamine under the action of ammonia hydroxylase, and the hydroxylamine is subsequently oxidised to nitrite ($E_0^{\prime}$ NO_2^-/NH_2OH 66mV) by hydroxylamine cytochrome $\underline{c}$ reductase. The latter is a large and rather complicated enzyme which is located, like the ammonia hydroxylase, on the periplasmic side of the membrane; it contains multiple haem $\underline{c}$ and haem P460 centres, and transfers electrons both to the original mono-oxygenation reaction and to the respiratory chain. The subsequent reduction of molecular oxygen via cytochrome oxidase allows the generation of a protonmotive force, probably via a redox arm mechanism.

Interestingly a number of parallels can be drawn between the oxidation of methane and ammonia by bacteria, including the ability of methanotrophs to oxidise ammonia and of nitrifiers to oxidise methane, which implies that methane mono-oxygenase and ammonia hydroxylase may, at least originally, have been closely related.

The further evolution of nitrification was predicated on the appearance of a nitrite oxidoreductase. As found in *Nitrobacter*, this enzyme contains molybdenum and iron-sulphur centres, and is closely associated with a non-autoxidisable cytochrome a_1 which transfers electrons to cytochrome $\underline{c}$. Since it is located on the cytoplasmic side of the membrane, probably for good thermodynamic reasons, appropriate transport systems must also be present to import nitrite and export the resultant nitrate. The relatively high standard redox potential of the nitrate/nitrite couple (E_0' 420mV) again limits energy conservation to site 3, proton translocation occurring either by some form of redox loop or proton-pumping cytochrome oxidase ($\rightarrow H^+/O$ 1-2), or perhaps by a combination of the two ($\rightarrow H+/O > 2$) (30).

If nitrite-oxidising bacteria evolved from ancestors of extant ammonia oxidisers by losing the capacity to oxidise ammonia and hydroxylamine, and acquiring the ability to oxidise nitrite, the existence of species capable of catalysing the complete oxidation of ammonia to nitrate might be predicted. As yet, however, no such organisms have been isolated, suggesting either that they are extinct or that direct descent did not occur.

NITRATE RESPIRATION AND DENITRIFICATION

The advent of aerobic chemolithotrophs capable of oxidising reduced nitrogen compounds led eventually to the accumulation of nitrate, and heralded the last major development during the evolution of bacterial respiration, viz. anaerobic respiration to nitrate (dissimilatory nitrate reduction) and the subsequent reduction of nitrite, nitric oxide and nitrous oxide to yield dinitrogen (denitrification). The latter process, in marked contrast to nitrification, is an extremely widespread phenomenon, having been reported to occur in facultative phototrophs (e.g. *Rps. sphaeroides*, *Rps.capsulata*), Gram negative chemoheterotrophs (e.g. *Alcaligenes*, *Chromobacterium*, *Rhizobium*, *Paracoccus* and many species of *Pseudomonas*), Gram positive chemoheterotrophs (e.g. *Bacillus*, *Propionibacterium*) and chemolithotrophs (e.g. *Th.denitrificans*); in addition, many other species of bacteria catalyse only nitrate respiration, of which *E.coli* and *K.pneumoniae* have been most extensively investigated.

This widespread occurrence of nitrate respiration and denitrification in organisms capable of aerobic respiration, together with the well-conserved nature of the nitrate reductases so far investigated and the paucity of nitrate in the early biosphere, suggest that nitrate respiration has probably evolved via interspecies transfer between aerobes and facultative anaerobes of the operon responsible for the nitrate reductase apoprotein rather than via an accessory oxidant-dependent fermentation route analogous to that proposed for fumarate reductase. Furthermore, the presence of high redox potential molybdenum and iron-sulphur centres in both nitrate reductase and nitrite oxidoreductase, together with the known reversibility of the latter enzyme, is compatible with the idea that the former may have evolved from the latter in a manner analogous to the evolution of succinate dehydrogenase from fumarate reductase. It is likely, therefore, that nitrate respiration appeared almost simultaneously in many different species perhaps as a result of gene duplication in a nitrifier, followed by gene transfer, then the insertion of the nitrate reductase enzyme into an existing aerobic respiratory chain and finally the modification of a b-type cytochrome to facilitate electron transfer to the novel reductase.

Although both respiration and denitrification are widely distributed in extant bacteria, nitrite and nitrous oxide are both much stronger oxidising agents than nitrate (e.g. $E_\theta^!$ N_2O/N_2 = 1355mV). Denitrification is therefore a significantly higher redox potential process than nitrate respiration and hence is only found in organisms which are capable of synthesising a high redox potential cytochrome c. Indeed, it is now well established that in *Pc.denitrificans* and several species of *Pseudomonas* both nitrite reductase (cytochrome cd_1) and nitrous oxide reductase (a large multi-centred copper protein) are loosely attached to the periplasmic side of the membrane and receive electrons from azurin (a small copper protein; E_m 230mV) and/or cytochrome c_{551} respectively (27).

In spite of the fact that cytochrome cd_1 exhibits slight

cytochrome oxidase activity, it probably evolved independently of cytochrome oxidase $\underline{d}$. Indeed, the activity of some species of *Rhodopseudomonas* and many species of *Pseudomonas* to denitrify suggests that this process may have evolved initially in parallel with aerobic respiration in these organisms, only later being acquired by many other species via gene transfer. It may thus be envisaged that cytochrome $\underline{cd}_1$ arose via the ability of a relatively primitive, facultative phototroph to synthesise haem d and tether it to a $\underline{c}$-type cytochrome to yield an autoxidisable haemoprotein which was capable of reducing nitrite; cytochrome $\underline{c}_{551}$ and azurin probably appeared later for kinetic reasons, the azurin perhaps as the result of gene transfer from primitive cyanobacteria. Nitrite respiration was probably followed by the appearance in some, but not all, denitrifiers of nitrous oxide reductase, which thus allowed further energy conservation.

The ubiquitous presence of copper proteins in denitrification is interesting since it is compatible with their presence in some forms of aerobic respiration and in oxygenic photosynthesis, and confirms that the role of these proteins is to catalyse high redox potential electron transfer. Indeed, geological evidence indicates that copper, unlike iron, remained trapped in ocean sediments as highly insoluble cuprous sulphide until, during the later stages of evolution, the sulphide was oxidised to sulphate and the soluble copper ions became available for incorporation into novel redox proteins (iron, of course, was available during the early stages of evolution as soluble magnetite, Fe_3O_4, and was captured, with various sulphides, during the formation of ferredoxin at the other end of the redox scale).

The external location of both these reductases is particularly noteworthy since significantly higher $\rightarrow H^+/2e^-$ quotients would have been possible had these enzymes catalysed their protolytic reactions on the cytoplasmic side of the membrane. It must be concluded, therefore, that the spatial organisation of the terminal electron transfer system was determined principally by the external location of pre-existing $\underline{c}$-type cytochromes (and possibly also copper proteins) prior to their modification for use as reductases.

Finally, it should be noted that several species of non-denitrifying, obligate and facultative anaerobes (e.g. *V. succinogenes*, *D. desulfuricans*, *E. coli*) contain a nitrite oxidoreductase which catalyses the reduction of nitrite to ammonia rather than to nitrous oxide. This enzyme is a large $\underline{c}$-type cytochrome which contains up to 6 haems $\underline{c}$ per molecule and is loosely bound to the membrane probably on the periplasmic side. It thus shows a striking resemblance to the hydroxylamine oxidoreductase present in some nitrifying bacteria (e.g. *Nitrosomonas*). Indeed, it is tempting to speculate that it may have evolved from hydroxylamine oxidoreductase in a manner analogous to the evolution of nitrate reductase from nitrite oxidoreductase.

ACKNOWLEDGEMENTS

The author is indebted to Dr D. Jones for useful discussions, Mrs Amelia Dunning for her careful typing of the manuscript, and SERC for financial support.

REFERENCES

1. Mitchell, P. (1977). Vectorial Chemiosmotic processes. *Ann. Rev. Biochem.* 46, 998-1004.
2. Garland, P.B. (1977). Energy transduction in microbial systems. In "Microbial Energetics" (Eds B.A. Haddock and W.A. Hamilton). Soc. Gen. Microbiol. Symp. 27, pp. 1-21. Cambridge University Press, Cambridge.
3. Jones, C.W. (1979). Energy metabolism in aerobes. In "International Review of Biochemistry: Microbial Biochemistry" (Ed J.R. Quayle) Vol. 21, pp. 49-84. University Park Press, Baltimore.
4. Ferguson, S.J. and Sorgato, M.C. (1982). Proton electrochemical gradients and energy transduction processes. *Ann. Rev. Biochem.* 51, 185-217.
5. Thauer, R.K., Jungermann, K. and Decker, K. (1977). Energy conservation in chemotrophic anaerobic bacteria. *Bact. Revs.* 41, 100-180.
6. Hamilton, W.A. (1977). Energy coupling in substrate and group translocation. In "Microbial Energetics" (Eds B.A. Haddock and W.A. Hamilton). Soc. Gen. Microbiol. Symp. 27, pp. 185-216. Cambridge University Press, Cambridge.
7. Michels, P.A.M., Michels, J.P.J., Boonstra, J. and Konings, W.N. (1979). Generation of an electrochemical proton gradient in bacteria by the excretion of metabolic end products. *FEMS Microbiol. Letts.* 5, 357-364.
8. Broda, E. (1975). "The evolution of the bioenergetic processes." Pergamon, Oxford.
9. Broda, E. and Peschek, G.A. (1979). Did respiration or photosynthesis come first? *J. Theoretical Biol.* 81, 201-212.
10. Gest, H. (1980). The evolution of biological energy-transducing systems. *FEMS Microbiol. Letts.* 7, 73-77.
11. Gest, H. and Schopf, J.W. (1983). Biochemical evolution of anaerobic energy conversion: the transition from

fermentation to anoxygenic photosynthesis. In "Earth's earliest biosphere: its origin and evolution" (Ed J.W. Schopf) pp. 135-150. Princeton University Press, Princeton.
12. Rao, K.K. and Cammack, R. (1981). The evolution of ferredoxin and superoxide dismutase in microorganisms. In "Molecular and Cellular Aspects of Microbial Evolution" (Eds M.J. Carlile, J.F. Collins and B.E.B. Moseley). Soc. Gen. Microbiol. Symp. 32, pp. 175-213. Cambridge University Press, Cambridge.
13. Gest, H. (1981). Evolution of the citric acid cycle and respiratory energy conversion in prokaryotes. *FEMS Microbiol. Letts.* 12, 209-215.
14. Garland, P.B. (1981). The evolution of membrane-bound bioenergetic systems: the development of vectorial oxidoreductions. In "Molecular and Cellular Aspects of Microbial Evolution" (Eds M.J. Carlile, J.F. Collins and B.E.B. Moseley). Soc. Gen. Microbiol. Symp. 32, pp. 273-283. Cambridge University Press, Cambridge.
15. Gest, H. (1983). Evolutionary roots of anoxygenic photosynthetic energy conservation. In "The phototrophic bacteria: anaerobic Life in the Light" (Ed J.G. Ormerod). Studies in Microbiology, volume 4, pp. 215-233. Blackwell Scientific Publications, Oxford.
16. Wood, P.M. (1983). Why do c-type cytochromes exist? *FEBS Letters* 164, 223-226.
17. Hauska, G., Hurt, E., Gabellini, N. and Lockau, W. (1983). Comparative aspects of quinol-cytochrome c/plastocyanin oxidoreductases. *Biochim. Biophys. Acta* 726, 97-133.
18. Dickerson, R.E. (1980). Cytochrome c and the evolution of energy metabolism. *Sci. Amer.* 242, 98-110.
19. Dickerson, R.E. (1980). Evolution and gene transfer in purple photosynthetic bacteria. *Nature (London)* 283, 210-212.
20. Cole, S.T. (1982). Nucleotide sequence coding for the flavoprotein subunit of the fumarate reductase of *Escherichia coli*. *Eur. J. Biochem.* 122, 479-484.
21. LeGall, J., DerVartanian, D.V. and Peck, H.D. (1979). Flavoproteins, iron proteins and haemoproteins on electron transfer components of the sulphate-reducing bacteria. *Curr. Top. Bioenergetics* 9, 237-265.
22. Jones, C.W. (1977). Aerobic respiratory systems in bacteria. In "Microbial Energetics" (Eds B.A. Haddock and W.A. Hamilton). Soc. Gen. Microbiol. Symp. 27, pp. 23-59. Cambridge University Press, Cambridge.
23. Jones, C.W. (1980). Cytochrome patterns in classifica-

tion and identification including their relevance to the oxidase test. In "Microbiological Classification and Identification" (Eds M. Goodfellow and R.G. Board). Soc. Appl. Bacteriol. Symp. 8, pp. 127-138. Academic Press, London.
24. Poole, R.K. (1983). Bacterial cytochrome oxidases: a structurally and functionally diverse group of electron transfer proteins. *Biochim. Biophys. Acta* 726, 205-243.
25. Haddock, B.A. and Jones, C.W. (1977). Bacterial respiration. *Bact. Revs.* 41, 47-99.
26. Collins, M.D. and Jones, D. (1981). Distribution of isoprenoid quinone structural types in bacteria and their taxonomic implications. *Microbiol. Revs.* 45, 316-354.
27. Stouthamer, A.H. (1980). Bioenergetic studies on *Paracoccus denitrificans*. *TIBS* 5, 164-166.
28. Kelly, D.P. (1982). Biochemistry of the chemolithotrophic oxidation of inorganic sulphur. *Phil. Trans. Roy. Soc. London (Series B)* 298, 499-528.
29. Ingledew, J. (1982). *Thiobacillus ferro-oxidans*: the bioenergetics of an acidophilic chemolithotroph. *Biochim. Biophys. Acta* 683, 89-117.
30. Ferguson, S.J. (1982). Is a proton-pumping cytochrome oxidase essential for energy conservation in *Nitrobacter*? *FEBS Letters* 146, 239-243.

EVOLUTION OF CHEMOLITHOAUTOTROPHY

Bärbel Friedrich

Institut für Mikrobiologie
Universität Göttingen
Göttingen, Federal Republic of Germany

INTRODUCTION

Autotrophs are defined as organisms which are able to synthesize their cell constituents from carbon dioxide. There are three assimilation pathways known that account for the net synthesis of organic carbon from CO_2: The reductive penthose phosphate (Calvin) cycle (5), the reductive tricarboxylic acid cycle (23,34), and the acetyl CoA pathway (33). Of these three routes the Calvin cycle is possibly not the most ancient but certainly the most widely distributed mechanism of CO_2 fixation. It operates in plants, algae, photosynthetic and most aerobic chemolithoautotrophic bacteria.

The Calvin cycle constitutes a highly endergonic sequence of reactions. Synthesis of 1 mol triose phosphate from CO_2 requires 9 mol ATP and 6 mol $NAD(P)H_2$. Consequently, not only a new set of assimilatory enzymes had to be developed for autotrophic CO_2 fixation but a concurrent ability to efficiently convert chemical energy was necessary. Quayle and Ferenci (72) concluded that organic substrate-linked phosphorylation would not be sufficient in itself because there would be no selective advantage for an organism to develop a metabolism based on autotrophy when reduced carbon substrates were available. Energy production coupled to exergonic reactions of inorganic compounds or to photosynthetic phosphorylation could have provided the thrust towards autotrophy. As the concentration of nutrients in the environment decreased due to their assimilation by heterotrophic organisms, a selective advantage would have conferred to any bacterium that adapted to less reduced substrates (65).

The metabolism of chemolithoautotrophic bacteria strictly depends on respiration. It could appear only after oxygen was available. Although some oxygen in the ancient atmosphere may have been formed through ultraviolet photolysis of water vapor, the prevailing view is that the bulk of oxygen arose from biological photosynthesis (79) which started approximately 2×10^9 years ago, presumably by organisms resembling the present-day cyanobacteria (86).

The characteristic feature of chemolithoautotrophic bacteria is their ability to produce metabollically useful energy from the oxidation of reduced or incompletely oxidized forms of sulfur, nitrogen, and ferrous iron or of carbon monoxide and hydrogen (81). The concept of chemolithotrophy arose from Winogradsky's studies of *Beggiatoa* (98) and *Lepthothrix* (99) almost hundred years ago. He interpreted the oxidation of reduced sulfur and iron compounds as the physiological equivalent of respiration of organic molecules.

The present survey will primarily focus on energy metabolism of chemolithoautotrophy. For a detailed discussion of autotrophic carbon metabolism the reader is referred to an article in this symposium by G. Fuchs.

CLASSIFICATION OF CHEMOLITHOAUTOTROPHIC BACTERIA

The group of aerobic chemolithoautotrophic bacteria is physiologically defined and comprises taxonomically diverse species of microorganisms. On the basis of their electron donor five categories of chemolithoautotrophs can be differentiated (Table 1).

The first three groups, the nitrifying and the sulfur- and iron-oxidizing bacteria are mostly obligately bound to the lithoautotrophic mode of existence. They fail to grow heterotrophically although the incorporation of organic compounds has been demonstrated in several cases (55). The nature of obligate chemolithoautotrophy is far from being understood. It seems as if a complex combination of metabolic lesions, a low transport and respiration capacity for organic substrates, and a deficiency for regulatory mechanisms is responsible for this highly specialized way of life (52). Apparently, this group of lithoautotrophs represents the end of a line of extreme evolutionary specialization.

EVOLUTION OF CHEMOLITHOAUTOTROPHY

TABLE 1

Lithoautotrophic bacteria

Electron donor	Group
NH_4^+, NO_2^-	Nitrifying bacteria
$S_2O_3^{2-}$, S^{2-}, S^o	Sulfur-oxidizing bacteria
Fe^{2+}	Iron-oxidizing bacteria
CO	Carboxydotrophic bacteria
H_2	Hydrogenotrophic bacteria

The carboxydotrophic bacteria and the hydrogenotrophic bacteria are in general facultative lithoautotrophs. Their lithoautotrophic abilities can be considered as a bypath of normal organotrophic energy conversion, advantageous during transient exhaustion of organic nutrients (81). In their natural habitats these bacteria presumably encounter both type of substrates simultaneously. Thus, the majority of facultative lithoautotrophs utilize mixtures of both organic and inorganic compounds concomitantly, referred to as mixotrophic growth. This potential may be of benefit under conditions of nutrient limitation which is probably the case with most of the natural environments where these bacteria thrive (55).

Aerobic chemolithoautotrophic bacteria share characteristics which are specific for photolithotrophic bacteria (14). This relationship is also obvious from studies of the 16S RNA of these organisms (88). In fact, species of the purple sulfur bacteria are capable of using thiosulfate or hydrogen in the dark for autotrophic of mixotrophic growth under conditions of reduced oxygen partial pressure (45,49).

ENERGY-CONVERTING SYSTEMS

Electron transport and reverse electron flow

Chemolithoautotrophic bacteria rely on ATP synthesis by oxidative phosphorylation via an electron transport chain, in which flavins, quinons, and cytochromes are involved (Fig.1). The redox carriers are basically similar to those found in other aerobic bacteria and in mitochondria (71,70).

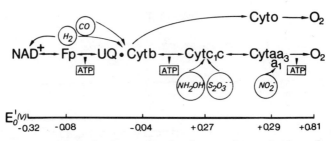

Fig. 1 A simplified, hypothetical version of the electron transport chain of lithoautotrophic bacteria. Fp, flavoprotein; UQ, ubiquinone; Cyt, cytochrome; E_o', standard redox potential.

It has been emphasized before that ATP and NAD(P)H_2 are in high demand for autotrophic CO_2 fixation. Thus, aerobic chemolithotrophs have to meet two specific requirements: (i) The conversion of energy from mostly high potential electron donors (Table 2), and (ii) the production of reducing power (NADH) by energy-driven reverse electron transport. Although it is thermodynamically feasable that the direct reduction of NAD by some sulfur compounds (19) or by H_2 and CO could occur, it is clear that the oxidation of sulfide, ammonia, nitrite or ferrous iron does not result in NADH production. In fact, with the exception of a few hydrogen bacteria the lithoautotrophs are the only aerobic microorganisms for which reverse electron transport is a necessity.

Wheelis (96) concluded from a thermodynamic analysis that in chemoautotrophs a respiratory electron transport generates an electrochemical potential across the cytoplasmic membrane which in turn drives both ATP synthesis and reverse electron transport.

According to Bose and Gest (10) photosynthetic phosphorylation is the primary event in bacterial photosynthesis, and reducing power is generated by ATP-driven electron flow. Thus, from the point of evolution reverse electron transfer may have been the original function of the electron transport chain (81).

TABLE 2

Redox potentials of substrates important in lithoautotrophic metabolism

Couple	E_o' (mV)
NH_4^+/NH_2OH	+ 899
NH_2OH/NO_2^-	+ 660
NO_2^-/NO_3^-	+ 433
$S_2O_3^{2-}/HS^- + HSO_3^-$	− 402
$S_2O_3^{2-}/2SO_4^{2-}$	− 240
HS^-/S^o	− 270
Fe^{2+}/Fe^{3+}	+ 772
H_2/H^+	− 414
CO/CO_2	− 540
$NAD(P)H/NAD(P)$	− 320

The values are taken from Wheelis (96)

Nitrification

Oxidation of ammonia. The oxidation of NH_4^+ to NO_3^- is carried out by two physiologically and morphologically distinct groups of lithoautotrophs. The first reaction to NO_2^- is a two step process with hydroxylamine as a free intermediate (91). It is catalyzed by ammonia monooxygenase (39) and by hydroxylamine oxidoreductase (42).

$$NH_3 + O_2 + AH_2 \longrightarrow NH_2OH + A + H_2O$$

$$NH_2OH + A + 0.5\ O_2 \longrightarrow HNO_2 + AH_2$$

Hollocher et al. (39) have shown that the oxygen in NH_2OH produced from NH_3 derives from dioxygen. Suzuki (91) has indicated that cytochrome c-554 may be involved as an electron donor in the oxygenation of NH_3. The reaction of NH_2OH to HNO_2 is a dehydrogenation, mediated by hydroxylamine oxidoreductase. This enzyme has been extensively studied, and some of its properties are listed in Table 3. Olson and Hooper (66) reported that hydroxylamine oxidoreductase is located in the periplasmic space of *Nitrosomonas* cells. They proposed the following scheme for energy coupling:

$$NH_2OH \longrightarrow HNO + 2e^- + 2 H^+ \quad \text{(outside)}$$
$$2e^- + 2 H^+ + 0.5 O_2 \longrightarrow H_2O \quad \text{(inside)}$$

Electrons are released from NH_2OH via c-hems of the oxidoreductase and pass to the membrane-bound terminal oxidase on the cytoplasmic side resulting in the reduction of O_2 in a proton-utilizing reaction. This may contribute to the total transmembrane potential generated for ATP synthesis. Hooper et al. (41) suggested that this scheme for the generation of the proton gradient generally applies to the oxidation of small molecules such as H_2, H_2O (photosynthesis), Fe^{2+}, NO_2^-, CO and reduced sulfur compounds.

In the context of evolution it is noteworthy that the oxidation of ammonia is analogous to methane oxidation. An oxygenase step ($CH_4 \longrightarrow CH_3OH$) is followed by a water-utilizing proton-yielding dehydrogenase (6). In fact, methane is oxidized by cells of *Nitrosomonas*, and ammonia is used as an alternate substrate by methane monooxygenase (44). Despite the similarities, ammonia oxidizers can not grow on methane as the sole carbon and energy source, and methane oxidizers fail to assimilate CO_2 with ammonia as the electron donor (91).

Oxidation of nitrite Nitrite oxidation is initiated by a membrane-bound nitrite oxidoreductase (Table 3) (95). The source of oxygen in nitrite-derived nitrate is water (53). As indicated in Fig. 1, there is evidence that nitrite enters the *Nitrobacter* respiratory chain at cytochrome a_1 (3). In addition to the cytochromes c and aa_3, cytochrome a_1 is a major component of electron transport particles from *Nitrobacter* cells (91).

Nitrite oxidoreductase, whose properties are listed in Table 3, contains iron, molybdenum, sulfur, and copper. Moreover, it exhibits a higher affinity to nitrate than to nitrite. From these data it was concluded that nitrite oxidoreductase is a nitrate reductase type of protein and may have developed from a nitrate reductase of a photosynthetic organism (90). A close phylogenetic relationship between *Rhodopseudomonas palustris* and *Nitrobacter* has been established (56).

TABLE 3

Properties of nitrifying enzymes

Property	Enzyme	
	Hydroxylamine[a] oxidoreductase	Nitrite[b] oxidoreductase
Location	Periplasm	Membrane
Molecular weight	220.000	390.000
Structure	Hexamer ($\alpha_3\beta_3$)	Pentamer ($\alpha_2\beta_2\gamma$)
Subunit composition	α : 63.000	α : 116.000
	6 c-type hemes	β : 65.000
	1 P-460	
	β : c-type cytochrome	γ : c-type cytochrome
Substrates		NO_2^-, NO_3^-
Trace elements	Fe	Fe, Mo, Cu

[a] from *Nitrosomonas* (66)
[b] from *Nitrobacter hamburgensis* (90)

Sulfur oxidation

When thiosulfate serves as electron donor, its oxidation starts with the cleavage of the bond between the sulfane and the sulfone sulfurs (Fig. 2). The reaction may be catalyzed by rhodanese. The ubiquitous rhodanese, a sulfur transferase, capable of transferring sulfur to various acceptors was shown to be present in *Thiobacillus denitrificans*, *Ferrobacillus ferrooxydans* and also in the photosynthetic bacterium *Chromatium* (91). Alternatively, a thiosulfate reductase may exist for catalysis of the first step (67).

Lu and Kelly (54) recently described a thiosulfate-cleaving enzyme complex (Table 4) of *Thiobacillus versutus* (A2). This bacterium oxidizes thiosulfate completely to sulfate without the accumulation of intermediates such as sulfur or tetrathionate.

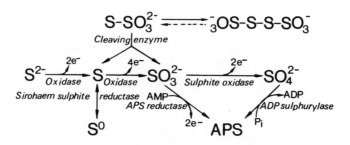

Fig. 2 Inorganic sulfur oxidation and enzymes involved in this process. APS, adenosine-5'-phosphosulfate.

The present view is that the oxygen in the product sulfate derives from water and not from dioxygen. The high level of c-type cytochrome in thiosulfate-grown cells of *T.versutus* indicates the importance of cytochromes in thiosulfate oxidation. The thiosulfate-oxidizing enzyme complex (thiosulfate: cytochrome c oxidoreductase) is supposed to act in the overall process, initiated by hydration, followed by dehydrogenation and coupled to electron transport-dependent oxygen reduction as follows:

$$S_2O_3^{2-} + 5H_2O \longrightarrow 2SO_4^{2-} + 10\,H^+ + 8e^-$$

$$8H^+ + 8e^- + 2O_2 \longrightarrow 4H_2O$$

The last step in this reaction is analogous to the dehydrogenation of NH_2OH as described in the preceding section.

The precise function of the colorless enzymes A and B in thiosulfate oxidation is still unknown. Both proteins contain neither iron, nickel nor copper. Sulfite oxidase is an integral part of the multienzyme complex (Table 4) and appears to be linked to cytochrome c_{551}. It resembles sulfite oxidase of *Thiobacillus novellus* which revealed a molecular weight of 40,000 and the presence of molybdenum (94). The fact that none of the five constituents of the multienzyme complex exhibited rhodanese or APS (adenosine-5'-phosphosulfate) reductase activity led to the conclusion that these proteins are not involved in thiosulfate oxidation by *T. versutus* (54).

TABLE 4

Structure of the thiosulfate-oxidizing multienzyme system from Thiobacillus versutus

Component[a]	Native	Subunit
Enzyme A	16,000	
Enzyme B	64,000	32,000
Cytochrome c-551	260,000	43,000
Cytochrome c-552.5	56,000	29,000
Sulfite oxidase	43,000	

[a]The values are taken from Lu and Kelly (54)

The presence of the AMP-dependent energy-conserving APS pathway (Fig. 2) in addition to sulfite oxidase (sulfite:cytochrome c oxidoreductase) has been reported for *Thiobacillus thioparus* (46,67). Peck (67) postulated that APS reductase may have an evolutionary priority originating from ancient, sulfate-reducing bacteria. In fact, the FAD-, iron-, and labile sulfide-containing APS reductases from *T. denitrificans* and *T. thioparus* were found to be similar to the enzyme from *Desulfovibrio desulfuricans*. Organisms that contain only sulfite oxidase, such as *Thiobacillus intermedius*, may be evolutionary newcomers adapted to strictly aerobic environments. The cytochrom c-coupled enzyme resembled liver sulfite oxidase (91).

When sulfur or sulfide serve as electron donors, they enter the pathway as indicated in Fig. 2. Suzuki and Silver (92) demonstrated the presence of a sulfur-oxidizing enzyme in *T. thioparus* and *Thiobacillus thiooxydans* producing sulfite from elemental sulfur by using reduced glutathione as a cofactor. However, the present view favors the assumption that this oxidase is unimportant and that sulfite is produced by a sirohaem sulfite reductase (78). Thus, the conversion of sulfur, sulfide, and the sulfane sulfur involves similar mechanisms which function in reverse direction in sulfate-reducing bacteria for the production of sulfide.

Carbon monoxide-oxidizing enzyme system

Carbon monoxide: acceptor oxidoreductase (CO dehydrogenase) is the key enzyme in CO oxidation by carboxydotrophic bacteria catalyzing the following reaction:

$$CO + H_2O + X_{ox} \longrightarrow CO_2 + X_{red}$$

CO dehydrogenase has three functions. (i) It transfers two electrons from CO into a CO-insensitive branch of the respiratory chain for oxidative phosphorylation; (ii) it provides CO_2 as a substrate for autotrophic growth, and (iii) it supplies electrons for an ATP-driven reverse electron flow to generate NADH (62). The enzyme is highly specific for CO, it uses only NADH as an additional substrate. CO dehydrogenase is loosely attached to the inner aspect of the cytoplasmic membrane and released into the cytoplasm when the cells enter the stationary phase of growth (61).

The CO-oxidizing enzymes thus far isolated from aerobic carboxydotrophic bacteria seem to be closely related on the basis of their catalytic, physical and immunochemical properties. They represent a new type of molybdo iron-sulfur flavoprotein (molybdenum hydroxylase) which appears to be strikingly similar to the mammalian xanthine-oxidizing enzymes (62). The presence of a molybdenum cofactor, common to nitrate reductase and xanthine dehydrogenase (63) was first recognized by Meyer (59). It contains molybdenum and a reduced form of molybdopterin as a prosthetic group. Selenium, another metal present in CO dehydrogenase from *Pseudomonas carboxydovorans*, is covalently bound to the protein. There is evidence that it stimulates exclusively the activity of the cytoplasmic enzyme species (60).

From the viewpoint of evolution the question emerges of whether there are any similarities between the CO dehydrogenation of aerobic carboxydotrophic bacteria and CO-metabolizing enzymes present in groups of anaerobic organisms such as the acetogens, methanogens, phototrophs and sulfate-reducers (Table 5). As suggested by Meyer and Fiebig (60), the current status indicates that aerobes and anaerobes have developed different types of CO-oxidizing enzymes. This concept is supported by a comparison of the molecular and catalytic properties of CO dehydrogenases from various bacterial sources, summarized in Table 5.

Instead of a molybdo iron-sulfur flavoprotein the anaerobic species, represented by *Clostridium thermoaceticum* (73), *Methanosarcina barkeri* (51) and *D. desulfuricans* (60), contain a flavin-free, nickel iron-sulfur CO dehydrogenase whose cofactor composition appears to be less complex than that of carboxydotrophs. Moreover, the enzyme from aerobes has a high affinity to its substrate (K_m: 50 to 60 µM) and a well defined physiological role of CO consumption (61). CO dehydrogenase from methanogenic and acetogenic bacteria has a low affinity to CO (K_m: 3 to 5 mM). Diekert and Thauer (20) proposed its dual function in acetate catabolism as well as in acetate synthesis from one carbon substrates.

TABLE 5

Properties of CO dehydrogenases from aerobic and anaerobic bacteria

Species	Mol wt	Subunit No x mol wt (kDa)	Metal content Mo Ni Fe Zn	Ref
Pseudomonas carboxydovorans	300,000	2 x 86 2 x 34 2 x 17	+ − + +	61
Clostridium thermoaceticum	440,000	3 x 78 3 x 71	− + + +	73
Methanosarcina barkeri	232,000	2 x 92 2 x 18	nk + nk nk	51
Desulfovibrio desulfuricans	180,000	nk	nk + + nk	60

nk, not known; +, present; −, absent; kDa, kilodaltons

Structure and function of hydrogenase

Hydrogenase catalyzes the reversible activation of molecular hydrogen:

$$H_2 \longleftrightarrow 2H^+ + 2e^-$$

Its existence has been first demonstrated in *Escherichia coli* about fifty years ago. Since then, hydrogenases have been found in a wide variety of bacterial and algal species (82).

Despite considerable differences in molecular composition and catalytic properties, all hydrogenases so far known, reveal one common feature, all are iron-sulfur proteins (Table 6). The non-heme iron and the acid labile sulfur are arranged in one or more clusters similar to those found in ferredoxins. For a long time these metal clusters were considered as the only redox-active components in hydrogenase.

The observation that the formation of active hydrogenase in the hydrogenotrophic bacterium *Alcaligenes eutrophus* requires nickel ions (25), led to the subsequent discovery that almost all bacterial hydrogenases are nickel iron-sulfur proteins (50, and references therein).

Isotopic substitution analyses and/or paramagnetic resonance spectroscopy have unambigously shown that nickel is an integral part of the hydrogenase protein (93,36,32,89,2). The studies have implicated that nickel undergoes a one electron reduction, it may represent the primary binding site for H_2 (93). It is of interest that the two hydrogenases of *Clostridium pasteurianum* have proven to be nickel-free (1,18). Moreover, the recently isolated hydrogenase from *Desulfovibrio vulgaris* revealed no nickel content (43). The authors emphasized that the latter enzyme exhibits a high specific activity in the hydrogen evolution assay, and resembles with that respect the clostridial hydrogenase rather than the nickel-containing enzymes from *Desulfovibrio gigas* (Table 6) and *D.desulfuricans*.

Hydrogen bacteria are characterized by the possession of two very distinct hydrogenase proteins, both of which are present in *A. eutrophus*. The enzymes differ in location, molecular structure cofactor content, and catalytic properties (Table 6).

TABLE 6

Properties of hydrogenases from bacterial sources

Species	Mol mass (kDa)	Subunits (kDa)	Prosthetic groups	Natural electron carrier	Location
Alcaligenes eutrophus	98 205	1x 67,31 1x 63,56,30,26	Ni,Fe-S Ni,Fe-S,FMN	nk NAD	Membrane Cytoplasm
Chromatium vinosum	62	1x 62	Ni,Fe-S	nk	Membrane
Desulfovibrio gigas	89	1x 62,26	Ni,Fe-S	Cytochrome-c_3	Periplasm
Methanobacterium thermoautotrophicum	170 nk	2x 40,31 1x 26 52	Ni,Fe-S,FAD Ni,Fe-S	F_{420} nk	Cytoplasm nk
Clostridium pasteurianum	60 53	1x 60 1x 53	Fe-S Fe-S	Fd Fd	Cytoplasm Cytoplasm

nk, not known; Fd, ferredoxin, kDa, kilodaltons
References (from top to bottom): 84,83; 85, 2, 93, 36, 48, 1, 18.

The NAD-linked hydrogenase which also exists in strains of a very diverse genus, namely *Nocardia* (82), has the most complex structure of all bacterial hydrogenases so far studied.

It shares some molecular properties with the F_{420} (8-hydroxy-5-deazaflavin)-reducing hydrogenase of *Methanobacterium thermoautotrophicum* (36,48). Flavin, present in both enzymes, supposedly acts as a two electron switch to NAD or F_{420}, respectively.

Recently, Schneider et al. (83) succeeded in dissociating the NAD-linked tetrameric protein of *Nocardia opaca* in two dimeric species. Reactivity and structural analyses led to the postulation of a model presented in Fig. 3. The authors concluded that the large dimer contains FMN but no nickel. It has no hydrogenase but NADH-oxidizing activity (diaphorase). The small flavin-free dimer contains nickel, is diaphorase-inactive, displays no NAD-reducing activity with H_2 but instead reduces artificial electron carriers with H_2 as the electron donor. Thus, the small dimer, the nickel iron-sulfur protein, seems to be the basic H_2-activating complex which is similar to the more simple structure of the particulate type of hydrogenase (Table 6). The large dimer, an iron-sulfur flavoprotein, resembles the highly developed NADH dehydrogenase from bovine heart mitochondria (83).

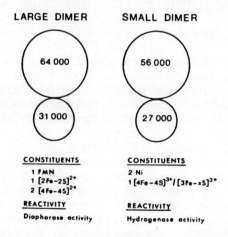

Fig. 3 Model of the NAD-linked hydrogenase of Nocardia opaca as proposed by Schneider et al. (83).

The fact that there is no immunological relatedness between the small dimer of the NAD-linked hydrogenase and the particulate enzyme of *A. eutrophus* (85) suggests that these two protein species are at least not closely related. It has to be considered that in general the particulate hydrogenases show a considerable diversity (Table 6). Of interest is the observation that the particulate hydrogenase of *A. eutrophus* is immunochemically similar to that of *Pseudomonas*, *Azotobacter*, *Aquaspirillum* and *Rhodopseudomonas* species (80). However, it is serologically unrelated to the hydrogenase of the thermophile *Bacillus schlegelii* (68), the nitrogen-fixing *Xanthobacter autotrophicus* (12) and the thiosulfate-oxidizing (31) *Paracoccus denitrificans* (80). On the other hand, the complex structure of the NAD-linked enzyme seems to be well conserved in the two diverse species in which this protein has been recognized (85).

On the basis of our current knowledge it is reasonable to assume the development of at least two types of hydrogenases: (i) a H_2-evolving, nickel-free system, and (ii) a H_2-consuming nickel-containing enzyme. The question of whether the latter, the so called "uptake" hydrogenase became further modified in the course of evolution to yield a structure of complexity as the NAD- or F_{420}-reducing flavin-containing proteins remains as yet unresolved. The possibility can not be excluded that these two proteins represent independent lines of evolution.

LOCATION AND EXPRESSION OF "LITHOAUTOTROPHIC" GENES

Occurrence of plasmids in lithoautotrophic bacteria

Although our knowledge on the genetics of lithoautotrophic systems is very limited, several experimental approaches have begun to explore the genetic origin of lithoautotrophic characters. Investigations conducted in a number of different laboratories indicate that extrachromosomal DNA is widely distributed among lithoautotrophic bacteria. Plasmids have been detected in thiosulfate-, iron-, CO- and H_2-oxidizing bacteria (Table 7). The extensive survey of Gerštenberg et al. (35) revealed a considerable diversity in number and sizes of plasmids, present in autotrophic, predominantly hydrogenotrophic bacteria. One of the most striking observations was the discovery of extremely large plasmid molecules, termed megaplasmids according to giantly sized plasmid species in *Rhizobium* (77).

The largest plasmids (600 to > 700 kilobase (kb) pairs) are resident in the H_2- and/or thiosulfate-oxidizing bacteria *P. denitrificans* (31), *Thiosphaera pantotropha* (76) and *T. versutus* (Fig. 4). The size of these plasmids exceeds one tenth of the *E. coli* chromosome. All of these three strains harbor a second medium-sized plasmid. So far no phenotypic function can be assigned to this extrachromosomal DNA. As indicated in Table 7, most of the plasmids can only be characterized as cryptic.

TABLE 7

Plasmids in lithoautotrophic bacteria

Species	Plasmid			Ref
	Number	Size (kb)	Function	
Thiobacillus versutus (A2)	2	106;>450	Cryptic	35
	2	5.3;86	Cryptic	7
Thiobacillus ferrooxidans	3	6.7;16;>23	Cryptic	40
Pseudomonas carboxydovorans strain OM5	1	129	Cryptic	35
Alcaligenes eutrophus H16	1	450	H_2-oxidation	27
			CO_2-fixation	11
			NO_3^--respiration	a
A. eutrophus CH34	2	166;219	Metal resistance	35,b
Nocardia opaca 1b	2	17.4;140	Tl resistance	c

a, D. Römermann and B. Friedrich; b, D. Nies, M. Mergeay and H.G. Schlegel; c, M. Reh and C. Sensfuß; unpublished results.

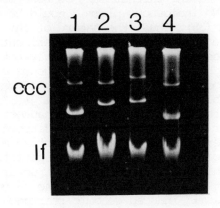

Fig. 4 Plasmid pattern of Paracoccus denitrificans (DSM 65), lane 1; Thiosphaera pantotropha, lane 2; Thiobacillus versutus, lane 3; and Alcaligenes eutrophus, carrying plasmid pHG1 (450 kb) and RP4 (54 kb), lane 4. ccc, covalently closed circular DNA; lf, linear fragments.

Genetic transfer of hydrogen autotrophy

The existence of plasmids is not extraordinary. They are found in virtually all bacterial species (15). What is the evidence that extrachromosomal elements are involved in lithoautotrophic metabolism?

Mutants of *Nocardia opaca* 1b impaired in hydrogen autotrophy have been found relatively frequently. These mutants regained the ability to grow autotrophically with H_2 after mating with the wild type. Moreover, the lithoautotrophic character could be transferred via conjugation from *N. opaca* 1b to non-autotrophic recipients such as *Rhodococcus erythropolis* and *Corynebacterium hydrocarboclastus* (74,75). There is accumulating evidence, however, that none of the plasmids resolved in lysates of the donor (Table 7) carries lithoautotrophic genes. M. Reh and coworkers favor the assumption that these genes are located on the chromosome and mobilized by a fertility factor (personal communication).

The transfer of large pieces of chromosomal DNA including "autotrophic" genes has been reported for *Xanthobacter autotrophicus* GZ29. This transfer requires cell contact, and surprisingly both mating partners act as donors as well as recipients (97). No indigenous plasmid could be detected in this bacterium (35). Thus, the nature of this gene transfer is still unclear. Despite several experimental efforts we have no evidence yet to support the early notion of Pootjes (69) that hydrogenase-specifying genes of *Pseudomonas facilis* are located extrachromosomally (J.Warrelmann and B.Friedrich, unpublished result). Cantrell *et al.* (17) examined strains of the nitrogen-fixing lithoautotrophic *Rhizobium japonicum* for plasmid content. They discovered two plasmids in a non-revertible hydrogenase-defective mutant; however, no plasmid could be resolved in the parent strain. From these data the authors concluded a structural rearrangement of a hypothetical megaplasmid or of the chromosomal genome. In fact, plasmid-induced genomic rearrangements in *R. japonicum* have been reported recently (9).

The only plasmid-encoded lithoautotrophic trait, so far conclusively identified, is H_2-oxidizing ability (Hox) in some *Alcaligenes* spp. (4,27). A large, conjugative plasmid, designated pHG, is involved in the transmission of Hox between different strains and species of *Alcaligenes* (27,38). Curing of this plasmid is strictly correlated with an irreversible loss of Hox activity which can only be recovered by backtransfer of the megaplasmid. Recent studies have implicated that plasmid pHG1 (450 kb) of *A.eutrophus* H16 confers two additional phenotypic traits on its host. (i) The capability to derepress the formation of CO_2-fixing (Cfx) enzymes under specific heterotrophic conditions of growth (11), and (ii) nitrate-respiring ability (Nitd), (D.Römermann and B.Friedrich, unpublished result).

The data reviewed in this section clearly indicate that there exist several modes of gene exchange between lithoautotrophs based on cell contact. They may have facilitated the spread of lithoautotrophic traits in nature. The following questions arise: How does a species like *A. eutrophus* maintain its identity? Is there a flow of lithoautotrophic genes between heterologous strains? Does the plasmid DNA interact with euchromosomal DNA?

Maintenance and host range of Hox-encoding plasmids

The finding that Hox in *A.eutrophus* is a plasmid-encoded character was unexpected, since hydrogen autotrophy is an extremely stably maintained metabolic trait of this host. We have shown that pHG megaplasmids are not essential for the "everyday" metabolism of this lithoautotroph (27,38). Nevertheless, the replication of these large molecules must constitute at least some metabolic burden to the cell. Their long-term survival may indicate that these plasmids earn to be kept. The effort to perpetuate a plasmid of this size should be balanced by some positive contribution to the organism's phenotype. Thus, the aquisition of hydrogen autotrophy should convey a selective advantage to the plasmid-harboring strain. Campbell (15) suggested that especially large conjugative plasmids have achieved a structural and regulatory complexity indicating a long successful evolution. He concluded that their ability to replicate and disseminate themselves must itself be of some value to the host.

Plasmid pHG1 can be transferred among different strains and species of *Alcaligenes* (24,27). Initial attempts to convert non-hydrogenotrophs such as *Pseudomonas oxalaticus* or *T. versutus* to hydrogen bacteria by mating with *A. eutrophus* failed. Thus, we improved the selection for and the mobilization of plasmid pHG1 by introducing the kanamycin (Km) resistance-encoding tranposon Tn5 into pHG1 in addition to a mobilizing DNA sequence (Mob) derived from the broad-host range plasmid RP4 (84). The pHG1::Tn5-Mob containing *A. eutrophus* donor gave rise to Km-resistant transconjugants of the non-hydrogen-utilizing soil bacterium JMP222 whose parent strain is known for its herbicide-degrading ability (21). Most of the Km-resistant transconjugants of JMP222 could grow lithoautotrophically and expressed hydrogenase activity (B.Friedrich and A.Kamienski, unpublished result). The presence of plasmid pHG1::Tn5-Mob in addition to the indigenous cryptic megaplasmid of JMP222 is illustrated in Fig. 5.

The spread of pHG plasmids among heterologous strains is certainly a rare event. The host range of these extrachromosomal elements seems to be limited. However, the result described above, provides evidence that a heterologous transmission of these plasmids is possible and may have evolutionary significance in generating phenotypes with new lithoautotrophic traits.

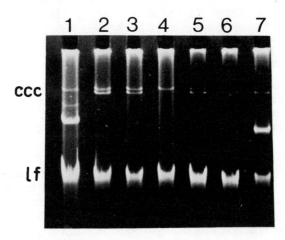

Fig. 5 Plasmid pattern of Hox^+ transconjugants (lanes 2,3,5) isolated after transfer of plasmid pHG1::Tn5-Mob (lane 6) to the recipient JMP222 (lane 4), a derivative of strain JMP134 (21), (lane 1). Size standards (lane 7) are the same as in Fig. 4.

pHG1-encoded lithoautotrophic genes and their interaction with the chromosome

The observation that plasmid-cured Hox^- mutants or those carrying deletant plasmids are incapable of forming the key enzymes of the Calvin cycle, phosphoribulokinase and ribulose bisphosphate carboxylase (RuBPCase) under heterotrophic growth with gluconate or fructose (11) led to the conclusion that megaplasmid pHG1 is not only involved in the expression of Hox but also carries determinants for autotrophic CO_2 fixation (Cfx). This result was completely unexpected, since plasmid-free mutants are still able to grow on formate (27) which depends on the operation of the Calvin cycle (29). Behki et al. (8) also observed this apparent descrepancy.

Plasmid pHG1 must at least encode Cfx regulatory functions which are necessary for the expression of Cfx enzymes under certain heterotrophic and possibly hydrogen-dependent autotrophic conditions. We have cloned a 12.4-kb EcoRI fragment of plasmid pHG1, reintroduced this piece of DNA via the broad-host range shuttle vector pVK101 (47) from *E. coli* to plasmid-free or plasmid-deleted mutants of *A. eutrophus* and found that contrary to the recipients the transconjugants were enhanced in RuBCase activity when grown on gluconate (G.Eberz and B.Friedrich, unpublished result). Some of the data are summarized in Table 8, they clearly indicate that the recombinant plasmid complements the Cfx regulatory phenotype.

TABLE 8

Activity of ribulose bisphosphate carboxylase in wild type, mutants and recombinant DNA-carrying strains of Alcaligenes eutrophus

Strain[a]	Plasmid[b]	Specific activity mU/mg of protein
H16	pHG1	54.0
HF23	pHG1⁻	8.4
HF219	pHG1⁻, pGE1	45.2
HF47	ΔpHG1	12.0
HF220	ΔpHG1, pGE1	31.5

[a] The cells were grown on gluconate-containing minimal medium as described (11)

[b] pGE1 is a recombinant plasmid composed of the vector pVK101 (47) and a 12.4-kb EcoRI fragment of plasmid pHG1, ΔpHG1 carries a deletion (26).

Determinants for hydrogenase activity are genetically linked in a strain of *Rhizobium leguminosarum* to a non-self transmissible plasmid of an approximate size of 285 kb (13). Distinct *hox* genes have not been conclusively identified. Our studies mainly focused on the location of *hox* genes on plasmid pHG1 of *A. eutrophus* H16. It has been emphasized before that the two hydrogenases common to three species of *Alcaligenes*, namely *A. eutrophus*, *A. hydrogenophilus* (64) and *A. ruhlandii* are very similar if not identical (12).

However, there are clear differences among *Alcaligenes* wild types concerning the location and expression of *hox* genes. The metal-resistant hydrogen bacterium *A. eutrophus* CH34 (58) seems to carry lithoautotrophic genes on the chromosome rather than on its indigenous plasmids (Table 7), (D.Nies and H.G.Schlegel, unpublished result). We have shown that even similarly sized pHG plasmids from closely related *A. eutrophus* strains exhibit distinct differences in the restriction maps (37). Two examples are demonstrated in Fig. 6.

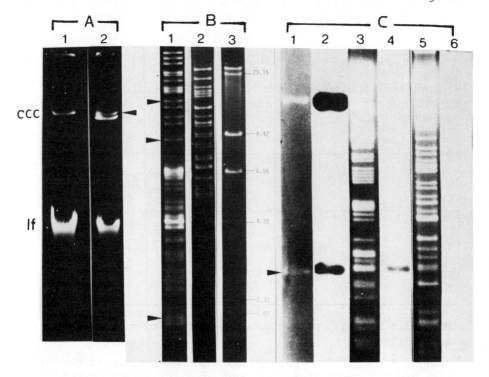

Fig. 6 Gel photographs of pHG plasmids.
A: *cccDNA of plasmids pHG1 of A. eutrophus H16 (lane 1) and pHG21-a (indicated by arrow) of A. hydrogenophilus (lane2)*
B: *EcoRI restriction analysis of plasmids pHG1 (lane 1), and pHG21-a (lane 2). Arrows mark hox gene encoding fragments. Hind III digest of λ DNA (lane 3)*
C: *EcoRI cleavage fragments of pCH 102 (lane 1) composed of pSUP202 (87)(upper band) and a 2-kb fragment (lower band) of pHG1, and pHG1 and pHG21-a (lanes 3 and 4). Autoradiograms of the same digests after Southern transfer and hybridization with ^{32}P-labeled pCH102 (lanes 2,4, and 6)*

The *Eco*RI restriction map of plasmid pHG1 from *A. eutrophus*
H16 (B1) is comparatively illustrated to that of plasmid
pHG21-a derived from *A. hydrogenophilus* (B2). The latter is
host of two megaplasmids (A2). By means of Tn5-induced mutants, marker rescue and molecular cloning we have identified
a 13.3-kb and a 1-kb *Eco*RI fragment as loci for NAD-linked
hydrogenase genes. We have also cloned a Hox regulatory region of plasmid pHG1 consisting of a contigously arranged
9-kb and a 2-kb EcoRI restriction fragment (26, C.Hogrefe,
G.Eberz and B.Friedrich, unpublished result). The regulatory DNA sequence encodes gene(s) required for the temperature-sensitive (30), energy-dependent (28) expression of hox genes
in *A. eutrophus*. As shown by Southern hybridization (Fig.6)
the 2-kb *Eco*RI fragment is absent from the *A. hydrogenophilus*
plasmid (C 5,6). The cloned fragment (C 1,2) clearly hybridized to the DNA of plasmid pHG1 (C 3,4). This result is interesting since *A. hydrogenophilus* has developed a different
regulation of *hox* gene expression which is also plasmid-borne.
Hydrogenase synthesis is subject to induction by hydrogen and
not thermosensitive (30).

Cangelosi and Wheelis (16) described an oxygen-sensitive
Hox regulation in *A. eutrophus* 177o7 which appears to be determined by a single plasmid-borne genetic locus. Synthesis
of hydrogenase in *R. japonicum* is also negatively controlled
by oxygen (57). In both cases mutants are easily selectable
that escape from oxygen repression. The authors proposed that
oxygen-tolerant regulatory systems as found in most of the
A. eutrophus strains may be the product of laboratory selection (16).

Hox regulation gains considerably more complexity by the
identification of chromosomally located mutation(s) that
lead to reduction or loss of both hydrogenase activities
(38). This observation has been consistently reported only
recently (16). However,the pleiotropic character of this
class of mutants (M.Lohmeyer, B. Friedrich and C.F.Friedrich)
suggests that they are affected in a general control system
comparable to that of nitrogen control or catabolite repression. Although our data are still preliminary, Hox regulation in *A. eutrophus* shows striking analogies to the model
proposed for the regulation of nitrogen fixation *(nif)*genes
(22). Thus, hydrogen autotrophy may prove to be a useful system for elucidating the evolution of complex regulatory mechanisms.

CONCLUSIONS

The prevalent view at the beginning of this centuary regarded chemolithoautotrophs as primitive first forms of life on earth. The evidence presented in this article demonstrates that these microorganisms represent metabolically highly developed prokaryotes which belong to one of the most biosynthetically complete groups of bacteria. In support of this notion I point to the following facts:
(i) The energy-converting systems of lithoautotrophs are biochemically related to ancient pathways of anaerobic bacteria, particularly phototrophs. It seems as if reductive metabolic routes became modified to function physiologically in the reverse direction in concert with oxygen-based respiration.
(ii) The evolution of extremely sophisticated regulatory mechanisms in facultatively lithoautotrophic bacteria may serve to mediate more stringent control on the expression of energy-intensive biochemical processes, to enhance the response to a constantly changing environment, and to allow the concomitant function of autotrophic and heterotrophic metabolism.
(iii) The most likely beneficial role of megaplasmids in lithoautotrophic metabolism is to facilitate gene transfer among different strains and species of phylogenetically even diverse bacterial hosts. Moreover, chromosomes retain their general gene order whereas plasmids offer many opportunities for rearrangements and incorporation of new genetic information. The question of whether *hox* genes owe their plasmid location to pure accident or because natural selection has favored that location awaits further genetic exploration.

ACKNOWLEDGEMENTS

I thank my collaborators and my colleagues for making available unpublished information, and H.G.Schlegel for stimulating discussion. Parts of the work were supported by a grant from the Deutsche Forschungsgemeinschaft.

REFERENCES

1. Adams, M.W.W. and Mortenson, L.E. (1984). The physical and catalytic properties of hydrogenase II of *Clostridium pasteurianum*. A comparison with hydrogenase I. *J.Biol.Chem.* 259, 7045-7055.
2. Albracht, S.P.J., Kalkman, M.L. and Slater, E.C. (1983). Magnetic interaction of nickel (III) and iron-sulphur cluster in hydrogenase from *Chromatium vinosum*. *Biochim. Biophys. Acta* 724, 309-316.
3. Aleem, M.I.H. and Sewell, D.L. (1981). Mechanism of nitrite oxidation and oxidoreductase systems in *Nitrobacter agilis*. *Curr.Microbiol.* 5, 267-272.
4. Andersen, K., Tait, R.C. and King, W.R. (1981). Plasmids required for utilization of molecular hydrogen by *Alcaligenes eutrophus*. *Arch.Microbiol.* 129, 384-390.
5. Bassham, J.A., Benson, A.A., Kay, L.D., Harris, A.Z., Wilson, A.T., and Calvin M. (1954). The path of carbon in photosynthesis XXI. The cycle regeneration of carbon dioxide acceptor. *J.Am.Chem.Soc.* 76, 1760-1770.
6. Beardmore-Gray, M., O'Keefe, D.T. and Anthony, C. (1983). The methanol:cytochrome c oxidoreductase activity of methylotrophs. *J.Gen.Microbiol.* 129, 923-933.
7. Bednarska, M., Jagusztyn-Krynicka, E.K., Popowski, J. and Wlodarczyk, M. (1983). Extrachromosomal DNA in *Thiobacillus* A2. *FEMS Microbiol.Lett.* 16, 183-185.
8. Behki, R.M., Selvaraj, G. and Iyer, V.N. (1983). Hydrogenase and ribulose-1,5-bisphosphate carboxylase activities of *Alcaligenes eutrophus* ATCC 17706 associated with an indigenous plasmid. *Can.J.Microbiol.* 29, 767-774.
9. Berry, J.O. and Atherly, A.G. (1984). Induced plasmid-genome rearrangements in *Rhizobium japonicum*. *J.Bacteriol.* 157, 218-224.
10. Bose, S.K. and Gest, H. (1962). Electron transport systems in purple bacteria. *Nature* 195, 1168-1171.
11. Bowien, B., Friedrich, B. and Friedrich, C.G. (1984). Involvement of megaplasmids in heterotrophic derepression of the carbon dioxide assimilation enzyme system in *Alcaligenes* spp. *Arch.Microbiol.* In Press.
12. Bowien, B. and Schlegel, H.G. (1981). Physiology and biochemistry of aerobic hydrogen-oxidizing bacteria. *Ann.Rev. Microbiol.* 35, 405-452.
13. Brewin, N.J., DeJong, T.M., Phillips, D.A. and Johnston, A.W.B. (1980). Co-transfer of determinants for hydrogenase activity and nodulation ability in *Rhizobium leguminosarum*. *Nature* 288, 77-79.

14. Broda, E. (1975)."The Evolution of Bioenergetic Processes." Pergamon, Press, Oxford.
15. Campbell, A. (1981). Evolutionary significance of accessory DNA elements in bacteria. Ann.Rev.Microbiol. 35, 55-83.
16. Cangelosi, G.A. and Wheelis, M.L. (1984). Regulation by molecular oxygen and organic substrates of hydrogenase synthesis in Alcaligenes eutrophus. J.Bacteriol. 159, 138-144.
17. Cantrell, M.A., Hickok, R.E. and Evans, H.J. (1982).Identification and characterization of hydrogen uptake positive and hydrogen uptake negative strains of Rhizobium japonicum. Arch.Microbiol. 131, 102-106.
18. Chen, J.S. and Blanchard, D.K. (1984). Purification and properties of the H_2-oxidizing (uptake)hydrogenase of the N_2-fixing anaerobe Clostridium pasteurianum W5. Biochem. Biophys.Res.Commun. 122, 9-16.
19. Cobley, J.G. and Cox, J.C. (1983). Energy conservation in acidophilic bacteria. Microbiol. Rev. 47, 579-595.
20. Diekert, G. and Thauer, R.K. (1982). The effect of nickel on carbon monoxide dehydrogenase formation in Clostridium thermoaceticum and Clostridium formicoaceticum. FEMS Lett. 7, 187-189.
21. Don, R.H. and Pemberton, J.M. (1981). Properties of six pesticide degradation plasmids isolated from Alcaligenes eutrophus. J.Bacteriol. 145, 681-686.
22. Drummond, M., Clements, J., Merrick, M. and Dixon, R. (1983). Positive control and autogenous regulation of the nif LA promoter in Klebsiella aerogenes. Nature 301, 302-313.
23. Evans, M.C.W., Buchanon, B.B. and Arnon, D.I. (1966). A new ferredoxin-dependent carbon reduction cycle in a photosynthetic bacterium. Proc.Natl.Acad.Sci.U.S.A.55, 928-934.
24. Friedrich, B., Friedrich. C.G., Meyer, M. and Schlegel, H.G. (1984). Expression of hydrogenase in Alcaligenes spp. is altered by interspecific plasmid exchange. J.Bacteriol. 158, 331-333.
25. Friedrich, B., Heine, E., Finck, A. and Friedrich, C.G. (1981). Nickel requirement for active hydrogenase formation in Alcaligenes eutrophus. J. Bacteriol. 152,42-48.
26. Friedrich, B. and Hogrefe, C. (1984). Genetics of lithoautotrophic metabolism in Alcaligenes eutrophus. In "Microbial Growth on C_1 Compounds" (Eds. R.L.Crawford and R.S.Hanson) pp. 244-247. ASM, Washington DC.
27. Friedrich, B., Hogrefe, C. and Schlegel, H.G. (1891). Naturally occurring genetic transfer of hydrogen-oxidizing

ability between strains of *Alcaligenes eutrophus*.
J.Bacteriol. 147, 198-205.
28. Friedrich, C.G. (1982). Derepression of hydrogenase during limitation of electron donors and derepression of ribulosebisphosphate carboxylase during carbon limitation of *Alcaligenes eutrophus*. *J.Bacteriol.* 149, 203-210.
29. Friedrich, C.G., Bowien, B. and Friedrich, B. (1979). Formate and oxalate metabolism in *Alcaligenes eutrophus*. *J.Gen.Microbiol.* 115, 185-192.
30. Friedrich, C.G. and Friedrich, B. (1983). Regulation of hydrogenase formation is temperature sensitive and plasmid coded in *Alcaligenes eutrophus*. *J.Bacteriol.* 153, 176-181.
31. Friedrich, C.G. and Mitrenga, G. (1981). Oxidation of thiosulfate by *Paracoccus denitrificans* and other hydrogen bacteria. *FEMS Microbiol.Lett.* 10, 209-212.
32. Friedrich, C.G., Schneider, K. and Friedrich, B. (1982). Nickel in the catalytically active hydrogenase of *Alcaligenes eutrophus*. *J.Bacteriol.* 152, 42-48.
33. Fuchs, G. and Stupperich, E. (1890). Acetyl-CoA, a central intermediate of autotrophic CO_2 fixation in *Methanobacterium thermoautotrophicum*. *Arch.Microbiol.* 127, 267-272.
34. Fuchs, G., Stupperich, E. and Eden, G. (1980). Autotrophic CO_2 fixation in *Chlorobium limicola*. Evidence for the operation of a reductive tricarboxylic acid cycle in growing cells. *Arch.Microbiol.* 128, 64-71.
35. Gerstenberg, C., Friedrich, B. and Schlegel H.G. (1982). Physical evidence for plasmids in autotrophic, especially hydrogen-oxidizing bacteria. *Arch.Microbiol.* 133, 90-96.
36. Graf, E.G., and Thauer, R.K. (1981). Hydrogenase from *Methanobacterium autotrophicum*, a nickel-containing enzyme. *FEBS Lett.* 136, 165-169.
37. Hogrefe, C. and Friedrich, C. Isolation and characterization of megaplasmid DNA from lithoautotrophic bacteria. *Plasmid*, In Press.
38. Hogrefe, C., Römermann, D. and Friedrich, V. (1984). *Alcaligenes eutrophus* hydrogenase genes (Hox). *J.Bacteriol.* 158, 43-48.
39. Hollocher, T.C., Tate, M.E. and Nicholas, D.J.D. (1981). ^{18}O-idation of ammonia by *Nitrosomonas europaea*. Definitive ^{18}O-tracer evidence that hydroxylamine formation involves a monooxygenase. *J.Biol.Chem.* 256, 10834-10836.
40. Holmes, D.S., Lobos, J.H., Bopp, L.H. and Welch, G.C. (1984). Cloning of a *Thiobacillus ferrooxidans* plasmid in *Escherichia coli*. *J.Bacteriol.* 157, 324-326.
41. Hooper. A.B., DiSpirito, A.A., Olsen, T.C., Andersson,

K.K., Cunningham, W. and Taaffe, L.R. (1984). Generation of the proton gradient by a periplasmic dehydrogenase. In "Microbial Growth on C_1 Compounds" (Eds. R.L.Crawford and R.S.Hanson) pp. 53-58. ASM, Washington, DC.

42. Hooper, A.B., Maxwell, P.C. and Terry, K.R. (1978). Hydroxylamine oxidoreductase from *Nitrosomonas:* absorption spectra and content of heme and metal. *Biochemistry* 17, 2984-2989.
43. Huynh, B.H., Czechowsk, M.H., Krüger, H.-J., Der Vartanian, D.V., Peck, Jr., H.D. and LeGall, J. (1984). *Desulfovibrio vulgaris* hydrogenase: A non heme iron enzyme lacking nickel that exhibits anomalous EPR and Mössbauer spectra. *Proc.Natl.Acad.Sci. U.S.A.* 81, 3728-3732.
44. Jones, R.D. and Morita, R.X. (1983). Methane oxidation by *Nitrosococcus oceanus* and *Nitrosomonas europaea. Appl. Environ.Microbiol.* 45, 401-410.
45. Kämpf, C. and Pfennig, N. (1980). Capacity of Chromatiaceae for chemotrophic growth. Specific respiration rates of *Thiocystis violacea* and *Chromatium vinosum. Arch. Microbiol* 127, 125-135.
46. Kelly, D.P. (1982). Biochemistry of chemolithotrophic oxidation of inorganic sulphur. *Phil.Trans.Roy.Soc.London Ser. B.* 298, 499-528.
47. Knauf, V.C. and Nester, E.W. (1982). Wide host range cloning vectors. A cosmid clone bank of an *Agrobacterium* Ti plasmid. *Plasmid* 8, 45-54.
48. Kojima, N.J., Fox, A., Hausinger, R.P., Daniels, L., Orme-Johnson, W.H. and Walsh, C. (1983). Paramagnetic centers in the nickel-containing, deazaflavin-reducing hydrogenase from *Methanobacterium thermoautotrophicum. Proc.Natl.Acad.Sci. U.S.A.* 80, 378-382.
49. Kondratieva, E.N., Zhukov, V.G., Ivanovsky, R.N., Petrushkova, Yu.P. and Monosov, E.Z. (1976). The capacity of phototrophic sulfur bacterium *Thiocapsa roseopersicina* for chemosynthesis. *Arch.Microbiol.* 108, 287-292.
50. Krüger, H.-J., Huynh, B.H., Ljungdahl, P.O., Xavier, A.V., Der Vartanian, D.V., Moura, I., Peck, Jr., H.D., Teixeira, M., Moura, J.J.G. and LeGall, J. (1982). Evidence for nickel and three-iron center in the hydrogenase of *Desulfovibrio desulfuricans. J.Biol.Chem.* 257, 14620-14623.
51. Krzycki, J.A. and Zeikus, J.G. (1984). Characterization and purification of carbon monoxide dehydrogenase from *Methanosarcina barkeri. J.Bacteriol.* 158, 231-237.
52. Kuenen, J.G. and Beudeker, R.F. (1982). Microbiology of thiobacilli and other sulphur-oxidizing autotrophs, mixotrophs and heterotrophs. *Phil.Trans.Roy.Soc.London,Ser.B.*

298, 473-497.
53. Kumar, S., Nicholas, D.J.D. and Williams, E.H. (1983). Definite ^{15}N NMR evidence that water serves as source of 'O' during nitrite oxidation by *Nitrobacter agilis*. *FEBS Lett.* 152, 71-74.
54. Lu, W.-P. and Kelly, D.D. (1984). Oxidation of inorganic sulphur compounds by thiobacilli. In Microbial Growth on C$_1$ Compounds" (Eds. R.L.Crawford and R.S.Hanson) pp.34-41. ASM, Washington DC
55. Matin, A. (1978). Organic nutrition of chemolithotrophic bacteria. *Ann. Rev. Microbiol.* 32, 433-468.
56. Mayer, H., Bock, E. and Weckesser, J. (1983). 2,3-Diamino-2,3-dideoxyglucose containing lipid A in the *Nitrobacter* strain X14. *FEMS Microbiol.Lett.*17, 93-96.
57. Merberg, D., O'Hara, E.B. and Maier, R.J.(1983). Regulation of hydrogenase in *Rhizobium japonicum*: analysis of mutants altered in regulation by carbon substrates and oxygen. *J.Bacteriol.* 156, 1236-1242.
58. Mergeay, M., Houba, G. and Gerits, J. (1978). Extrachromosomal inheritance controlling resistance to cadmium, cobald, copper and zinc ions: evidence from curing in a *Pseudomonas*. *Arch.Int.Physiol.Biochem.*86, 440-441.
59. Meyer, O. (1982). Chemical and spectral properties of carbon monoxide: methylene blue oxidoreductase, the molybdenum-containing iron-sulphur flavoprotein from *Pseudomonas carboxydovorans*. *J.Biol.Chem.* 257, 1333-1341.
60. Meyer, O. and Fiebig, K. (1984). Enzymes oxidizing carbon monoxide. In "Gas Enzymology" (Eds. R.P.Cox and H. Degn). D.Reidel Publishing Comp., Dordrecht, In Press
61. Meyer, O. and Rohde, M. (1894). Enzymology and bioenergetics of carbon monoxide-oxidizing bacteria. In "Microbial Growth on C$_1$ Compounds" (Eds R.L.Crawford and R.S. Hanson) pp. 26-33. ASM Washington, DC
62. Meyer, O. and Schlegel, H.G. (1983). Biology of aerobic carbon monoxide-oxidizing bacteria.*Ann. Rev.Microbiol.* 37, 277-310.
63. Nason, A., Lee, K.-Y., Pan, S.-S., Ketchum, P.A., Lamberti, A. and DeVries, J. (1971). In vitro formation of assimilatory reduced nicotinamide adenosine dinucleotide phosphate: nitrate reductase from a *Neurospora crassa* mutant and a component of molybdenum enzymes. *Proc.Natl. Acad.Sci.U.S.A.* 68, 3242-3246.
64. Ohi, K., Takada, N., Komemushi, S., Okazaki, O. and Miura, Y. (1979). A new species of hydrogen-utilizing bacterium. *J.Gen.Appl.Microbiol.* 25, 53-58.
65. Olson, J.M. (1970). The evolution of photosynthesis.

Science 438-446.
66. Olson, T.C. and Hooper, A.B. (1983). Energy coupling in the bacterial oxidation of small molecules: an extracytoplasmic dehydrogenase in *Nitrosomonas*. *FEMS Microbiol. Lett.* 19, 47-50.
67. Peck, H.D.Jr. (1968). Energy-coupling mechanisms in chemoautotrophic bacteria. *Ann.Rev.Microbiol.* 22, 489-514.
68. Pinkwart, M., Schneider, K. and Schlegel, H.G. (1983). The hydrogenase of a thermophilic hydrogen-oxidizing bacterium. *FEMS Microbiol. Lett.* 17, 137-141.
69. Pootjes, C.F. (1977). Evidence for plasmid coding of the ability to utilize hydrogen gas by *Pseudomonas facilis*. *Biochem.Biophys.Res.Commun.* 76, 1002-1006.
70. Porte, F. and Vignais, P.M. (1980). Electron transport chain and energy transduction in *Paracoccus denitrificans* under autotrophic growth conditions. *Arch.Microbiol.* 127, 1-10.
71. Probst, I. (1980). Respiration in hydrogen bacteria. In "Diversity of Bacterial Respiratory" (Ed C.J.Knowles) pp. 159-181. CRC Press Boca Raton
72. Quayle, J.R. and Ferenci, T. (1978). Evolutionary aspects of autotrophy. *Microbiol.Rev.* 42, 251-273.
73. Ragsdale, S.W., Ljungdahl, L.G. and Der Vartanian, D.V. (1983). Isolation of carbon monoxide dehydrogenase from *Acetobacterium woodii* and comparison of its properties with those of the *Clostridium thermoaceticum* enzyme. *J.Bacteriol.* 155, 1224-1237.
74. Reh, M. and Schlegel, H.G. (1975). Chemolithoautotrophie als eine übertragbare, autonome Eigenschaft von *Nocardia opaca* 1b. Nachr.Akad.Wiss. Göttingen Math.-Phys.Kl.2 12, 207-216.
75. Reh, M. and Schlegel, H.G. (1981). Hydrogen autotrophy as a transferable genetic character of *Nocardia opaca* 1b. *J.Gen.Microbiol.* 126, 327-336.
76. Robertson, L.A. and Kuenen, J.G. (1983). *Thiosphaera pantotropha* gen.nov. sp. nov., a facultatively anaerobic, facultatively autotrophic sulphur bacterium. *J.Gen.Microbiol.* 129, 2847-2855.
77. Rosenberg, C., Casse-Delbart, F., Dusha, I., David, M., and Boucher, C. (1982).Megaplasmids in the plant-associated bacteria *Rhizobium meliloti* and *Pseudomonas solanacearum*. *J.Bacteriol.* 150, 402-406.
78. Schedel, M. and Trüper, H.G. (1979). Purification of *Thiobacillus denitrificans* siroheme sulfite reductase and investigation of some molecular and catalytic properties. *Biochim.Biophys.Acta* 568, 454-467.
79. Schidlowski, M. (1976). Archaen atmosphere and evolution

of the terrestrial oxygen budget. *In* "The Early History of the Earth" (Ed B.Windley) pp. 525-535. John Wiley & Sons, New York.
80. Schink B. and Schlegel, H.G. (1980). The membrane-bound hydrogenase of *Alcaligenes eutrophus*: II. Localization and immunological comparison with other hydrogenase systems. *Antonie van Leeuwenhoek J.Microbiol.Serol.* 46,1-14.
81. Schlegel, H.G. (1975). Mechanisms of chemoautotrophy. *In* "Marine Ecology II" (Ed O.Kine) Part I, pp.9-60. John Wiley & Sons, London.
82. Schlegel, H.G. and Schneider, K. (1978). Introductory report: distribution and physiological role of hydrogenase in microorganisms. *In* "Hydrogenases: Their Catalytic Activity, Structure and Function" (Eds H.G.Schlegel and K. Schneider) pp. 15-44. Erich Goltze, KG Göttingen.
83. Schneider, K., Cammack, R. and Schlegel, H.G. (1984).Content and localization of FMN, Fe-S clusters and nickel in the NAD-linked hydrogenase of *Nocardia opaca* 1b. *Eur.J.Biochem.* 142, 75-84.
84. Schneider, K., Patil, D.S. and Cammack, R. (1983). ESR properties of membrane-bound hydrogenase from aerobic hydrogen bacteria. *Biochim.Biophys.Acta* 748, 353-361.
85. Schneider, K., Schlegel, H.G. and Jochim, K. (1984). Effect of nickel on activity and subunit composition of purified hydrogenase from *Nocardia opaca* 1b. *Eur.J.Biochem.* 138, 533-541.
86. Schopf, J.W. (1974). Paloebiology of the precambrian: The age of blue-green algae. *Evol.Biol.* 7, 1-43.
87. Simon, R.,Priefer, U. and Pühler, A. (1983). A broad host range mobilization system for in vivo genetic engineering: transposon mutagenesis in gram-negative bacteria. *Biotechnology* 1, 784-790.
88. Stackebrandt, E. and Woese, C.R. (1979). Primärstruktur der ribosomalen 16s RNA - ein Marker der Evolution der Prokaryonten. *Forum Microbiol.* 2, 183-190.
89. Stults, C.W., O'Hara, E.B., and Maier, R.J. (1984). Nickel is a component of hydrogenase in *Rhizobium japonicum*. *J.Bacteriol.* 159, 153-158.
90. Sundermeyer-Klinger, H., Meyer, W., Warninghoff, B. and Bock, E. Membrane-bound nitrite oxidoreductase of *Nitrobacter*; evidence for a nitrate reductase type enzyme. *Arch.Microbiol.* In Press.
91. Suzuki, I. (1974). Mechanism of inorganic oxidation and energy coupling. *Ann. Rev.Microbiol.* 28, 85-101.
92. Suzuki, I. and Silver, M. (1966). The initial product and properties of the sulphur-oxidizing enzyme of thiobacilli. *Biochim.Biophys.Acta* 122, 22-33.

93. Teixeira, M., Moura, I., Xavier, A.V. Der Vartanian, D. V., LeGall, J., Peck, H.D., Jr., Huynh, B.H. and Moura, J.J.G. (1983). *Desulfovibrio gigas* hydrogenase: redox properties of the nickel and iron-sulphur centers. *Eur. J.Biochem.* 130, 481-484.
94. Toghrol, F. and Southerland, W.M. (1983). Purification of *Thiobacillus novellus* sulfite oxidase. *J.Biol.Chem.* 258, 6762-6766.
95. Tsien, H.C., Lambert, R. and Laudelout, H. (1968). Fine structure and the localization of the nitrite oxidizing system in *Nitrobacter winogradskyi*. *Antonie van Leeuwenhoek J.Microbiol.Serol.* 34, 483-494.
96. Wheelis, M. (1984). Energy conservation and pyridine nucleotide reduction in chemoautotrophic bacteria: a thermodynamic analysis. *Arch.Microbiol.* 138, 166-169.
97. Wilke, D. (1980). Conjugational gene transfer in *Xanthobacter autotrophicus* GZ 29. *J.Gen.Microbiol.* 117,431-436.
98. Winogradsky, S. (1887). Über Schwefelbakterien. *Bot.Ztg.* 45, 489-600, 606-616.
99. Winogradsky, S. (1888). Über Eisenbakterien. *Bot.Ztg.* 46, 261-276.

EVOLUTION OF AUTOTROPHIC CO_2 FIXATION

G. Fuchs and E. Stupperich

Abteilung für Angewandte Mikrobiologie
Universität Ulm
Ulm, Federal Republic of Germany

INTRODUCTION

If one thinks about living nature in quantitative terms two processes can be recognized which are dominating all the others. All the energy is derived from the sunlight by photosynthetic energy transformation. All the cell material is derived from CO_2 assimilated by the autotrophs. Since life cannot exist for a longer period of time without one of these two processes, they must have evolved early in history of life. Primitive life forms probably dependend on preformed organic compounds. However, it is also safe to assume that, once life became succesful, soon a local shortage of the organic material preformed by chemical processes resulted, followed by a global shortage.

There is general agreement that the element carbon, which represents approximately 50% of cell material (dry matter), existed under anoxic conditions not only as CO_2, but partially in form of different reduced one-carbon compounds such as carbon monoxide, methane, methanol, formaldehyde, cyanide, among others. For biological purposes their ambient concentrations must not have been high as illustrated by the fact that life presently depends on 300 ppm CO_2.

Hence, primitive autotrophs were forced to develop catalysts, - coenzymes and enzymes -, enabling them to form all their cell constituents from these carbon compounds. Far too little is known about the conditions prevailing 3-4 billions of years ago. Therefore, we are hesitating to present ideas how this could have happened; but we do show how the existing autotrophs assimilate inorganic carbon. These pathways certainly are highly developed and adapted to the present conditions; therefore, they should be regarded more as models for rather than as copies of ancient paths.

The concept of autotrophy has often been disputed in the past (1-5). For our purpose we define a carbon autotroph as an organism which is able to derive most or all of its cell carbon from CO_2, which is the prevalent form of inorganic carbon of today. An ancient autotroph should have been able to derive most or all of its cell carbon from the then prevalent molecular form(s) of the element carbon.

CRITERIONS FOR AN "ANCIENT" CO_2 FIXATION MECHANISM

If the assumption is valid that autotrophic carbon assimilation in a more general sense has evolved early, one can postulate conditions under which it might have evolved. Some of the following points are purely hypothetical.

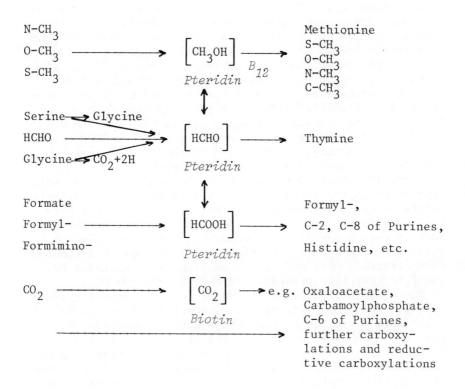

Fig. 1 *Some coenzymes and enzymes involved in biosynthesis using one-carbon compounds.*

(1) The atmosphere was anaerobic, and organic compounds were limiting. (2) Archaebacteria and Eubacteria might not have been separated yet. (3) The primitive autotrophs probably were energy limited. Chemolithotrophic energy metabolism was not based on electron acceptors such as oxygen, nitrate, and probably also not on sulfate. Photosynthesis, if existing, was not highly developed. (4) CO_2 as well as reduced one-carbon compounds were available. (5) Common coenzymes and enzymes of intermediary metabolism were already existing.

If a CO_2 fixation pathway persisted which kept some characters of such an ancient enzymic machinery it should fulfil most of the following criterions deduced from the above postulates. (1) It is likely to occur in anaerobes. (2) It might occur in Archaebacteria and in Eubacteria as well. (3) The pathway should require a minimum of energy. (4) It should enable the organism to use not only CO_2, but other one-carbon compounds such as carbon monoxide, formaldehyde, methanol as well. (5) It should use ubiquitous coenzymic structures.

If such a common pathway occurs in quite distantly related groups its enzymes should exhibit homology, though of low degree; this is an important point. If they are not homologous they resulted from convergent developments. If they are highly homologous they might have been acquired later by lateral gene transfer mechanisms. Some of the more common enzymes and coenzymes involved in one-carbon metabolism in autotrophs and heterotrophs are presented in Fig. 1.

CO_2 FIXATION PATHWAYS OPERATING IN AUTOTROPHIC ORGANISMS

Presently we known of three different ways how autotrophs can assimilate all their cellular carbon from CO_2; there might be even more. By comparison one then has to answer the question which of the pathways is more likely to be derived from an ancient CO_2 fixation mechanism. For one-carbon assimilation see the reviews and books on this topic (6-23).

(1) CO_2 fixation via the reductive pentosephosphate cycle (Calvin-Bassham-Benson Cycle)

Distribution This is the most important pathway of CO_2 assimilation in autotrophs, occuring both in chloroplasts of the green plants and in bacteria from different groups. Interestingly enough it has not been found in Archaebacteria and in chemolithotrophic anaerobes studied so far. Since the Calvin cycle has been the only accepted autotrophic CO_2 fixation mechanism until recently, earlier treatises on evolution of autotrophy centred on this path (12, 13, 15).

Pathway For details the reader is referred to the excellent reviews which appeared on this topic (7, 12, 13, 15, 16). The Calvin cycle is schematically presented in Fig. 2. The only CO_2 fixation reaction in this pathway is the carboxylation of ribulose 1,5-bisphosphate to give two molecules of glycerate 3-phosphate, catalyzed by ribulose 1,5-bisphosphate carboxylase (6). It has been pointed out that there is a close similarity to the hexulosemonophosphate path of formaldehyde assimilation in aerobic methylotrophic bacteria; formaldehyde has been proposed being a likely carbon source of the primitive environment. Formaldehyde assimilation pathways therefore might have preceeded CO_2 assimilation paths and functioned as autotrophic templates (15).

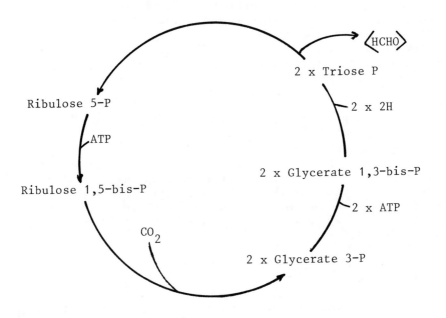

Fig. 2 Autotrophic CO_2 fixation via the reductive pentose-phosphate cycle (Calvin cycle).

Key enzymes and energy demand The key enzymes of this pathway are ribulose 5-phosphate kinase and ribulose 1,5-bisphosphate carboxylase. Per mol of triosephosphate (representing the average oxidation state of cell carbon) synthesized from 3 CO_2, 9 ATP equivalents are required, i.e. 3 ATP/CO_2.

(2) CO_2 fixation via the reductive citric acid cycle

Distribution This pathway was first proposed to occur in the phototrophic green sulfur bacterium *Chlorobium limicola* by Evans et al.(24). Because of inconsistency of data it has been questioned by many workers in the field (25, 26, 27). Recent evidence provided by Ivanovski et al.(28) and by our group (29, 30) is in favour of the operation of this path in *Chlorobium* (Fig. 3).

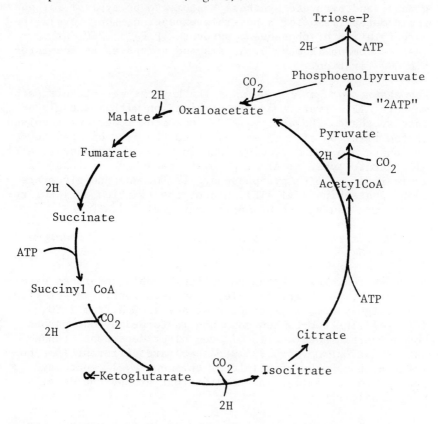

Fig. 3 Autotrophic CO_2 fixation via the reductive citric acid cycle.

Its distribution in the *Chlorobiaceae* and *Chloroflexaceae* remains to be studied. Recent work by Minoda and associates (31, 32; personal communication by Dr. Y. Igarashi) suggests that a similar pathway might occur in *Hydrogenobacter thermophilus*, an aerobic, obligately chemolithoautotrophic hydrogen-oxidizing bacterium. It is a gram-negative, non-motile, rod-shaped thermophile isolated from hot springs in Japan.

Pathway The organism uses reverse reactions of the citric acid cycle in order to form citrate starting from oxaloacetate and two CO_2 (24, 27). As a result of citrate cleavage by ATP citrate lyase (28, 33), oxaloacetate is reformed and the CO_2 fixation product acetyl CoA is released. The synthesis of C_3- and C_4-compounds requires further carboxylations. The fixation of CO_2 by reductive carboxylations catalyzed by three enzymes - α-ketoglutarate synthase, isocitrate dehydrogenase, and pyruvate synthase - seems to be typical for anaerobic metabolism (for a possible exception, see above). The carboxylation of phosphoenolpyruvate (PEP) by PEP carboxylase is common in many organisms and serves as an anaplerotic CO_2 fixation.

Key enzymes and energy demand The key enzymes of this pathway are ATP citrate lyase in combination with α-ketoglutarate synthase. The energy (ATP) consuming reactions involved in the fixation of 3 molecules of CO_2 to give one triose phosphate are the following: Succinyl CoA synthetase (1 ATP), ATP citrate lyase (1 ATP), pyruvate phosphate dikinase (2 equivalents of ATP; pyrophosphate is formed) and glycerate 3-phosphate kinase (1 ATP). Hence, the fixation of 1 CO_2 requires the equivalent of approximately 1 2/3 ATP.

(3) CO_2 fixation via a total synthesis of acetyl coenzyme A (activated acetic acid pathway)

Distribution Most of the chemolithotrophic anaerobes examined so far, which are capable of an autotrophic way of life, use a highly complex synthesis of acetyl CoA from 2 CO_2 (Fig. 4). The two-carbon compound is formed by one-carbon to one-carbon addition (34). The sulfur-dependent Archaebacteria (*Thermoproteus, Sulfolobus*) are different (see below). Although the principle of this non-cyclic pathway appears to be similar in the different groups the details vary considerably.

Pathway The process has been studied in methanogenic bacteria (21, 35, 36), acetogenic bacteria (34, 37, 38), and sulfate-reducing bacteria (39). In contrast to the other CO_2-fixation mechanisms it principally allows not only the assimilation of carbon from CO_2, but also from methanol, methylamines, formaldehyde, formate, and carbon monoxide. This does not necessarily mean that each individual species has the capability to do so.

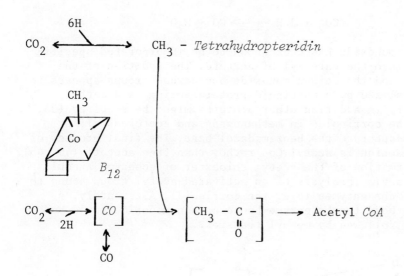

Fig. 4 CO_2 fixation via a total synthesis of acetyl coenzyme A from 2 CO_2 (ativated acetic acid pathway).

Origin of the methyl group of acetate   One CO_2 is reduced to a methyl group bound to a tetrahydropteridin which gives rise to the methyl of acetate. In acetogenic bacteria, formate is a free intermediate formed via formate dehydrogenase; it is subsequently activated to formyl tetrahydrofolate and then reduced by several enzymes to methyl tetrahydrofolate (40).

In methanogenic bacteria formate is not a free intermediate; rather, CO_2 appears at first to be bound as a carbamate to the coenzyme methanofuran (41) and then to be reduced to the formyl state. After being transferred to tetrahydromethanopterin (a tetrahydrofolate counterpart in methanogens), the one-carbon is reduced to methyl tetrahydromethanopterin (42-44). These reactions are shared by CO_2 reduction to methane (45, 46).

In sulfate-reducing bacteria details about the intermediates and enzymes involved are not yet known, although the principle clearly is the same (39).

Origin of the carboxyl group of acetate The carboxyl group of acetate is synthesized from carbon monoxide (34, 47), which is formed in a second CO_2 reduction catalyzed by the nickel-enzyme carbon monoxide dehydrogenase (48).

$$CO_2 + 2H \rightleftharpoons CO + H_2O$$

Carbon monoxide has been shown to be incorporated specifically into the carboxyl of acetate. The addition of the methyl and the carbon monoxide one-carbon groups appears to be catalyzed by a corrinoid protein acting as a methyl carrier; in addition other proteins might be required (34, 40). The corrinoids in methanogens and acetogens differ with respect to the benzimidazol base. The final product of the reaction is acetyl CoA rather than free acetate (40, 45). The formation of the acetyl thioester of coenzyme A could proceed via thiolysis of an activated acetyl group bound to one of the enzymes (e.g. the corrinoid). Alternately, a sulfur bond to the enzyme-bound carbon monoxide could be formed followed by methyl transfer.

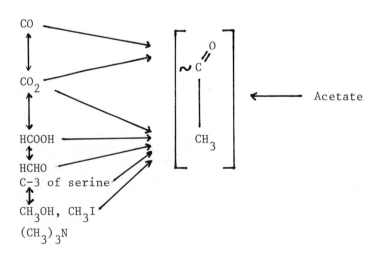

Fig. 5 Incorporation of various one-carbon compounds and acetate into the central intermediate acetyl CoA in autotrophic methanogenic, acetogenic and sulfate-reducing bacteria.

Assimilation of reduced one-carbon compounds and acetate

This path also allows the assimilation of reduced one-carbon compounds (for CO see above) and of acetate. Methanol and methylamine(?)metabolism in methanogenic and acetogenic bacteria appears to involve corrinoid proteins (49, 50) allowing the conversion of these one-carbon compounds to methane and acetate, respectively, as well as their assimilation into the methyl of acetate (Fig. 5).

Formaldehyde reacts spontaneously with tetrahydrofolate and tetrahydromethanopterin (44); it is incorporated into the methyl of acetate or oxidized to the formyl level. Formate is directly incorporated into the methyl of acetate (in acetogenic and sulfate reducing bacteria), and/or is oxidized to CO_2.

Under hydrogen limitation all reduced one-carbon compounds mentioned can be disproportionated, i.e. part of them is reduced to methane or incorporated into acetate at the cost of another variable part which is oxidized to CO_2. The reversal of the activated acetic acid pathway, therefore, could bring about a complete oxidation of acetyl CoA to CO_2.

Incorporation of acetyl CoA

The incorporation of acetyl CoA into cellular compounds does not proceed via a glyoxylic acid cycle. Rather, acetyl CoA is reductively carboxylated to pyruvate. The synthesis of phosphoenolpyruvate and oxaloacetate are accomplished by different sets of enzymes. The synthesis of ∝-ketoglutarate proceeds via citrate or via an incomplete reductive carboxylic acid cycle (Fig. 6).

Key enzymes and energy demand

It should be pointed out that the activated acetic acid pathway shows considerable variations, depending on the organism. The nickel-enzyme carbon monoxide dehydrogenase in combination with the acetyl CoA synthesizing multienzyme complex are unique to this pathway. Much more information is needed in order to define the key enzymes.

The energy demand of this path cannot been specified yet, but it appears to be rather low. The synthesis of acetyl CoA from CO_2, 4 H_2, and coenzyme A is exergonic even under cellular conditions and possibly does not require a net ATP consumption. However, there has to be an energetic coupling of the endergonic reduction of CO_2 to the formaldehyde state of oxidation with the following reduction of the formaldehyde state of oxidation. Also, CO_2 reduction to CO appears to be energy driven (48). However, autotrophic acetogenic bacteria can grow at the expense of acetate formation from 2 CO_2 and 4 H_2. One ATP per acetate formed is synthesized via phosphotransacetylase and acetate kinase. Since the growth yields indicate the gain of approximately 1 mol of ATP per mol of acetate formed (51), the synthesis of acetyl CoA from 2 CO_2 must be energy-neutral or should require less than 1 ATP. Evidently, the energetics of methane and acetate formation remains puzzling. Two steps in triose phosphate formation require ATP, PEP synthetase (2 equivalents of ATP) or pyruvate phosphate dikinase (2 equivalents of ATP, pyrophosphate is formed), and glycerate 3-phosphate kinase (1 ATP). Given that acetyl CoA synthesis requires less than 1 ATP, triose phosphate synthesis from 3 CO_2 costs less than 4 ATP, or less than 1 1/3 ATP per CO_2.

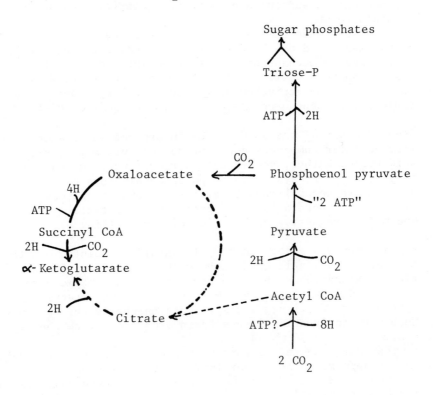

Fig. 6 Assimilation of acetyl CoA into cell compounds by methanogenic, acetogenic and sulfate-reducing bacteria. Note that there are two possibilities to synthesize α-ketoglutarate (⎯⎯→ , ----→).

(4) CO_2 fixation via a reductive carboxylic acid pathway

The sulfur-dependent Archaebacteria *Thermoproteus sp.* and *Sulfolobus sp.* are chemolithotrophic bacteria, which grow at the expense of anaerobic sulfur reduction to H_2S and aerobic sulfur oxidation to sulfate, respectively (52). Although recognized as autotrophs the species *Thermoproteus neutrophilus* and *Sulfolobus brierleyi* do not seem to contain the key enzymes of one of the other CO_2 fixation paths. Pulse-labelling studies with *Sulfolobus* using $^{14}CO_2$ indicate the presence of a novel path tentatively designated reductive carboxylic acid path (53); similarly *Thermoproteus* exhibits an unusual labelling pattern. It will be interesting to learn more about how these and other groups live autotrophically.

COMPARISON OF THE KNOWN CO_2 FIXATION PATHWAYS IN PROKARYOTES

Table 1 compares the different paths of autotrophic CO_2 fixation. The question is which of them is more likely to represent a good model for an "ancient" CO_2 assimilation path, as defined before.

From an energetic point of view the Calvin cycle is not an attractive model for an anaerobe growing with e.g. hydrogen or reduced sulfur compounds as electron donors. In addition, it seems not to occur in Archaebacteria. On the other hand little additional genetic information is required to transform a heterotroph in an organism capable of autotrophic growth using the Calvin cycle. Since plasmids in chemolithoautotrophs carry some information for autotrophy (see the contribution by B. Friedrich, this volume), a lateral gene transfer could have been a means for distribution. Another reason for its wide distribution in Eubacteria could be that the Calvin cycle allows a high flexibility of metabolism; it fits to any kind of energy metabolism. Many bacteria and most plants are limited by factors other than energy; therefore, the high energy demand of the Calvin cycle is not of great weight.

The reductive tricarboxylic acid cycle appears to be fairly restricted to a few Eubacteria. It represents a most simple, energetically favorable device for CO_2 assimilation. More information is needed in order to evaluate its distribution and its evolutionary position.

The total synthesis of acetyl CoA fulfils most of the criterions postulated for an ancient pathway. Its distribution in only distantly related anaerobes (Archaebacteria and Eubacteria, (gram (+) and gram (-)) and its unusual biochemistry are noteworthy. It requires the lowest amount of ATP. It is a most versatile one-carbon and two-carbon assimilation path. It will be interesting to determine the degree of homology of its key enzymes from different sources.

TABLE 1

Comparison of the different autotrophic CO_2 fixation pathways

Pathway	Key enzymes	ATP required per CO_2 in Triose phosphate
1. Calvin cycle	- Ribulose 1,5-bis-phosphate Carboxylase plus - Ribulose 5-phosphate Kinase	9/3
2. Reductive Citric Acid Cycle	- α-Ketoglutarate Synthase plus - ATP Citrate Lyase	5/3
3. Activated Acetic Acid Pathway	- CO Dehydrogenase (Ni) plus - Acetyl CoA synthesizing enzyme complex	4/3
4. Reductive Carboxylic Acid Pathway	?	?

CONCLUSIONS

The previous discussions on evolution of autotrophy were based on the understanding that the Calvin cycle is the universal path of CO_2 assimilation. While these studies are highly interesting with respect to the evolution of this cycle and of related paths, they cannot be considered comprehensive with respect to autotrophy in general. At least two additional CO_2 fixation pathways exist. It seems unlikely that these different carbon assimilation paths can be deduced from one common ancient mechanism of CO_2 fixation. Therefore, they have developed independently at different times to meet different needs. Unfortunately, historical records such as carbon isotope fractionation in the oldest sediments cannot be interpreted unambiguously (54). For theoretical reasons the total synthesis of acetyl CoA represents an attractive model for an ancient C_1 fixation path. However, an understanding of evolution of autotrophy requires more information, (i) a comprehensive knowledge of pathways and enzymes occuring in all autotrophs, (ii) a deeper understanding of the phylogeny, and (iii) a more precise picture of the conditions existing in the early biosphere.

ACKNOWLEDGEMENTS

The authors work quoted herein was supported by the Deutsche Forschungsgemeinschaft.

REFERENCES

1. Rittenberg, S.C. (1969). The roles of exogenous organic matter in the physiology of chemolithotrophic bacteria. *Adv Microb Physiol* 3, 159-195.
2. Kelly, D.P. (1971). Autotrophy: concepts of lithotrophic bacteria and their organic metabolism. *Annual Review of Microbiology* 25, 177-210.
3. Rittenberg, S.C. (1972). The obligate autotroph - the demise of a concept. *Antonie van Leeuwenhoek* 38, 457-478.
4. Whittenbury, R., Kelly, D.P. (1977). Autotrophy: a conceptual phoenix. *In* "Microbial energetics" (Eds B.A. Haddock, W.A. Hamilton) 27. Symposium Society for general microbiology, pp. 121-150. Cambridge University Press, New York.
5. Smith, A.J., Hoare, D.S. (1977). Specialist phototrophs, lithotrophs, and methylotrophs: a unity among a diversity of procaryotes? *Bacteriol. Rev.* 41, 419-448.
6. Quayle, J.R., Fuller, R.C., Benson, A.A., Calvin, M. (1954) Enzymatic carboxylation of ribulose diphosphate. *J. Am. Chem. Soc.* 76, 3610-3611.
7. Calvin, M., Bassham, J.A. (1962). The photosynthesis of carbon compounds. W.A. Benjamin, Inc., New York.
8. Ljungdahl, L.G., Wood, H.G. (1969). Total synthesis of acetate from CO_2 by heterotrophic bacteria. *Ann. Rev. Microbiol.* 23, 515-538.
9. Ribbons, D.W., Harrison, J.E., Wadzinski, A.M. (1970). Metabolism of single carbon compounds. *Ann. Rev. Micro-*

biol. <u>24</u>, 135-158.
10. Wood, H.G., Utter, M.F. (1965) The role of CO_2 fixation in metabolism. *In* "Essays in biochemistry" (Eds P.N. Campbell, G.D. Greville) Vol. 1, pp. 1-27.
11. Quayle, J.R. (1972). *In* "Advances in Microbial Physiology"(Eds A.H. Rose, D.W. Tempest), Vol. 7, pp. 119-203.
12. McFadden, B.A. (1973). Autotrophic CO_2 assimilation and the evolution of ribulose diphosphate carboxylase. *Bacteriol. Rev.* <u>37</u>, 289-319.
13. McFadden, B.A., Tabita, F.R. (1974). D-ribulose-1,5-diphosphate carboxylase and the evolution of autotrophy. *Bio. Systems* <u>6</u>, 93-112.
14. Schlegel, H.G., Gottschalk, G., Pfennig, N (Eds) (1976) Symposium on microbial production and utilization of gases (H_2, CH_4, CO). Akademie der Wissenschaften zu Göttingen. E. Goltze KG, Göttingen.
15. Quayle, J.R., Ferenci, T. (1978). Evolutionary aspects of autotrophy. *Microbiological reviews* <u>42</u>, 251-273.
16. McFadden, B.A. (1978). Assimilation of one-carbon compounds. *In* "The bacteria" (J.C. Gunsalus, L.N. Ornston, J.R. Sokatch, Eds), Vol. 6, pp. 219-304. Academic Press, New York, San Francisco, London.
17. Matin, A. (1978).Organic nutrition of chemolithotrophic bacteria. *Ann. Rev. Microbiol.* <u>32</u>, 433-68.
18. Colby, J., Dalton, H., Whittenbury, R. (1979). Biological and biochemical aspects of microbial growth on C_1 compounds. *Ann. Rev. Micriobiol.* <u>33</u>, 481-517.
19. Dalton, H. (Ed) (1981). Microbial growth on C_1 compounds. Proceedings of the third international symposium, Sheffield, 1980, Heyden & Son, London.
20. Anthony, C. (1982). The biochemistry of methylotrophs. Academic Press, London.
21. Zeikus, J.G. (1983). Metabolism of one-carbon compounds by chemotrophic anaerobes. *Adv. Microbial. Physiol.* <u>24</u>, 215-299.
22. Large, P.J. (1983). Methylotrophy and methanogenesis. American Society for Microbiology, Washington DC.
23. Crawford, R. L., Hanson, R.S. (1984). Microbial growth on C_1 compounds. Proceedings of the 4th international symposium. American Society for Microbiology, Washington DC.
24. Evans, M.C.W., Buchanan, B.B., Arnon, D.I. (1966). A new ferredoxin dependent carbon reduction cycle in a photosynthetic bacterium. *Proc. Natl. Acad. Sci. USA* <u>55</u>, 928-934.
25. Tabita, F.R., McFadden, B.A., Pfennig, N. (1974). D-ribulose-1,5-bisphosphate carboxylase in *Chlorobium*

Thiosulfatophilum Tassajara. *Biochim. Biophys. Acta* 341 187-194.
26. Buchanan, B.B., Sirevåg, R. (1976). Ribulose-1,5-diphosphate carboxylase and *Chlorobium thiosulfatophilum*. *Arch. Microbiol.* 109, 15-19.
27. Buchanan, B.B. (1979). Ferredoxin-linked carbon dioxide fixation in photosynthetic bacteria. In "Photosynthesis II. Encyclopedia of plant physiology" (Eds M. Gibbs, E. Latzko) New Series 6, pp. 416-424. Springer Verlag Berlin, Heidelberg, New York.
28. Ivanovsky, R.N., Sintsov, N.V., Kondratieva, E.N. (1980). ATP-linked citrate lyase activity in the green sulfur bacterium *Chlorobium limicola* forma thisulfatophilum. *Arch. Microbiol.* 128, 239-241.
29. Fuchs, G., Stupperich, E., Jaenchen, R. (1980). Autotrophic CO_2 fixation in *Chlorobium limicola*. Evidence against the operation of the Calvin cycle in growing cells. *Arch. Microbiol.* 128, 56-63.
30. Fuchs, G., Stupperich, E., Eden, G. (1980). Autotrophic CO_2 fixation in *Chlorobium limicola*. Evidence for the operation of a reductive tricarboxylic acid cycle in growing cells. *Arch. Microbiol.* 128, 64-71.
31. Shiba, H., Kawasumi, T., Igarashi, Y., Kodama, T., Minoda, Y. (1982). The deficient carbohydrate metabolic pathways and the incomplete tricarboxylic acid cycle in an obligately autotrophic hydrogen-oxidizing bacterium. *Agric. Biol. Chem.* 46, 2341-2345.
32. Kawasumi, T., Igarashi, Y., Kodama, T., Minoda, Y.(1980). Isolation of strictly thermophilic and obligately autotrophic hydrogen bacteria.*Agric.Biol.Chem.*44,1985-1986.
33. Antranikian, G., Herzberg, C., Gottschalk, G. (1982). Characterization of ATP citrate lyase from *Chlorobium limicola*. *J. Bacteriol.* 152, 1284-1286.
34. Wood, H.G., Drake, H.L., Hu, S.J. (1982). Studies with *Clostridium thermoaceticum* and the resolution of the pathway used by acetogenic bacteria that grow on carbon monoxide or carbon dioxide and hydrogen. In "Proc. Biochem. Symp." (Ed E.E. Snell), pp. 29-56. Annual Reviews Inc., Palo Alto.
35. Fuchs, G., Stupperich, E. (1982). Autotrophic CO_2 fixation pathway in *Methanobacterium thermoautotrophicum*. In "Archaebacteria"(Ed O. Kandler)pp. 277-288, Gustav Fischer, Stuttgart, New York.
36. Fuchs, G., Stupperich, E. (1983). CO_2 fixation pathways in bacteria. *Physiol. Veg.* 21, 845-854.
37. Eden, G., Fuchs, G. (1982). Total synthesis of acetyl coenzyme A involved in autotrophic CO_2 fixation in

Acetobacterium woodii. *Arch. Microbiol.* 133, 66-74.
38. Eden, G., Fuchs, G. (1983). Autotrophic CO_2 fixation in *Acetobacterium woodii*. II. Demonstration of enzymes involved. *Arch. Microbiol.* 135, 68-73.
39. Jansen, K., Thauer R.K., Widdel, F., Fuchs, G. (1984). Carbon assimilation pathways in sulfate reducing bacteria. Formate, carbon dioxide, carbon monoxide, and acetate assimilation by *Desulfovibrio baarsii*. *Arch. Microbiol.* 138, 257-262.
40. Ljungdahl, L.G., Wood, H.G. (1982). Acetate biosynthesis. In "B_{12}" (Ed D. Dolphin), Vol. 2, pp. 165-202. John Wiley & Sons, Inc.
41. Leigh, J.A., Rinehart, K.L., Wolfe, R.S. (1984). Structure of methanofuran, the carbon dioxide reduction factor of *Methanobacterium thermoautotrophicum*. *J. Amer. Chem. Soc.* 106, 3636-3640.
42. VanBeelen, P., Van Neck, J.W., deCock, R.M., Vogels, G.D., Guijt, W., Haasnoot, C.A.G. (1984). 5,10-Methenyl-5,6,7,8-tetrahydromethanopterin, a one-carbon carrier in the process of methanogenesis. *Biochemistry* 23, 4448-4454.
43. VanBeelen, P., Stassen, A.P.M., Bosch, J.W.G., Vogels, G.D., Guijt, W., Haasnoot, C.A.G. (1984). Elucidation of the structure of methanopterin, a coenzyme from *Methanobacterium thermoautotrophicum*, using two-dimensional nuclear-magnetic-resonance techniques. *Eur. J. Biochem.* 138, 563-571.
44. Escalante-Semerena, J.C., Rinehart, K.L., Wolfe, R.S. (1984). Tetrahydromethanopterin, a carbon carrier in methanogenesis. *J. Biol. Chem.* 259, 9447-9455.
45. Stupperich, E., Fuchs, G. (1984a). Autotrophic synthesis of activated acetic acid from two CO_2 in *Methanobacterium thermoautotrophicum*. I. Properties of in vitro system. *Arch. Microbiol.* 139, 8-13.
46. Stupperich, E., Fuchs, G. (1984b). Autotrophic synthesis of activated acetic acid from two CO_2 in *Methanobacterium thermoautotrophicum*. II. Evidence for different origins of acetate carbon atoms. *Arch. Microbiol.* 139, 14-20.
47. Stupperich, E., Hammel, K.E., Fuchs, G., Thauer, R.K. (1983). Carbon monoxide fixation into the carboxyl group of acetyl coenzyme A during autotrophic growth of *Methanobacterium*. *FEBS Letters* 152, 21-23.
48. Eikmanns, B., Fuchs, G., Thauer, R.K. (1984). Carbon monoxide formation from CO_2 and H_2 by *Methanobacterium thermoautotrophicum*. *Eur. J. Biochem.* In press.
49. Van der Meijden, P., Jansen, L.P., Drift, C.van der,

Vogels, G.D. (1983). Involvement of corrinoids in the methylation of coenzyme M (2-mercaptoethanesulfonic acid) by methanol and enzymes from *Methanosarcina barkeri*. *FEMS Microbiol. Lett.* 19, 247-251.
50. Naumann, E., Fahlbusch, K., Gottschalk, G. (1984). Presence of a trimethylamine: HS-coenzyme M methyltransferase in *Methanosarcina barkeri*. *Arch. Microbiol.* 138, 79-83.
51. Tschech, A., Pfennig, N. (1984).Growth yield increase linked to caffeate reduction in *Acetobacterium woodii*. *Arch. Microbiol.* 137, 162-167.
52. Zillig, W., Stetter, K.O., Schäfer, W., Janekovic, D., Wunderl, S., Holz, I., Palm, P. (1981). Thermoproteales: A novel type of extremely thermoacidophilic anaerobic archaebacteria isolated from icelandic solfataras. *Zbl. Bakt. Hyg., I. Abt. Orig. C 2*, 205-227.
53. Kandler, O, Stetter, K.O. (1981).Evidence for autotrophic CO_2 assimilation in *Sulfolobus brierleyi* via a reductive carboxylic acid pathway. *Zbl.Bakt. Hyg., I. Abt. Orig. C 2*, 111-121.
54. Schidlowski, M. (1983). Biologically mediated isotope fractionations: Biochemistry, geochemical significance and preservation in the earth's oldest sediments. *In* "Cosmochemistry and the origin of life" (Ed. C. Ponnamperuma) pp. 277-322. D. Reidel Publishing Co.

EVOLUTION IN THE CITRIC ACID CYCLE

P. D. J. Weitzman

Department of Biochemistry
University of Bath
Bath, England

1. INTRODUCTION

All students of the life sciences encounter the citric acid cycle (CAC), also referred to as the Krebs or tricarboxylic acid cycle (Fig.1). This central metabolic pathway is portrayed as a virtually ubiquitous pathway throughout Nature and may be hailed as a shining example of the unity of biochemistry. It is also emphasised that the cycle fulfils a dual role, making a crucial contribution both to energy metabolism and to biosynthesis. Having "mastered" the sequence of steps in the CAC, the would-be life scientist sets the pathway to the back of his mind as a biochemical device which all cells make use of in their respiratory metabolism. It is a fundamental, but not particularly exciting, piece of cellular activity which holds no unexpected novelties or surprises for the modern investigator.

How far from the truth is this dismissive view! For one thing, although the chemical steps of the cycle are preserved intact throughout Nature, diverse organisms make diverse use of this potential chemistry, in some cases using only selected portions of the cycle and in others even regearing their enzymic "hardware" to operate portions or all of the cycle in reverse. At another level, the regulatory mechanisms operating to control the activities of the enzymes of the cycle differ significantly between diverse organisms. Indeed, one would *a priori* anticipate this to be the case, as different organisms would be expected to make different demands on the several functions of the cycle and thus to control the cycle in different ways in accord with their individual metabolic "life-styles". It seems reasonable to

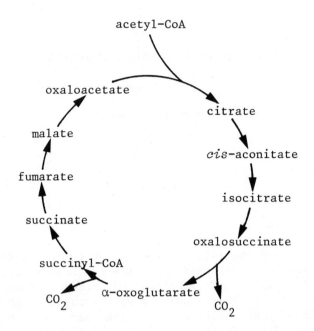

Fig. 1 The citric acid cycle

assume that the evolutionary paths to different organisms have been accompanied by the evolution of distinctive regulatory and other individual functional features in the CAC. Indeed, Krebs himself argued for a teleological viewpoint in studying metabolism (1) and this case has been further championed by Atkinson (2). It is amongst prokaryotes that metabolic diversity is particularly displayed, and it is therefore appropriate that some attention to the CAC be paid at this symposium.

I shall address myself to two questions:
1) How might the CAC pathway have evolved? (evolution *of* the cycle).
2) What differences have evolved in the fine details of the operation of the CAC in diverse organisms? (evolution *in* the cycle).

2. EVOLUTION *OF* THE CYCLE

Reference was made above to the dual, or multifunctional, role of the cycle in both energy and biosynthetic metabolism. It oxidises 2-carbon (acetyl) fragments to CO_2 and reduced nucleotides, NADH and NADPH, these electron sources acting respectively to drive the formation of ATP by oxidative phosphorylation and to effect reductive biosynthesis. In addition, the cycle generates a number of metabolites which serve as starting materials for a range of biosynthetic pathways. Thus oxaloacetate and α-oxoglutarate are the respective precursors of aspartate and glutamate and the related families of amino acids, while succinyl-CoA is utilised in both porphyrin and amino acid biosynthesis.

Clearly, the CAC must have evolved in a complex stepwise manner, which would make guessing at its evolutionary route a daunting task. However, perhaps the multifunctional aspect of the cycle renders such guessing at its evolution just a little less formidable. The fact that present-day organisms display such diversity, using different segments of the cycle for their individual needs, provides some possible clues and facilitates intelligent guessing about the stages in the evolution of the cycle. In view of the central position of the CAC in *aerobic* metabolism, it may be ironic that we should seek the origins and evolutionary development of the cycle in the more primitive *anaerobic* organisms of the past. However, the basic scheme of intermediary metabolism is essentially anaerobic, and reactions dependent on the presence of oxygen are later evolutionary additions to an already complete framework (3).

A strictly biosynthetic evolutionary route has been proposed by Dillon (4) in which the gradual build-up of the cycle is based on the functional contribution of each new step to the synthesis of a product required for survival. This is the same as the idea of stepwise retrograde evolution put forward by Horowitz (5). As primitive organisms drew on the content of the primeval seas, key biochemicals were depleted and it became essential to gain the ability to convert other organic compounds into those depleted substances. Organisms which acquired (by whatever molecular or mutational process) the enzyme catalysts to effect such conversion would obviously be selected for survival. By sequential repeat of this process, a metabolic pathway may be constructed in a backward manner.

The scheme proposed by Dillon (4) centres on three key amino acids - glycine, aspartate and glutamate, and describes the evolution of the CAC in concert with the glyoxylate bypass. As these amino acids became depleted from the environment, primitive organisms evolved to produce enzymes which could yield these compounds by the amination of

glyoxylate, oxaloacetate and α-oxoglutarate respectively.
The next steps in the scheme introduced the conversion of
malate to oxaloacetate, in order to maintain the availability
of the latter, and the formation of malate from glyoxylate
and an activated acetyl group. This drain on glyoxylate may
have favoured the acquisition of an enzyme capable of cleaving isocitrate to glyoxylate and succinate (Fig.2).

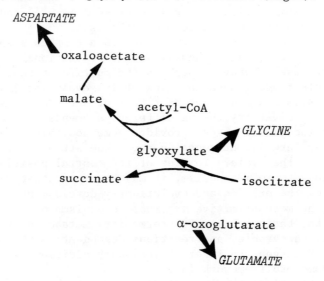

Fig. 2 Early stage in a scheme for the evolution of the CAC

In the next stage, the ability to hydrate fumarate to malate
was acquired as well as the conversion of succinate to
succinyl-CoA. The next steps involved the formation first
of isocitrate from *cis*-aconitate and then of *cis*-aconitate
from citrate. At this time, the pathway developed from
succinyl-CoA and glycine leading to porphyrins and haems
(for cytochrome and chlorophyll production). Evolution of
a succinate-fumarate interconversion, of the oxidation of
isocitrate to α-oxoglutarate, and of the condensation of
oxaloacetate with acetyl-CoA to form citrate produced a
nearly complete CAC identical with that found in present-day
cyanobacteria (6-8) (Fig.3).

In the final stage, the evolution of the multienzyme
complex oxidising α-oxoglutarate to succinyl-CoA completed
the cycle of reactions and finally presented a pathway that
could function, not only as a biosynthetic device, but also
for oxidative metabolism. This would have coincided with the
appearance of an oxidising atmosphere created by the oxygenic

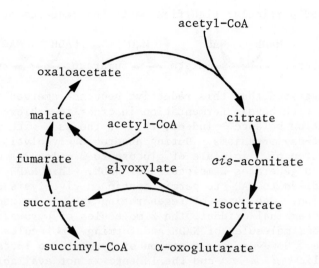

Fig. 3 *Nearly complete citric acid cycle*

photosynthesis of the cyanobacteria. It may be noted that Dillon (4) considered the final adjustment in the cycle to be the succinate thiokinase reaction,

$$\text{succinate} + \text{CoA} + \text{ATP} = \text{succinyl-CoA} + \text{ADP} + P_i$$

replacing the alternative reaction,

$$\text{succinate} + \text{acetoacetyl-CoA} = \text{succinyl-CoA} + \text{acetoacetate}$$

However, we have shown (9) that despite its reported absence (7), succinate thiokinase is present in cyanobacteria.

This fully developed cyclic pathway has then persisted throughout the subsequent evolution of eukaryotic and multicellular organisms, the enzymes of the glyoxylate bypass having been lost in these higher organisms (though see (10, 11) for evidence of glyoxylate cycle activity in animal tissues). Dillon (4) attributed no respiratory function to the pathway until after the evolution of cyanobacteria and the generation of an oxygen-containing atmosphere, when the cycle could be regeared to serve the additional oxidising role. However, he offered no suggestions as to the fate of the reduced coenzymes generated in the oxidising dehydrogenase reactions.

A different approach was adopted by Gest (12). He

attached particular significance to the sequence of reactions

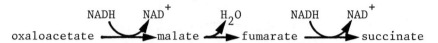

and suggested that this reductive sequence emerged in association with hexose fermentation in order to achieve oxidation-reduction balance. Indeed, this is the case with many present-day organisms. During anaerobic glycolysis, the conversion of 1 molecule of glucose to 2 molecules of pyruvate also generates 2 molecules of NADH. This NADH must be re-oxidised to NAD^+ to permit continued glycolysis and ATP formation. One means of regenerating NAD^+ is to use pyruvate as the terminal oxidant, the 2 molecules of pyruvate oxidising the 2 molecules of NADH and forming 2 molecules of lactate. However, the reaction of pyruvate to lactate is a metabolic *cul de sac* and the lactate is not available for biosynthesis. On the other hand, if one of the 2 molecules of pyruvate is carboxylated to oxaloacetate and the 2 molecules of NADH are re-oxidised by the sequence from oxaloacetate to succinate, the second molecule of pyruvate is spared and may be used for biosynthetic metabolism. The sequence from oxaloacetate to succinate thus supplies an "electron sink" and the evolution of such a sequence would confer a selective metabolic advantage on an organism. The use of fumarate as a terminal electron acceptor relies on the presence of a fumarate reductase, which is distinct from succinate dehydrogenase which operates in the oxidative direction. It is also to be noted that the pathway from oxaloacetate to succinate fulfils a biosynthetic role, as the succinate may be converted to succinyl-CoA for porphyrin and amino acid biosynthesis.

In anaerobic bacteria, pyruvate may be metabolised by a ferredoxin-linked enzyme to produce acetyl-CoA, CO_2 and H_2. Thus the formation of acetyl-CoA from pyruvate is likely to have evolved in early anaerobic organisms and could therefore have been used in the evolution of that branch of the CAC leading from oxaloacetate to α-oxoglutarate and furnishing a biosynthetic route to glutamate: oxaloacetate + acetyl-CoA → citrate → *cis*-aconitate → isocitrate → α-oxoglutarate → glutamate. A primitive anaerobic cell could thus possess two "arms" of the CAC (Fig.4), an arrangement which is still encountered in present-day organisms, e.g. *E. coli* growing anaerobically (13).

To convert these two sequences into a CAC capable of

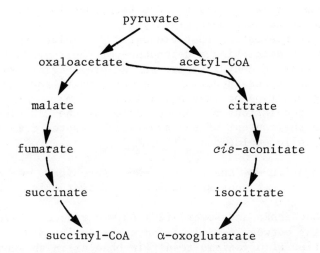

Fig. 4 *"Split"* citric acid cycle in anaerobic organisms

undertaking a bioenergetic role would then have required the filling of the gap between α-oxoglutarate and succinyl-CoA. This was achieved by the appearance of α-oxoglutarate dehydrogenase, an enzyme complex which could have evolved from the analogous pyruvate dehydrogenase. In addition, it would have been necessary to regear the enzymes of the C_4 branch to operate only in the oxidative direction. It is noteworthy that present-day organisms such as *E. coli* can produce both fumarate reductase and succinate dehydrogenase, the former being repressed under aerobic conditions and the latter under anaerobic conditions (14). Succinate dehydrogenase may well have evolved from fumarate reductase to provide the correct directionality for an oxidative cycle. Indeed, Cole (15) has found strong resemblance between the amino acid compositions of the flavoprotein subunits of fumarate reductase and succinate dehydrogenase as well as a common sequence of nine residues at the probable flavin binding site of the enzymes.

The driving force behind such developments may have been the appearance of oxygen in the atmosphere (as a result of oxygenic photosynthesis) and the elaboration of the electron transport system for phosphorylation of ADP. The opportunity to use oxygen as terminal electron acceptor and to derive high ATP yield by running the cycle in the manner with which we are now familiar would have conferred major selective advantages on emerging aerobic organisms. Thus the ATP yield is nearly 20 times greater when glucose is oxidised completely via the CAC than when its catabolism is restricted to anaerobic glycolysis.

It is interesting that present-day facultative anaerobes persist in displaying a dependence of their production of α-oxoglutarate dehydrogenase on the presence of oxygen; the enzyme is repressed during anaerobic growth and induced by the introduction of oxygen (13,16-19).

Any speculations on the evolution of the CAC should be made in the context of the generally accepted time sequence of metabolic evolution:

Fermentation ➔ Anaerobic photosynthesis ➔ Aerobic photosynthesis ➔ Oxidative respiration

As Hochachka and Somero (20) have stressed, there is interplay between organisms and their environment. The activities of organisms result in changes in the environment which then become exploited through the evolution of new metabolic capabilities. In such metabolic evolution, random mutations of the genome would have resulted in new enzymes and selection must have favoured those which were capable of exploiting the changed chemical environment.

The oxidation of pyruvate to acetyl-CoA presumably first evolved to use compounds other than NAD^+ as electron acceptor. Reference was made above to the ferredoxin-linked reaction encountered in present-day anaerobic bacteria (pyruvate: ferredoxin oxidoreductase). Some organisms use flavodoxin in place of ferredoxin (21) while the methanogens employ a deazaflavin derivative (22). If we assume that something akin to this ferredoxin-linked enzyme evolved in the primitive anaerobes, we can account for the acetyl-CoA required for the segment of the CAC leading from oxaloacetate (via citrate) to α-oxoglutarate. As photosynthetic mechanisms evolved, a ferredoxin-linked pyruvate oxidation could be driven backwards by photoreduced ferredoxin, i.e. the enzyme would act as a pyruvate synthase (an advantage over the NAD-linked pyruvate dehydrogenase which is irreversible). Referring back to Fig.4, depicting the branched pathway evolved in anaerobic organisms, if the carboxylation of acetyl-CoA to pyruvate is effected with reduced ferredoxin, a pathway is presented for the conversion of acetyl-CoA to succinyl-CoA. It is reasonable to postulate the evolution of an enzyme α-oxoglutarate synthase from the strictly analogous pyruvate synthase. The establishment of such an enzyme would close the gap between succinyl-CoA and α-oxoglutarate and permit the pathway to operate round to citrate. Evolution of the citrate cleavage enzyme ATP citrate lyase (perhaps by mutation from citrate synthase?) would then allow citrate to be converted to oxaloacetate plus acetyl-CoA. Fig.5 illustrates the pathway so constructed, and this is seen to be the reductive carboxylic acid cycle first proposed by Evans *et al.* (23) to account for the fixation of CO_2 in the green photosynthetic *Chlorobiaceae*. The operation of this cycle has been confirmed (24) and the presence of ATP citrate lyase, previously thought to occur only in eukaryotes (25), has recently been established (26,27).

It would thus appear that the advent of photosynthesis allowed the exploitation of metabolic reactions already established; recruitment and regearing of the appropriate enzymes resulted in a pathway for the assimilation of CO_2. This seems a plausible account of the evolution of the reductive carboxylic acid cycle.

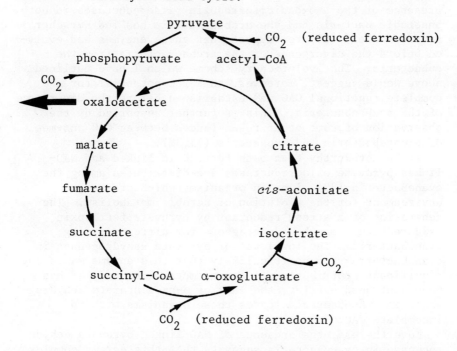

Fig. 5 The reductive carboxylic acid cycle

If the cycle shown in Fig.5 were now to be regeared again and operated in the oxidative direction (i.e. clockwise in the figure), in concert with a suitable electron-accepting system, the pathway would be essentially that of the complete CAC. This appears to be the situation in the archaebacteria. These organisms are capable of utilising the CAC in its oxidative mode but their pyruvate and α-oxoglutarate oxidoreductases are linked to ferredoxin (or similar, see above). No NAD-linked α-oxoacid dehydrogenase complexes have been detected in the archaebacteria (28,29). It is therefore likely that the NAD-linked pyruvate and α-oxoglutarate dehydrogenases evolved in eubacteria after the divergence of the archaebacteria. The α-oxoacid:ferredoxin oxidoreductases are considerably less complex enzymes than the NAD-linked multienzyme dehydrogenases (molecular weights around 300,000 as opposed to 3 million or more) and, unlike the latter, they do not contain lipoic acid as an integral cofactor, though Danson et al. (30) have recently established the presence of dihydrolipoamide dehydrogenase in halophilic archaebacteria - a finding which may indicate an as yet undisclosed role for lipoic acid in these organisms. The presence of the α-oxoacid:ferredoxin oxidoreductases in both anaerobic bacteria and the archaebacteria has led Kerscher and Oesterhelt (29) to suggest that these enzymes had evolved before the divergence of the archaebacteria from the eubacteria. The evolutionary scheme which I have outlined above would suggest, moreover, that the potential for a complete functional CAC was established before the divergence of the archaebacteria. This is further supported by the observation of some other resemblances between CAC enzymes of archaebacteria and eubacteria (31,32).

It is noteworthy that both ferredoxin-linked and NAD-linked pyruvate oxidoreductases are distributed among the cyanobacteria (33), i.e. in organisms which created an environment for the evolution of aerobic metabolism. The generation of a strong reductant by pyruvate:ferredoxin oxidoreductase may be advantageous for nitrogen-fixing cyanobacteria. The low level of pyruvate dehydrogenase in cyanobacteria renders it unlikely that this enzyme makes any significant contribution to energy metabolism and, as has been mentioned earlier, the lack of α-oxoglutarate dehydrogenase in cyanobacteria leaves these organisms with an incomplete CAC.

Nevertheless, the presence of NAD-linked pyruvate dehydrogenase in cyanobacteria suggests that this enzyme complex may have evolved in early aerobic photosynthetic organisms. The stage would then have been set for the final recruitment of this enzyme, and for the evolution of the analogous α-oxoglutarate dehydrogenase, to provide the oxidative directionality to the CAC in emerging non-photosynthetic aerobic organisms.

It is significant that present-day facultative anaerobes, e.g. *E. coli*, use pyruvate dehydrogenase during aerobic growth though, under anaerobic conditions, they may employ pyruvate oxidation linked to ferredoxin or flavodoxin (21). The development of efficient oxidative phosphorylation mechanisms in strict aerobes has resulted in the loss of ferredoxin-linked α-oxoacid oxidoreductases, a total reliance on NAD-linked α-oxoacid dehydrogenases and an obligatory dependence on the fully functional oxidative CAC.

In a recent review (34) I drew attention to another facet of the CAC which had not previously been commented on and which may have some bearing on the pathway of evolution of the cycle. Here, I would like to elaborate further on this idea. Rather than hypothesise about the gradual, step-by-step evolution of the cycle, this approach examines the chemical reaction sequence in the completed cycle and reveals a striking pattern of duplication. First, consider the following general scheme of six reactions:

α-oxoacid
↓
oxidise, decarboxylate and form an acyl-CoA
↓
"spend" free energy
↓
introduce a C=C double bond
↓
add H_2O across the double bond
↓
oxidise the CHOH group to C=O
↓
decarboxylate to yield α-oxoacid

If this general scheme is now applied to the CAC it may be seen that the reactions of the cycle fit this scheme in two consecutive groups. Thus, starting with pyruvate, acetyl-CoA is first produced, its free energy is "spent" in driving the condensation with oxaloacetate to form citrate, a double bond is introduced with the formation of *cis*-aconitate, hydration across this bond produces isocitrate, oxidation of the secondary alcoholic group gives oxalosuccinate and decarboxylation of the latter yields α-oxoglutarate. Carrying out a similar sequence of reactions on the product of the first "run", α-oxoglutarate, initially produces succinyl-CoA, "spending" its free energy brings about the phosphorylation of a nucleoside diphosphate to the triphosphate, introduction of a double bond yields fumarate from succinate, hydration produces malate, oxidation of the alcoholic group gives oxaloacetate and decarboxylation of the latter yields pyruvate, the starting α-oxoacid.

Combination of these two sets of reactions produces a cyclic pathway based on a two-fold passage through the same sequence of six reactions starting with, and returning to, pyruvate (Fig.6). Such a scheme involves a total of four decarboxylations, i.e. a molecule of the 4-carbon compound oxaloacetate is consumed per turn of the cycle. If we conserve this molecule of oxaloacetate by abandoning the final decarboxylation of oxaloacetate to pyruvate and allow the oxaloacetate to re-circulate round the pathway, then we are left with the familiar CAC in which pyruvate provides the "fuel" (acetyl-CoA) for respiration.

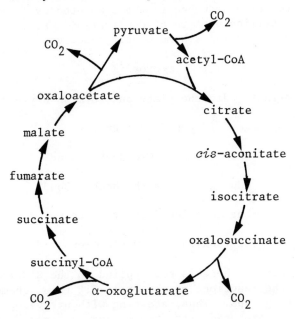

Fig. 6 Cyclic scheme of reactions from pyruvate

This economy of metabolic strategy is remarkable and invites the speculation that the two halves of the CAC bear some evolutionary relationship to each other. It is conceivable that the pairs of enzymes catalysing equivalent reactions in the two halves may have evolved by a process of gene duplication and diversification or, as Jensen (35) has proposed for metabolic pathway evolution, by the recruitment of enzymes with broad specificity, followed by subsequent refinement. Indeed, there are some distinct similarities within such enzyme pairs (34). I have discussed above how the cycle may have evolved in two "arms"; the matching pairs of reactions in the two arms suggest an interplay between their respective evolutions. Future detailed comparative studies on the enzymes and their genes may cast more light on this possibility.

Finally, mention should be made of Krebs' own final thoughts on the evolution of the CAC. In two publications shortly before his death (36,37) he presented an argument for the evolution of the cycle based on its greater efficiency for the combustion of acetate into ATP energy than feasible alternative pathways of acetate oxidation (e.g. via glycollate). The greater efficiency of the CAC route means that, in a competitive environment, the CAC allows an organism to make optimal use of available resources and thus enhances the chances of survival. Furthermore, because the more efficient pathway of acetate oxidation proceeds by means of a primary attachment of the acetate to the carrier molecule oxaloacetate (in citrate formation), this pathway must be a cyclic one in which the carrier is regenerated. If the pathway were not cyclic, the carrier would need constantly to be produced afresh and would give rise to stoichiometric quantities of a wasteful by-product.

3. EVOLUTION *IN* THE CYCLE

In this section I wish to consider some examples of enzymic diversity, at the structural, catalytic and regulatory levels, which presumably represent evolution within the established citric acid cycle pathway. It is probably safe to assume that activity-regulating mechanisms evolved *after* the evolution of pathways and their enzymes. Thus controls over CAC enzymes by metabolites signalling the energy state of the cell are meaningful only after the evolution of the pathway as an energy-producing device. The acquisition of regulatory sensitivity to metabolic needs would confer evolutionary advantages in terms of metabolic efficiency. I will not here consider further how such controls may have evolved, but this subject has recently been discussed in a stimulating article by Baumberg (38).

Several enzymes of the CAC display molecular, catalytic or regulatory diversity among different organisms (34). Here I shall present aspects of this diversity displayed by two enzymes of the cycle - citrate synthase and succinate thiokinase. These catalyse respectively the reactions:

acetyl-CoA + oxaloacetate + H_2O → citrate + CoA

and

succinyl-CoA + NDP + P_i = succinate + NTP + CoA

where NDP and NTP represent nucleoside di- and triphosphate. These reactions are the two "spend free energy" steps described above in the hypothetical scheme for CAC evolution.

Citrate Synthase (CS)

A cyclic pathway may not appear at first sight to have an obvious *initial* enzymic step or an *end product*. However, a little consideration reveals that CS fulfils the role of initial enzyme. It catalyses the only step in which carbon enters the cycle and is therefore expected to be an appropriate target for the exercise of regulatory effects. Moreover, insofar as the oxidative ATP-generating role of the CAC is achieved by the production of NADH, the latter compound may be considered an end product of the operation of the cycle.

In linear metabolic pathways we are familiar with the phenomenon of feedback control of an initial enzyme by the end product. A similar process appears to operate in the CAC; NADH has been found to act as a powerful and specific end-product inhibitor of CS activity (39). Remarkably. this inhibition occurs *only* with CS from Gram-negative bacteria; CS from Gram-positive bacteria or eukaryotes is insensitive to NADH (40). It has further been found that AMP can completely overcome the NADH inhibition of CS. This does not occur with all Gram-negative bacterial CSs, but only with those from strictly aerobic species (34,40). No explanation can currently be offered for the absence in Gram-positive bacteria of the eminently "sensible" feedback control of CS by NADH. However, the AMP effect may be rationalised in terms of the metabolic differences between organisms and the different energy-generating pathways they employ. Strict aerobes are absolutely dependent on the CAC for energy production and it is thus appropriate that they should regulate the key enzyme of the cycle, CS, in response to the energy state of the cell. Facultative anaerobes, on the other hand, can generate energy by fermentation, then using the CAC for biosynthesis only; regulation of CS by AMP would be a futile mechanism and is hence absent.

Reference was made earlier to the branched variant of the CAC which operates during the anaerobic growth of certain bacteria and satisfies the biosynthetic demands for α-oxoglutarate and succinyl-CoA (Fig.4). Examination of this metabolic scheme shows that α-oxoglutarate is the end product of a short sequence of reactions initiated by CS; feedback inhibition of CS by α-oxoglutarate would thus constitute a plausible biosynthetic regulatory mechanism. Significantly, such a control mechanism is indeed observed and only with CS from Gram-negative facultatively anaerobic bacteria (34, 41). The fact that the inhibition is observed only in organisms which employ the split CAC pathway (Fig.4) strengthens the conclusion that this is a physiologically significant control process.

Another variant on the CAC was also mentioned earlier as operating in present-day cyanobacteria (Fig.3). These Gram-negative organisms lack α-oxoglutarate dehydrogenase and do not use the CAC for energy production. Inhibition of CS by NADH would thus be a redundant mechanism and has been found to be absent (42,43). Fig.3 shows that α-oxoglutarate is an end product of CS action, and cyanobacterial CSs are inhibited by this compound (42,43). Unlike the situation in the split pathway of facultative anaerobes, the cyanobacterial scheme in Fig.3 shows that, because of the operation of the glyoxylate cycle, succinyl-CoA is formed from succinate which itself is produced from isocitrate. Hence succinyl-CoA is also an end product of the initial action of CS and it might serve as an additional feedback inhibitor of CS. We have found (44) that this is indeed the case; cyanobacterial CS is distinctive in being sensitive to inhibition by both succinyl-CoA and α-oxoglutarate.

The existence of enzyme variants of CS which are controlled by NADH, NADH plus AMP, α-oxoglutarate, and α-oxoglutarate plus succinyl-CoA presents a striking example of the evolution of regulatory diversity to fit the metabolic life-style of the host organism.

It is remarkable that the diverse regulatory effects just described are restricted to the CSs of Gram-negative bacteria and it is conceivable that there is a difference in metabolic organisation (spatial distribution of enzymes?) between Gram-negative and Gram-positive bacteria. The greater functional complexity of Gram-negative bacterial CSs suggests that there may be an accompanying greater complexity at the molecular structural level. This is so, as evidenced quite simply by the difference in molecular sizes and sub-unit compositions between CSs from Gram-negative bacteria on the one hand and Gram-positive bacteria and eukaryotes on the other. All CSs fall into two groups - "large" and "small" (34,45,46), and these may readily be determined by gel filtration with appropriate marker proteins. "Large" CS is encountered exclusively in Gram-negative bacteria, whereas CSs from Gram-positive bacteria and eukaryotes are all of the "small" type. The "large" enzymes have molecular weights in the region of 250,000 and appear to be composed of six subunits; the "small" CSs have molecular weights around 90,000 and contain two subunits. In all cases the subunits of a particular CS appear to be similar and no evidence has been found for the existence of distinct catalytic and regulatory subunits. The larger, hexameric structure of Gram-negative bacterial CS presumably confers the opportunities for subtle conformational rearrangements in response to interactions of the enzymes with the various allosteric metabolite effectors; such rearrangements lead to appropriate modulation of enzymic function (34,47,48).

There are some exceptions to the correlation between CS behaviour and bacterial taxonomy. Thus, species of *Acetobacter* and some other genera, although Gram-negative, possess CS which is of the "large" type but is quite insensitive to inhibition by NADH (34).

It is desirable to know what are the particular molecular design features which confer the various regulatory sensitivities and which determine the extent of subunit association (to dimers or to hexamers) in the different variants of CS. One approach to this goal which we have pursued has been to generate and isolate mutant forms of CS with altered properties (34). The strategy adopted was a two-stage one. First, CS-deficient bacterial mutants were produced and, secondly, revertants were selected which had regained CS activity. It was hoped that amongst such revertants there would be some that contained CS enzyme which, though active, might have altered molecular and regulatory properties. Such mutant CS enzymes were indeed produced (49-51). Starting with *E. coli*, for example, three distinct types of revertant CS were identified. One type was "large" with kinetic and regulatory properties very similar to those of the wild-type *E. coli* CS. A second type was "large", but lacked the regulatory properties normally associated with "large" CS, i.e. resembled the *Acetobacter* enzyme. The third type was "small" and had properties typical of small CS, i.e. resembled the Gram-positive bacterial enzyme. Hence, starting with a CS-deficient mutant of *E. coli*, a group of mutants may be isolated containing CSs resembling those occurring naturally in diverse bacteria. The genetic relatedness of these mutant enzymes strengthens the possibility that the evolution of the functional diversity and distinctiveness of bacterial CSs may have resulted from relatively minor genetic alterations.

Recent work in my laboratory has revealed yet further complexity and diversity among bacterial CSs (52,53). It now emerges that within the pseudomonads some species contain both "large" *and* "small" CS and that there is a measure of variation in the relative proportions of the two types dependent on the stage of growth of the bacterial culture. These new findings emphasise the degree of novelty and unexpectedness that attends the study of the enzymes of the citric acid cycle and highlights the immense metabolic diversity which has evolved in the prokaryotes.

Succinate Thiokinase (STK)

Considerable attention has been paid to the study of STK, largely because of its unique substrate phosphorylation role in the citric acid cycle and the complexity of the reaction catalysed. In particular, STKs from *E. coli* and pig heart have been investigated and a major difference in their molecular sizes has been revealed (54). Whereas the molecular weight of the *E. coli* enzyme is in the region of 140,000 - 150,000, that of the pig enzyme is one-half this value, i.e. 70,000 - 75,000. The *E. coli* enzyme is known to have an $\alpha_2\beta_2$ subunit composition, whereas the mammalian enzyme is of $\alpha\beta$ structure. It was implicitly assumed that these different molecular sizes characterised the bacterial and animal STKs and that other bacteria would have a similar STK to that of *E. coli*. However, in view of our discovery of the two classes of citrate synthase within the bacteria, it seemed conceivable that not all bacterial STKs would resemble the *E. coli* enzyme; there might well be a distinction between STKs from Gram-negative and Gram-positive organisms. Experiments confirmed this possibility (9,31,55). As with citrate synthases, STKs were found to fall into two groups - "large" (molecular weight ~150,000) and "small" (molecular weight ~75,000); the former are encountered *only* in Gram-negative bacteria, whereas the latter occur in Gram-positive bacteria and eukaryotes. The incidence of "large" and "small" STKs and CSs is strikingly parallel.

Another difference that has been noted between *E. coli* STK and the mammalian enzyme is that the latter appeared to be specific for GDP as the nucleoside diphosphate substrate, whereas the *E. coli* enzyme appeared to be specific for ADP. Once again, it was suggested that bacterial STKs might generally be specific for ADP. However, we had observed that STK from *Acinetobacter calcoaceticus* functioned more readily with GDP than with ADP, i.e. the K_m for GDP was considerably lower than that for ADP. This finding, together with the demonstration of molecular size differences among bacterial STKs, prompted an examination of the nucleotide utilisation of a range of STKs (56). Distinct patterns were revealed which show some correlation with bacterial classification. On the basis of the estimated STK K_m values for ADP and GDP, the organisms tested fell into four groups.

One group contained Gram-negative strict aerobes which, like *A. calcoaceticus*, had STKs with a very high K_m for ADP but a very much lower K_m for GDP. The second group also comprised Gram-negative aerobes, but these were the pseudomonads and related organisms; their STKs showed similar and low K_m for both ADP and GDP. The third group contained the Gram-negative facultative anaerobes, typified by *E. coli*; their STKs exhibited low K_m for ADP but significantly higher K_m for GDP. Finally, the fourth group contained Gram-positive bacteria; their STKs were active with ADP (low K_m) but appeared to be inactive with GDP.

In the case of the regulatory diversity of citrate synthases discussed above, we were able to identify particular control mechanisms with the metabolic individuality of the organisms concerned. In contrast, we cannot yet perceive any rationale for the marked diversity exhibited by succinate thiokinase. It may be assumed that this diversity represents evolutionary divergence and that its survival reflects as yet undiscovered physiological distinctiveness of, and benefit to, the different organisms.

4. CONCLUDING REMARKS

Despite our complete knowledge of the chemistry of the citric acid cycle pathway, there is clearly much we do not understand, or are not yet even aware of, concerning the more subtle aspects of the biological catalysis of these steps, their organisation and regulation, and their total integration into the complete fabric of cellular metabolism. Further exploration of Nature's evolutionary subtleties should prove most rewarding, and we should be encouraged by Dobzhansky's philosophy (57) - "Nothing in biology makes sense except in the light of evolution".

ACKNOWLEDGEMENTS

I wish to acknowledge my gratitude to Professor Howard Gest (Indiana University, USA) and Dr Dorothy Jones (Leicester University, UK) for particularly helpful and stimulating discussions. I also thank the Science and Engineering Research Council (UK) for supporting experimental work undertaken in my laboratory with research grants.

REFERENCES

1. Krebs, H.A. (1954). Excursion into the borderland of biochemistry and philosophy, *Bulletin of Johns Hopkins Hospital* 95, 45-51.
2. Atkinson, D.E. (1977). "Cellular Energy Metabolism and its Regulation". Academic Press, New York.
3. Wald, G. (1964). The origins of life, *Proceedings of the National Academy of Sciences of the USA* 52, 595-611.
4. Dillon, L.S. (1981). "Ultrastructure, Macromolecules and Evolution". Plenum Press, New York.
5. Horowitz, N.H. (1945). On the evolution of biochemical synthesis, *Proceedings of the National Academy of Sciences of the USA* 31, 153-157.
6. Hoare, D.S., Hoare, S.L. and Moore, R.B. (1967). The photoassimilation of organic compounds by autotrophic blue-green algae, *Journal of General Microbiology* 49, 351-370.
7. Pearce, J., Leach, C.K. and Carr, N.G. (1969). The incomplete tricarboxylic acid cycle in the blue-green alga *Anabaena variabilis*, *Journal of General Microbiology* 55, 371-378.
8. Smith, A.J. and Hoare, D.S. (1977). Specialist phototrophs, lithotrophs, and methylotrophs: a unity among a diversity of procaryotes? *Bacteriological Reviews* 41, 419-448.
9. Weitzman, P.D.J. and Kinghorn, H.A. (1980). Succinate thiokinase from cyanobacteria, *FEBS Letters* 114, 225-227.
10. Goodman, D.B.P., Davis, W.L. and Jones, R.G. (1980). Glyoxylate cycle in toad urinary bladder: possible stimulation by aldosterone, *Proceedings of the National Academy of Sciences of the USA* 77, 1521-1525.
11. Jones, C.T. (1980). Is there a glyoxylate cycle in the liver of the fetal guinea pig? *Biochemical and Biophysical Research Communications* 95, 849-856.

12. Gest, H. (1981). Evolution of the citric acid cycle and respiratory energy conversion in prokaryotes, *FEMS Microbiology Letters* 12, 209-215.
13. Amarasingham, C.R. and Davis, B.D. (1965). Regulation of α-ketoglutarate dehydrogenase formation in *Escherichia coli*, *Journal of Biological Chemistry* 240, 3664-3668.
14. Guest, J.R. (1981). Partial replacement of succinate dehydrogenase function by phage- and plasmid-specified fumarate reductase in *Escherichia coli*, *Journal of General Microbiology* 122, 171-179.
15. Cole, S.T. (1982). Nucleotide sequence coding for the flavoprotein subunit of the fumarate reductase of *Escherichia coli*, *European Journal of Biochemistry* 122, 479-484.
16. Gray, C.T., Wimpenny, J.W.T. and Mossman, M.R. (1966). Regulation of metabolism in facultative bacteria, *Biochimica et Biophysica Acta* 117, 33-41.
17. O'Brien, R.W. and Stern, J.R. (1969). Requirement for sodium in the anaerobic growth of *Aerobacter aerogenes* on citrate, *Journal of Bacteriology* 98, 388-393.
18. Keevil, C.W., Hough, J.S. and Cole, J.A. (1979). Regulation of 2-oxoglutarate dehydrogenase synthesis in *Citrobacter freundii* by traces of oxygen in commercial nitrogen gas and by glutamate, *Journal of General Microbiology* 114, 355-359.
19. Beatty, J.T. and Gest, H. (1981). Biosynthetic and bioenergetic functions of citric acid cycle reactions in *Rhodopseudomonas capsulata*, *Journal of Bacteriology* 148, 584-593.
20. Hochachka, P.W. and Somero, G.N. (1973). "Strategies of Biochemical Adaptation". W.B. Saunders Co., Philadelphia.
21. Blaschkowski, H.P., Neuer, G., Ludwig-Festl, M. and Knappe, J. (1982). Routes of flavodoxin and ferredoxin reduction in *Escherichia coli*, *European Journal of Biochemistry* 123, 563-569.
22. Zeikus, J.G., Fuchs, G., Kenealy, W. and Thauer, R.K. (1977). Oxidoreductases involved in cell carbon synthesis of *Methanobacterium thermoautotrophicum*, *Journal of Bacteriology* 132, 604-613.
23. Evans, M.C.W., Buchanan, B.B. and Arnon, D.I. (1966). A new ferredoxin-dependent carbon reduction cycle in a photosynthetic bacterium, *Proceedings of the National Academy of Sciences of the USA* 55, 928-934.

24. Fuchs, G., Stupperich, E. and Eden, G. (1980). Autotrophic CO_2 fixation in *Chlorobium limicola*. Evidence for the operation of a reductive tricarboxylic acid cycle in growing cells, *Archives of Microbiology* 128, 64-71.
25. Srere, P.A. (1975). The enzymology of the formation and breakdown of citrate, *Advances in Enzymology* 43, 57-101.
26. Ivanovsky, R.N., Sintsov, N.V. and Kondratieva, E.N. (1980). ATP-linked citrate lyase activity in the green sulfur bacterium *Chlorobium limicola* Forma *thiosulfatophilum*, *Archives of Microbiology* 128, 239-241.
27. Antranikian, G., Herzberg, C. and Gottschalk, G. (1982). Characterization of ATP citrate lyase from *Chlorobium limicola*, *Journal of Bacteriology* 152, 1284-1287.
28. Aitken, D.M. and Brown, A.D. (1969). Citrate and glyoxylate cycles in the halophil, *Halobacterium salinarium*, *Biochimica et Biophysica Acta* 177, 351-354.
29. Kerscher, L. and Oesterhelt, D. (1982). Pyruvate:ferredoxin oxidoreductase - new findings on an ancient enzyme, *Trends in Biochemical Sciences* 7, 371-374.
30. Danson, M.J., Eisenthal, R., Hall, S., Kessell, S.R. and Williams, D.L. (1984). Dihydrolipoamide dehydrogenase from halophilic archaebacteria, *Biochemical Journal* 218, 811-818.
31. Weitzman, P.D.J. and Kinghorn, H.A. (1983). Succinate thiokinase from *Thermus aquaticus* and *Halobacterium salinarium*, *FEBS Letters* 154, 369-372.
32. Danson, M.J., Black, S.C., Woodland, D.L. and Wood, P.A. (1984). Citric acid cycle enzymes of the archaebacteria: citrate synthase and succinate thiokinase, *FEBS Letters* in press.
33. Bothe, H. and Nolteernsting, U. (1975). Pyruvate dehydrogenase complex, pyruvate:ferredoxin oxidoreductase and lipoic acid content of microorganisms, *Archives of Microbiology* 102, 53-57.
34. Weitzman, P.D.J. (1981). Unity and diversity in some bacterial citric acid cycle enzymes, *Advances in Microbial Physiology* 22, 185-244.
35. Jensen, R.A. (1976). Enzyme recruitment in evolution of new function, *Annual Reviews of Microbiology* 30, 409-425.
36. Krebs, H.A. (1981). The evolution of metabolic pathways. *In* "Molecular and Cellular Aspects of Microbial Evolution" (Eds M.J. Carlile, J.F. Collins and B.E.B. Moseley), pp. 215-228. Cambridge University Press, Cambridge.

37. Baldwin, J.E. and Krebs, H.A. (1981). The evolution of metabolic cycles, *Nature* 291, 381-382.
38. Baumberg, S. (1981). The evolution of metabolic regulation. *In* "Molecular and Cellular Aspects of Microbial Evolution" (Eds M.J. Carlile, J.F. Collins and B.E.B. Moseley), pp. 229-272. Cambridge University Press, Cambridge.
39. Weitzman, P.D.J. (1966). Regulation of citrate synthase activity in *Escherichia coli*, *Biochimica et Biophysica Acta* 128, 213-215.
40. Weitzman, P.D.J. and Jones, D. (1968). Regulation of citrate synthase and microbial taxonomy, *Nature* 219, 270-272.
41. Weitzman, P.D.J. and Dunmore, P. (1969). Regulation of citrate synthase activity by α-ketoglutarate - metabolic and taxonomic significance, *FEBS Letters* 3, 265-267.
42. Taylor, B.F. (1973). Fine control of citrate synthase activity in blue-green algae, *Archives of Microbiology* 92, 245-249.
43. Lucas, C. and Weitzman, P.D.J. (1975). Citrate synthase from blue-green bacteria, *Biochemical Society Transactions* 3, 379-381.
44. Lucas, C. and Weitzman, P.D.J. (1977). Regulation of citrate synthase from blue-green bacteria by succinyl coenzyme A, *Archives of Microbiology* 114, 55-60.
45. Weitzman, P.D.J. and Dunmore, P. (1969). Citrate synthases: allosteric regulation and molecular size, *Biochimica et Biophysica Acta* 171, 198-200.
46. Weitzman, P.D.J. and Danson, M.J. (1976). Citrate synthase, *Current Topics in Cellular Regulation* 10, 161-204.
47. Rowe, A.J. and Weitzman, P.D.J. (1969). Allosteric changes in citrate synthase observed by electron microscopy, *Journal of Molecular Biology* 43, 345-349.
48. Mitchell, C.G. and Weitzman, P.D.J. (1983). Reversible effects of cross-linking on the regulatory cooperativity of *Acinetobacter* citrate synthase, *FEBS Letters* 151, 260-264.
49. Harford, S. and Weitzman, P.D.J. (1978). Mutant citrate synthases from *Escherichia coli*, *Biochemical Society Transactions* 6, 433-435.
50. Danson, M.J., Harford, S. and Weitzman, P.D.J. (1979). Studies on a mutant form of *Escherichia coli* citrate synthase desensitised to allosteric effectors, *European Journal of Biochemistry* 101, 515-521.

51. Weitzman, P.D.J., Kinghorn, H.A., Beecroft, L.J. and Harford, S. (1978). Mutant citrate synthases from *Acinetobacter* generated by transformation, *Biochemical Society Transactions* **6**, 436-438.
52. Solomon, M. and Weitzman, P.D.J. (1983). Occurrence of two distinct citrate synthases in a mutant of *Pseudomonas aeruginosa* and their growth-dependent variation, *FEBS Letters* **155**, 157-160.
53. Mitchell, C.G. and Weitzman, P.D.J. (1985). Molecular size diversity of citrate synthases from *Pseudomonas* species, *Archives of Microbiology* in press.
54. Bridger, W.A. (1974). Succinyl-CoA synthetase. *In* "The Enzymes" (Ed P.D. Boyer) 3rd edn, Vol. 10, pp. 581-606. Academic Press, New York.
55. Weitzman, P.D.J. and Kinghorn, H.A. (1978). Occurrence of 'large' or 'small' forms of succinate thiokinase in diverse organisms, *FEBS Letters* **88**, 255-258.
56. Weitzman, P.D.J. and Jaskowska-Hodges, H. (1982). Patterns of nucleotide utilisation in bacterial succinate thiokinases, *FEBS Letters* **143**, 237-240.
57. Dobzhansky, T. (1973). Nothing in biology makes sense except in the light of evolution, *American Biology Teacher* **35**, 125-129.

EVOLUTION OF ARGININE METABOLISM

V. Stalon

Laboratoire de Microbiologie
Faculté des Sciences
Université Libre de Bruxelles
Brussels, Belgium
and
Institut de Recherches du CERIA
Brussels, Belgium

INTRODUCTION

According to Oparin's ideas, the first living systems were heterotrophic and utilized organic compounds then present on the primitive earth as their source of energy and cellular material (1). The acquisition of biosynthetic functions occurred as these substrates were gradually depleted from the nutrient environment. Selection led to the establishment of complete biosynthetic sequences. In their respective ecological niches, modern bacteria encounter a broad spectrum of compounds able to serve as a carbon and/or nitrogen source. Highly elaborated mechanisms have evolved, enabling the organisms to utilize as efficiently as possible the nutrients available in their environment. In bacteria, the synthesis of enzymes required for the degradation of most carbon sources is regulated by induction and subject to catabolite repression exerted by a "good" carbon source. A complex situation arises when the carbon source also serves as a nitrogen source for the cells. Synthesis of the enzymes that play a rôle in utilization of such compounds is generally controlled by induction and both by carbon and nitrogen catabolite repression. Metabolic control must operate in such a way as to prevent carbon catabolite repression of enzymes involved in the catabolism of a nitrogen compound when the latter is the sole nitrogen source. In bacteria, the formation of a number of enzyme systems involved in nitrogen metabolism is regulated by the availability of ammonia in the growth medium. This nitrogen catabolite repression is generally assumed to be a manifestation of the selective use of high quality nitrogen over poorer nitrogen sources, and prevents the production of unnecessary enzymes and the uncontrolled breakdown of compounds that could be otherwise valuable for the cell.

The occurrence of competitive pathways such as biosynthesis and catabolism of an essential nitrogen compound, raises the problem of their balanced reciprocal exclusion under different growth conditions. Either the catabolic pathway involves intermediates entirely different than the biosynthetic route or some metabolites are common to both pathways. In the former case, reciprocal exclusion of the two metabolisms can be exerted by repression and/or inhibition of the biosynthetic route and induction of the degradative pathway. In the latter case, intermediate metabolites common to both anabolic and catabolic pathways, would become involved in a futile cycle detrimental to cellular economy. A selective pressure is thus created for the emergence of regulatory mechanisms that prevent the recycling of intermediates. In addition, more than one catabolic pathway may exist for a given compound.
The derivation of the metabolite drift into different catabolic pathways has often been recognized as a dilemna with respect to the control of enzymes. The establishment of hierarchies of signals and highly controlled systems enable the cells to coordinate the various segments of metabolism.

A multiplicity of examples illustrating the diversity of metabolic signals than can serve to mediate control functions have emerged from studies on the arginine metabolic systems in different bacteria. The existence of different options for catabolism of arginine seems to reflect significant phylogenic divergences and suggests an overall picture of pathway evolution in which modern organisms have adopted various solutions in response to their environmental constraints.

Several reviews have been dealing with aspects of arginine metabolism in procaryotes (2,3,4,5,6,7,8). This paper is restricted to a few selected topics from the larger field of arginine metabolism. The choice from available information is partial and dictated by the preference of the author.

ARGININE BIOSYNTHESIS

Prototrophic bacteria are able to make their own arginine from ammonia as a sole nitrogen source. This involves the transfer of ammonia into glutamate, glutamine and aspartate. Arginine biosynthesis commences with acetylation of glutamate which is converted in two steps to acetylglutamic semialdehyde. The second atom of nitrogen enters the pathway in

the form of glutamate through the action of acetylornithine
δ transaminase which catalyzes the transfer of the α amino
group of glutamate to glutamic semialdehyde to form acetylornithine. In Enterobacteria and *Bacillus* species, N-acetylornithine is hydrolyzed by acetylornithinase(Enzyme 5,Fig.1)

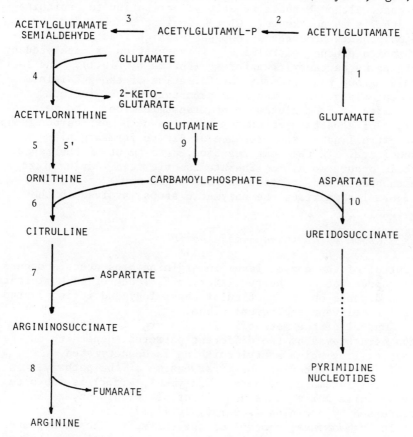

*Fig.1. Numbers 1 to 10 designate enzymes. 1, N-acetylgluta-
mate synthetase (EN 2.3.1.1); 2, N-acetylglutamate kinase
(EN 2.7.28); 3, N-acetylglutamylphosphate reductase (EN 2.1.
38); 4, N-acetylornithine transaminase (EN 2.6.1.11); 5,
acetylornithinase (EN 3.5.1.16); 5', ornithine acetyltransferase (EN 2.3.1.25); 6, ornithine carbamoyltransferase
(EN 2.1.3.3); 7, argininosuccinate synthetase (EN 6.3.4.5);
8, argininosuccinase (EN 4.3.2.1); 9, carbamoylphosphate
synthetase (EN 2.7.2.29); 10, aspartate carbamoyltransferase
(EN 2.1.3.2).*

to yield ornithine and free acetic acid whereas, in other microorganisms, N-acetylornithine donates its acetyl group to glutamic acid to yield free ornithine and acetylglutamate, a reaction catalyzed by an ornithine acetyltransferase (Enzyme 5', Fig.1). In *Pseudomonas*, both activities are present although their relative contribution to ornithine synthesis is not known. Glutamine is the donor for the biosynthesis of carbamoylphosphate which reacts with ornithine to produce citrulline. This reaction is catalyzed by ornithine carbamoyltransferase (Enzyme 6, Fig.1). The last amino group required for the formation of the guanidino group of arginine comes from aspartate which, in turn, acquires it from glutamate by transamination. The amino group of aspartate condenses with the carbamoyl-carbon group of citrulline in the presence of ATP to form argininosuccinate. In the next reaction, detachment of the aspartate chain occurs so as to liberate fumaric acid leaving nitrogen attached to the acceptor to form arginine.
Figure 1 summarizes the enzymatic steps involved in the pathway.

Regulation of Arginine Biosynthesis

Control of the enzyme level in arginine biosynthesis occurs by repression but the regulatory efficiency of each microbial type depends on its particular physiology and it is adapted to its relevant ecological niche.

Control of the metabolic flow in arginine biosynthesis can be achieved by two different patterns, depending on the mechanism by which acetylornithine is deacetylated.

In Enterobacteria, the first enzyme of the pathway, N-acetylglutamate synthetase (Enzyme 1, Fig.1) is inhibited by arginine while the synthesis of all eight enzymes is repressed by arginine (see ref. 10,11,12).

In *Pseudomonas*, control of the enzyme synthesis appears to be less extensive than in *Escherichia coli*. Only ornithine carbamoyltransferase (Enzyme 6, Fig.1) is significantly repressed by arginine (9,10,11). In *P. aeruginosa*, the main control exerted by end-products acts at the level of enzyme activity, the first and the second enzymes of the pathway being subjected to feedback inhibition (12,13). N-acetylglutamate synthetase (Enzyme 1, Fig.1) is synergistically inhibited by arginine, polyamines, and the reaction product N-acetylglutamate, whereas acetylglutamokinase (Enzyme 2, Fig.1) is inhibited by arginine as found for all microorganisms possessing an ornithine transacetylase (Enzyme 5', Fig.1) (14,15). Bacteria which possess an ornithine transacetylase synthesize arginine with less expenditure of energy.

They require enzyme 1 only to start the biosynthetic sequence since thereafter, N-acetylglutamate is produced by the transacetylase reaction.

Arginine biosynthesis encounters special problems because carbamoylphosphate, an intermediate of arginine biosynthesis, participates also in pyrimidine synthesis. This is a branch point in nitrogen metabolism and is also subject to regulation. The enteric carbamoylphosphate synthetase (Enzyme 9, Fig.1) is subject to cumulative repression by arginine and uracil, and feedback inhibited by UMP (16,17). In order to avoid depletion of arginine when there is an excess of pyrimidines, additional controls are required. Ornithine which would accumulate in absence of carbamoylphosphate antagonizes the UMP effect whereas inhibition of carbamoylphosphate utilization for pyrimidine synthesis results from the inhibition by CTP of aspartate carbamoyltransferase (Enzyme 10, Fig.1), the second step in pyrimidine biosynthesis (18). Carbamoylphosphate will thus be preferentially used by ornithine carbamoyltransferase (Enzyme 6, Fig.1) for the synthesis of citrulline. On the other hand, in the presence of an excess of arginine, the synthesis of ornithine is blocked, this allows the utilization of carbamoylphosphate for pyrimidine synthesis.

Carbamoylphosphate synthetase from *P. aeruginosa* is similar in many respects to the enzymes from enteric bacteria. However, the enzyme from *Pseudomonas* is also activated by N-acetylornithine. This signal was hypothesized to serve as an index of arginine limitation under conditions where ornithine is catabolized by the organism (19).

In another bacterial species, *Bacillus subtilis*, control of carbamoylphosphate synthesis has evolved to a quite different pattern. Ornithine is an intermediate in both the biosynthesis and catabolism of arginine and can thus not be used to control carbamoylphosphate synthesis. *Bacillus* solves this problem by elaborating two separate carbamoylphosphate synthetases, one involved in pyrimidine synthesis while the other belongs to the arginine biosynthetic pathway. The pyrimidine-specific carbamoylphosphate synthetase is repressed by uracil and inhibited by UMP while the other is repressed by arginine (20). Inhibition of ornithine carbamoyltransferase by arginase also allows the utilization of carbamoylphosphate for pyrimidine synthesis (21).

CATABOLISM OF ARGININE

By contrast with the remarkable constancy of the arginine biosynthetic pathway, various strategies are used by bacteria to utilize this nitrogen-rich compound.

The Arginase Pathway generally used by lower eucaryotes for arginine catabolism is restricted to a relatively few bacterial species such as *Bacillus* (22,23,24). *Agrobacterium*, *Rhizobium* (25), and among *Cyanobacteria*, *Aphanocapsa* (26) or *Anabaena variabilis* (27). It was claimed that the genus *Proteus* also possesses arginase activity (28). However, the enzyme was not characterized nor the pathway enzyme system. Moreover, we were unable to find any arginase activity in *Proteus vulgaris* of our collection. More data are required to assess the existence of arginase in this bacterium.

The first step involves the hydrolytic cleavage of arginine to ornithine and urea mediated by arginase (Enzyme 1 Fig.2). The δ nitrogen of ornithine is then transferred to 2-ketoglutarate by ornithine transaminase (Enzyme 2, Fig.2). The glutamate semialdehyde generated during this reaction is oxidized to glutamate by pyrroline-5-carboxylate dehydrogenase (Enzyme 3, Fig.2). In most *Bacillus* strains, urea is not metabolized and is excreted. In *Agrobacterium*, urea is

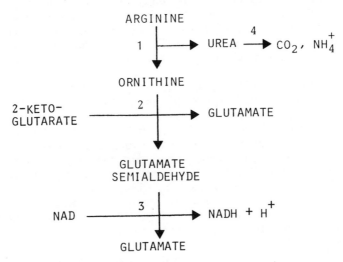

Fig.2. *The arginase pathway. 1, arginase (EN 3.5.3.1); 2, ornithine aminotransferase (EN 2.6.1.13); 3, Δ pyrroline-5-carboxylate dehydrogenase (EN 1.5.1.12); 4, urease (EN 3.5.1.5)*

is metabolized by urease to yield CO_2 and two ammonia (Enzyme 4, Fig.2).

Such a pathway would lead to the generation of a futile cycle if ornithine resulting from the cleavage of arginine by arginase is re-utilized by ornithine carbamoyltransferase involved in the biosynthesis of arginine. In *Agrobacterium*, ornithine carbamoyltransferase is indeed produced constitutively. This futile cycle is however abolished by specific inhibition of ornithine carbamoyltransferase by arginine (25). In *Bacillus subtilis*, the biosynthetic enzymes are repressed by arginine whereas the arginase pathway is induced. However, biosynthetic arginine precursors, ornithine and citrulline, have different effects on the two pathways in that they induce the catabolic enzymes but do not repress those of the biosynthesis (23). Futile cycle between ornithine and arginine is then prevented by specific interaction between ornithine carbamoyltransferase and arginase which results in the inhibition of ornithine carbamoyltransferase, arginase being unaffected (21).

Whereas *Bacillus subtilis* employs the arginase pathway to dissimilate arginine, some other *Bacilli* have been found to possess also a divergent array of enzyme sequences to catalyze arginine breakdown (29). The best studied organism is *Bacillus licheniformis* which has two pathways, the arginase and the arginine deiminase routes. *B. licheniformis* is a facultative aerobe rather than a stric aerobe as *B. subtilis* (24). The arginase pathway is only aerobic in *B. licheniformis* since it ceases to function under anaerobic conditions whereas the arginine deiminase route is used only under anaerobic conditions and does not operate under aerobic ones. In aerobic cultures grown on arginine as only carbon and nitrogen source, disappearance of arginine is accompanied by a stoechiometric formation of urea in the culture fluid thus indicating that all the arginine is converted by arginase.

One might expect that the inducible enzymes of the arginase pathway are subject to catabolite repression mediated by a signal of availability of both carbon and nitrogen sources. Glucose entails a 5-fold reduction of the arginase level and a 9-fold reduction of the transaminase activity. Neither ammonia nor glutamate seems to antagonize the induction of these enzymes but glutamine does it in the presence of a good carbon/energy source. In this organism, glutamine is a better nitrogen source than ammonia or glutamate since the generation time is increased from 50 min for the former case to 60 min the the latter.

In *B. subtilis*, the arginase pathway is not significantly affected by carbon catabolite repression, but is markedly repressed by various nitrogen sources, glutamine strongly repressing arginase induction by arginine (30). However, in other *Bacillus* species, nitrogen catabolic repression does not appear to be involved in the regulation of arginase synthesis (31).

The Arginine Deiminase Pathway. Arginine utilization by the arginine deiminase pathway is shown in Fig.3. Hydrolysis of the guanidino group of arginine is catalyzed by arginine deiminase giving citrulline and ammonia (Enzyme 1, Fig.3).

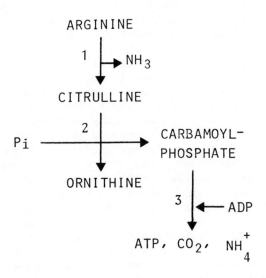

Fig.3. The arginine deiminase pathway. 1, arginine deiminase (EN 3.5.3.6); 2, ornithine carbamoyltransferase (EN 2.1.3.3); 3, carbamate kinase (EN 2.7.2.2)

The phosphorolysis of citrulline through the action of catabolic ornithine carbamoyltransferase (Enzyme 2, Fig.3) produces carbamoylphosphate and ornithine. In most bacteria having this pathway, conversion of carbamoylphosphate to carbon dioxide and ammonia is stoechiometrically coupled to the formation of ATP from ADP, catalyzed by carbamate kinase (Enzyme 3, Fig.3). In *Lactobacillus leichmanii*, carbamoyl-phosphate also donates its carbamoyl group to aspartate initiating the pyrimidine synthesis (32). In *Streptococcus faecium* lacking carbamate kinase, it has been shown by Pendey (33) that carbamoylphosphate transfers its phosphate group to glucose thereby initiating the first step of glycolysis.

Streptococcus faecalis ATCC 11700 is dependent on complex media for growth in addition to the energy source. Arginine can serve as an energy source since it no longer sustains growth in mutants affected in the arginine deiminase route (34). In the wild-type strain, ornithine and ammonia produced during arginine fermentation are found to be excreted thus indicating that only a fraction of the arginine carbon skeleton is used. The enzymes of the pathway are induced by arginine and are repressed by glucose, fumarate and oxalurate. Repression by glucose is increased in the presence of oxygen and haematin . Addition of haematin to the medium enables *S. faecalis* to synthesize porphyrin ; under these conditions, the organism develops a rudimentary electron transport system involving one or two cytochromes and becomes capable of oxidative phosphorylation. In other words, growth conditions which promote the energy status of the cell leads to reduction of the level of arginine deiminase pathway enzymes. Assuming that only one co-repressor is involved in repression, ATP is the only common product of the aerobic metabolism, glycolytic pathway, fumarate reduction and oxalurate utilization. It has thus been suggested that ATP or one of its derivatives, may control the formation of the arginine deiminase pathway (34).

Bacillus licheniformis. Of the two pathways for arginine breakdown, one, the arginase route, is expressed only in the presence of an electron acceptor, oxygen or nitrate, whereas the other, the arginine deiminase pathway, is implicated under anaerobic conditions (24). In anaerobiosis, no growth on arginine as the carbon, energy, and nitrogen source occurs, unless a suitable carbon source is added. When used as nitrogen source, arginine consumption is accompanied by the appearance of ornithine and ammonia in the growth medium. However, if arginine is growth-limiting, ammonia which accumulates during the early stage of growth, disappears as arginine concentration vanishes, but ornithine remains in the medium and is not used further. The presence of arginine is required for the induction of the arginine deiminase pathway whereas glucose causes repression of the induced enzymes. In the presence of a good energy source such as glucose, ammonia acts as a repressor but shows little effect when pyruvate is used as a carbon source instead of glucose. By contrast, glutamine, a better nitrogen source than ammonia for the aerobic *Bacillus* strain, causes, in anaerobiosis, a spectacular derepression of the deiminase pathway in the presence of glucose. In this respect, it should be noted that in anaerobiosis, glutamine impairs the utilization of glucose since the generation time increases two-fold when glutamine replaces ammonia as nitrogen source

(24). Besides degrading arginine as nitrogen source, the deiminase pathway fulfils another function. Indeed, no growth occurs on pyruvate and ammonia, but good growth is observed on pyruvate and arginine. In addition, a mutant displaying reduced carbamate kinase activity is unable to grow on this medium (Broman, Ph.D. Thesis, Brussels, 1978). These observations indicate that, in addition ot its nitrogen providing function, arginine can also serve as an energy source.

The expression of the arginine deiminase pathway in *B. licheniformis* seems controlled by at least three different circuits, one of them is responsible for turning the genes of the pathway on or off in the presence or in the absence of the inducer. The synthesis of the pathway is also governed by repression exerted by oxygen, nitrate and glucose. Again, assuming that only one co-repressor is involved in these regulations, ATP or one of its derivatives is the only common product to the oxygen and nitrate respiration and to the glycolytic pathway. A third circuit involves catabolite repression exerted by ammonia in the presence of a good carbon source.

Aeromonas formicans NCIB 9232. Catabolism of arginine in this bacterium can also proceed by different pathways (see next section) (35,36). As in *Bacillus licheniformis*, the deiminase pathway is induced by arginine and repressed by oxygen and/or glucose ; cAMP, a signal of energy deficit, overcomes these repressions. Arginine consumption is accompanied by ornithine excretion.

Pseudomonas. Derepression of the arginine deiminase pathway in Pseudomonads of the fluorescent group as well as in *P. mendocina*, occurs under particular growth conditions such as transition from high to low oxygen tension or depletion of carbon or phosphate source (36,37,38). Arginine is not required for induction although in *P. aeruginosa*, it boosts the induction process by a factor two (38). Oxygen represses enzyme induction whereas, in strains having an anaerobic mode of respiration, nitrate and nitrite attenuate the effect of enzyme induction (38). One enzyme of the pathway, the catabolic ornithine carbamoyltransferase is the target of activity regulation. The phosphorolytic cleavage of citrulline is inhibited by nucleotides triphosphate, a high energy level signal, and stimulated by nucleotides monophosphate, a signal of energy deficit (Legrain & Stalon, unpublished data).

An interesting point is that *P. aeruginosa*, classified as a strict aerobe, however is able to grow in anaerobiosis in the absence of electron acceptor, provided that arginine is supplied in the medium containing biosynthetic precursors.

Mutants that do not utilize arginine under anaerobic conditions have been isolated and were found to be mutationally affected in the arginine deiminase pathway enzymes (39). However, the physiology of *P. putida, P. fluorescens* or *P. mendocina* appear quite distinct from *P. aeruginosa*. Although having the arginine deiminase pathway, these bacteria cannot grow in anaerobiosis with arginine as an energy source.

The Two Ornithine Carbamoyltransferases of Pseudomonas and Properties of Homologous Enzymes. In *Pseudomonas* having the arginine deiminase pathway, biosynthetic and catabolic pathways may operate under mild conditions of aeration. Ornithine and citrulline being intermediates in both routes, a debiliting cycle could result in the absence of appropriate control. Citrulline biosynthesis and degradation are catalyzed by different, noticeably specialized ornithine carbamoyltransferases whose biosynthetic enzyme cannot function in the catabolic direction and vice-versa. The irreversibility properties of the biosynthetic enzyme result from its low affinity with citrulline as a substrate and from the inhibition of the enzyme by citrulline through the formation of a dead-end complex which strongly reduces the velocity of the reaction (40). In *Pseudomonas* of the fluorescent group, this enzyme has evolved to fulfil a biosynthetic rôle only. Inability of the catabolic ornithine carbamoyltransferase to catalyze the carbamoylation of ornithine *in vivo* is explained by the extreme cooperativity of carbamoylphosphate binding which rules out the use of this enzyme for biosynthetic purpose (41,42). That the catabolic enzyme cannot function biosynthetically *in vivo* is shown by the fact that a mutant lacking the biosynthetic activity is an arginine auxotroph (41,43). In both *P. putida* and *P. aeruginosa,* isolated arginine revertants were found to possess an altered catabolic ornithine carbamoyltransferase, with reduction or absence of cooperativity towards carbamoylphosphate allowing the mutant enzyme to fulfil an anabolic function *in vivo*. It is not unreasonable to see in this selection of mutants producing altered enzymes, a reversal of the selection process that had occurred in nature. These observations suggest that during evolution, duplication of an ornithine carbamoyltransferase gene may have occurred and that natural selection has resulted in these two distinct enzymes appropriately fitted for their metabolic rôle.

Conversely, it is possible to reverse the metabolic rôle of ornithine carbamoyltransferase fulfilling an anabolic function. *E. coli* K-12 lacking carbamoylphosphate synthetase (Enzyme 9, Fig.1) requires both uracil and arginine for growth. Cells of this strain plated on medium containing citrulline as a sole source of carbamoylphosphate could synthesize arginine by the final step of the arginine biosynthetic pathway but the only possibility they have to produce carbamoylphosphate for the synthesis of uracil would be the phosphorolysis of citrulline. Mutants having acquired this possibility were selected and were found to have very high level of ornithine carbamoylphosphate activity (44). These mutants are either constitutive, are a result of multiplication of ornithine carbamoyltransferase structural gene or a defect in argininosuccinate synthetase (Enzyme 7, Fig.1). In this latter condition, the conversion of citrulline is slow and a threshold repressing arginine concentration could not be reached *in vivo*.

In *E. coli*, they are two genes, *argI* and *argF*, encoding very similar polypeptides and since the enzyme is a trimer, a strain carrying both genes produces four isoenzymes made up of the two different subunits (45). Regulatory mutations allowing citrulline utilization as source of carbamoylphosphate may affect both genes (44). Duplication of gene followed by differentiation and specialization of one of the copies is often recognized as an effective way in evolutionary theories and this may explain the mechanism by which *Pseudomonas* of the fluorescent group have acquired the two specialized carbamoyltransferases.

The Pathway Initiated by Arginine Decarboxylase. Putrescine, a precursor of the polyamines, spermidine and spermine, is synthesized by two distinct routes. Ornithine, an intermediary metabolite of the arginine biosynthetic pathway, may be directly converted into putrescine by a decarboxylase (46). However, in arginine prototrophic strains, when arginine is added to the medium, ornithine formation is blocked by feedback inhibition of the first enzyme of the pathway. In those conditions, the conversion of arginine to putrescine remains the main route of polyamines synthesis. Two main strategies are used by bacteria to produce putrescine from arginine. In Enterobacteria, an arginine decarboxylase (Enzyme 1, Fig.4) converts arginine into carbon dioxide and agmatine which, in turn, is split into putrescine and urea mediated by agmatinase (46,47). (Enzyme 2, Fig.4). In *Pseudomonas* and *Aeromonas formicans*, agmatine produced by decarboxylation of arginine is deiminated into ammonia and N-carbamoylputrescine by agmatine deiminase (Enzyme 3, Fig.4). N-carbamoylputrescine, under the action of an hydrolase (Enzyme 4, Fig.4) produces carbon dioxide, ammonia and putrescine (35,36,37).

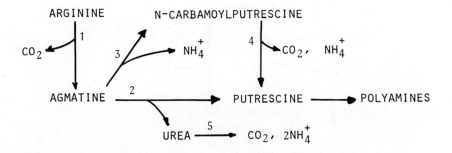

Fig.4. The arginine decarboxylase pathway. 1, arginine decarboxylase (EN 4.1.1.19); 2, agmatinase (EN 3.5.1.11); 3, agmatine deiminase (EN 3.5.3.11); 4, N-carbamoylputrescine hydrolase (EN 3.5.3.-); 4, urease (EN 3.5.1.5)

Lacking arginine decarboxylase activity, *Streptococcus faecalis* ATCC 11700 uses however agmatine as its sole energy source for growth (49,50). The pathway of agmatine utilization is analogous to the arginine deiminase route found in the same organism. N-carbamoylputrescine, produced by agmatine deiminase, is phosphorolyzed to give putrescine and carbamoylphosphate mediated by a putrescine carbamoyltransferase (49,51) which can react with ADP to yield ATP. Agmatine consumption by *Streptococcus* cells was characterized by excretion of putrescine which does not seem to be used further.

In other bacterial systems, putrescine can be diverted into polyamines or used as carbon and/or nitrogen source. For the different genus investigated, catabolism of putrescine is similar (Fig.5). Putrescine is first converted into 4-aminobutyraldehyde either by transamination of by oxidation (52,53)(Enzymes 1 and 1', Fig.5). 4-aminobutyraldehyde is oxidized to 4-aminobutyrate by a dehydrogenase (Enzyme 2, Fig.5), then a transaminase is responsible for the conversion of 4-aminobutyrate to succinate semialdehyde. A single transaminase seems to catalyze the transamination of putrescine and 4-aminobutyrate in *Pseudomonas* (53,54). The last step of the pathway is the oxidation of succinic semialdehyde to succinate catalyzed by a succinic semialdehyde dehydrogenase (Enzyme 4, Fig.5) (55).

In *P. stutzeri* or *P. pseudoalcaligenes* using arginine only as a nitrogen source, the arginine decarboxylase pathway may be a major route for the entry of ammonia in metabolism since their growth in the presence of arginine elicits both arginine decarboxylase and agmatine deiminase

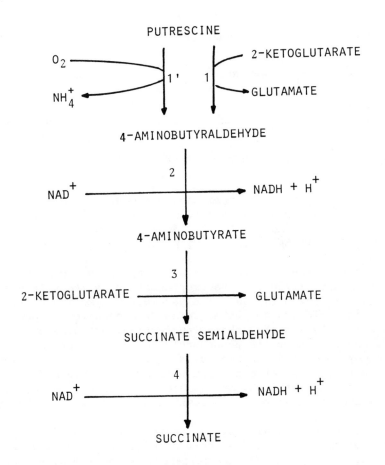

Fig.5. Pathway of putrescine degradation. 1, putrescine oxidase (EN 1.4.3.10); 1', putrescine transaminase (EN 2.6.1.29); 2, 4-aminobutyraldehyde dehydrogenase (EN 1.2.1.19); 3, 4-aminobutyrate aminotransferase (EN 2.6.1.19); 4, succinate semialdehyde dehydrogenase (EN 1.2.1.24).

pathways (36). Arginine decarboxylase appears to have little importance in P. aeruginosa where several pathways for arginine catabolism co-exist (see next section). Arginine elicits induction of arginine decarboxylase only, agmatine deiminase and N-carbamoylputrescine hydrolase being induced by agmatine or carbamoylputrescine and subject to carbon and nitrogen catabolite repression (48). Moreover, mutants of Pseudomonas aeruginosa affected in the agmatine deiminase pathway do not grow on agmatine but do so with arginine as a sole carbon and nitrogen source, thus indicating that only

a minor fraction of arginine is converted into agmatine.
The agmatinase of *Klebsiella aerogenes* is induced by
agmatine or arginine (57). This latter compound does not
induce the subsequent enzymes of putrescine utilization as
agmatine does it. Arginine utilization by the combined
action of arginine decarboxylase and agmatinase is likely
to fulfil the polyamine requirement when arginine is abundant.
In *E. coli*, the regulation of putrescine synthesis is
fairly complex. This bacterium exhibits two distinct
arginine decarboxylases (47). One form is induced by acid
culture : the physiological rôle of the resulting production
of amines is probably to buffer the bacterium against the
poisoning effects of acidic medium (47). The second form is
produced at neutral pH. It is synthesized by arginine,
negatively regulated by cAMP ; putrescine inhibits this
enzyme and represses its formation (47,48). *E. coli*
possesses an agmatinase which is induced by arginine or
agmatine and negatively controlled by cAMP ; agmatine acts
as an antagonist to cAMP (59). In *E. coli*, the arginine
decarboxylase route is the only pathway for arginine
dissimilation, as this organism produces no urease (60),
putrescine and/or 4-aminobutyrate transaminase are responsible for the entry of ammonia in the metabolism (61).

Arginine Catabolism via 2-Ketoarginine Formation. P. putida
and several other *Pseudomonas* species also use a catabolic
pathway by which arginine is first converted to 2-keto-
arginine by an arginine oxidase (Enzyme 1, Fig.6).
2-ketoarginine is decarboxylated by the action of a 2-keto-
arginine decarboxylase (Enzyme 2, Fig.6) giving 4-guanidino-
butyraldehyde which is converted to 4-guanidinobutyrate
mediated by the guanidinobutyraldehyde oxidoreductase
(Enzyme 3, Fig.6). Guanidinobutyrate is split into urea and
4-aminobutyrate by guanidinobutyrase (Enzyme 4, Fig.6).
Since no or low levels of urease appear by growing the cells
in the presence of arginine, 2-ketoarginine or guanidinobutyrate, urea   is accumulated in the medium (36,62).
4-aminobutyrate is converted into succinate ; this part of
the catabolism is common to the final step of putrescine
catabolism. 2-ketoarginine utilization is widely distributed
among *Pseudomonas* species (36) and it is found that their
guanidinobutyrase is induced by growing cells, either on
2-ketoarginine or guanidinobutyrate. Although *P.aeruginosa*,
P. fluorescens ATCC 13525 and *P. cepacia* have the entire
enzyme sequence enabling them to utilize 2-ketoarginine,
these organisms do not appear to use this pathway for
arginine dissimilation since their guanidinobutyrase is not
induced by arginine.

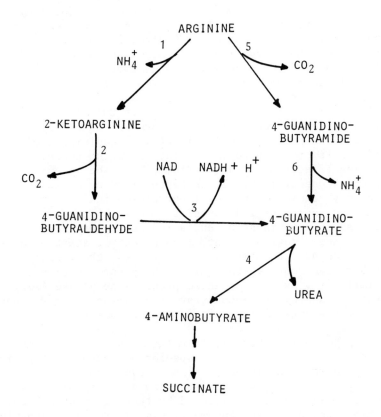

Fig.6. Arginine oxidase and arginine decarboxyoxidase pathways. 1, arginine oxidase (EN 1.4.3.-); 2, 2-ketoarginine decarboxylase (EN 4.1.1.-); 3, 4-guanidinobutyraldehyde oxidoreductase (EN 1.2.1.-); 4, guanidinobutyrase (EN 3.5.3.7); 5, arginine decarboxyoxidase (EN 1.13.12.1); 6, guanidinobutyramidase (EN 3.5.1.-).

Mutants of *P. putida* impaired in guanidinobutyrate catabolism do not grow on 2-ketoarginine nor on guanidinobutyrate but retain the ability to grow on arginine as only carbon and nitrogen source (Legrain & Stalon, unpublished data). By contrast, some mutants of *Pseudomonas aeruginosa* which were isolated as having lost the capacity to use arginine as a carbon source, but using it as a nitrogen source, have their guanidinobutyrase inducible by arginine (Vander Wauven, Ph.D. Thesis, Brussels, 1984). In these mutants, urea is accumulated in the medium in proportion to the amount of arginine utilized. This new metabolic activity may result from the reactivation of a silent gene, from a regulatory mutation conferring inducibility to a pre-existing structural gene, or from recruitment of a novel enzyme such as a transaminase known to have width of substrate specificity.

Apart from the genus *Pseudomonas*, arginine oxidase is also found in the cyanobacterium *Anacystis nidulans* (65). In this organism, the enzyme, a flavo-protein, catalyses both O_2 evolution in the photosystem II and also oxidization of arginine in the dark, playing a rôle similar to that of a dehydrogenase in aerobic organisms (66).

In *Streptomyces griseus*, Thoai found another arginine catabolic pathway which resembles the 2-ketoarginine pathway (64). The first step of this pathway is catalyzed by an arginine decarboxyoxidase producing 4-aminobutyramide which is hydrolyzed in 4-guanidinobutyrate. Catabolism of 4-guanidinobutyrate to succinate is similar to that found in *Pseudomonas* strains.

The Pathway Initiated by Arginine Succinyltransferase.
Although *P. cepacia* can efficiently use arginine, ornithine, citrulline, agmatine and 2-ketoarginine for growth, its arginine-grown cell extracts contain no arginase, arginine oxidase, decarboxylase nor deiminase actitivities. Arginine degradation in *P. cepacia* uses a novel arginine degradative pathway (Fig.7).

Degradation is initiated by an arginine succinyltransferase (Enzyme 1, Fig.7) which succinylates the α carbon bound nitrogen atom of arginine. Succinylarginine is then converted to succinylcitrulline by a succinylarginine deiminase (Enzyme 2, Fig.7) ; succinylcitrulline is hydrolyzed by an hydrolase giving succinylornithine. Succinylornithine gives its δ nitrogen atom by transamination to 2-ketoglutarate to form glutamate and succinylglutamate semialdehyde. This reaction is catalyzed by a succinylornithine transaminase (Enzyme 4, Fig.7). The aldehyde product of the transaminase reaction is oxidized to succinylglutamate which is split into succinate and glutamate by a succinylglutamate desuccinylase (Enzyme 6, Fig.7).

Induction of the arginine succinyltransferase pathway in *P. cepacia* is controlled by carbon and nitrogen repression exerted respectively by succinate and glutamine (Table 1). The distribution of this novel pathway among the bacterial kingdom has not yet been studied in depth. However, among bacterial species known to use arginine as only carbon and nitrogen source, its presence is attested in *Pseudomonas* of the fluorescent group, namely *P. aeruginosa* PAO,

P. fluorescens ATCC 13525, *P. putida* IRC 204 and *P. mendocina* NCIB 10541, and in the enterobacterial species *Aeromonas formicans* NCIB 9232 and *Klebsiella aerogenes* MK53 (Table 1 and unpublished results).

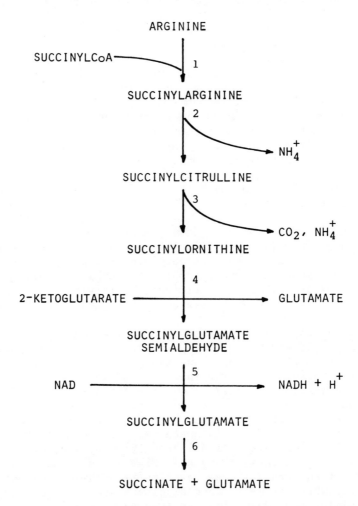

Fig.7. Arginine succinyltransferase pathway. 1, arginine succinyltransferase (EN 2.3.1.-); 2, succinylarginine deiminase (EN 3.5.3.-); 3, succinylcitrulline hydrolase (EN 3.5.3.-); 4, succinylornithine transaminase (EN 2.6.1.-); 5, succinylglutamate semialdehyde dehydrogenase (EN 1.2.1.-); 6, succinylglutamate desuccinylase (EN 3.5.1.-).

TABLE 1

Formation of the arginine succinyltransferase pathway in P. cepacia NCTC 10743

Growth Medium [a]	Arginine succinyl-transferase (sp.act.)	Deiminase + Hydrolase (sp.act.)	Succinyl-ornithine transaminase (sp.act.)	Succinyl-glutamate desuccinylase (sp.act.)
Succinate + ammonia	0.4	0.06	1.6	0.6
Arginine	22.0	26.0	64.0	19.5
Arginine + succinate	6.0	4.0	10.5	3.4
Arginine + ammonia	22.0	21.8	50.1	17.3
Arginine + glutamine	4.0	2.5	6.2	n.t.
Arginine + succinate + glutamine	1.0	2.2	6.2	n.t.

[a] Mineral medium 154 (9) was supplemented with carbon and nitrogen sources at a concentration of 20 mM each.
The enzyme specific activity is expressed as µmol of product formed (mg protein)$^{-1}$ h^{-1}.

Pseudomonas aeruginosa or *Pseudomonas putida* mutants blocked in enzyme activities of the arginine succinyltransferase pathway are also impaired in the utilization of arginine as a carbon source. For all bacterial systems reported in Table 2, it was observed that the high level of succinylornithine aminotransferase found in cells grown on arginine is always associated with the appearance of acetylornithine aminotransferase activity. The level of these activities are the same whatever the growth conditions are. However, the affinity of the transaminase for succinylornithine is always higher or equal to that observed for acetylornithine. It is likely that these two activities are carried out by the same molecule. This is supported by two facts. First, a partially purified preparation of acetylornithine transaminase of *Pseudomonas aeruginosa* used both acetyl- and succinylornithine as substrate (Haas, pers. com.) Second, mutants of *P. aeruginosa* or *P. putida* that have lost the inducible succinylornithine transaminase are also affected in the inducible acetylornithine transaminase activity (unpublished results). In previous works (67,68,69) it was assumed that inducible acetulornithine aminotransferase plays a major rôle in arginine dissimilation mainly by virtue of its ornithine aminotransferase activity. The data reported here indicate that succinylornithine is the actual intermediate in arginine catabolism via succinylarginine and succinylornithine formation.

There are also reasons to query the proposition of Friedrich and Magasanik (70) on the existence of an arginine transamidinase in *Klebsiella aerogenes*. Our results suggest that the induced catabolic acetylornithine transaminase in this strain is the same enzyme molecule as the succinylornithine transaminase involved in the arginine succinyltransferase pathway. The careful analysis of Friedrich *et al.* (69) on the formation of this enzymic activity, shows that this acetylornithine transaminase is subject to catabolite repression by glucose in the presence of ammonia. However, when the cells are deprived of ammonia, the enzyme escapes catabolic repression, thus indicating that this pathway plays a specific rôle in arginine utilization as carbon and nitrogen source.

TABLE 2

Formation of the Succinylarginine Transferase Pathway in Bacteria

Strains[a]	Growth Medium	Arginine succinyl-transferase (sp.act.)	Deiminase + Hydrolase (sp.act.)	Succinyl-ornithine transaminase (sp.act.)	Succinyl-glutamate desuccinylase (sp.act.)
K.aerogenes MK53	Glucose + ammonia Arginine	1.6 12.0	0.1 9.1	1.2 64.0	0.8 10.4
A. formicans NCIB 9232	Glucose + ammonia Arginine	1.0 4.0	2.2 12.5	2.3 15.0	2.5 11.0
P.aeruginosa PAO	Succinate + ammonia Arginine	< 1.0 28.0	0.2 27.0	1.5 41.0	0.9 12.4
P. putida IRC 204	Succinate + ammonia Arginine	< 0.2 13.0	0.9 16.7	1.0 3.4	2.4 19.0
P. fluorescens ATCC 13525	Succinate + ammonia Arginine	0.6 40.0	0.2 2.6	1.8 18.0	2.4 19.6
P. mendocina NCIB 1051	Succinate + ammonia Arginine	< 0.5 7.5	0.7 28.0	1.0 18.0	0.7 7.4

[a]Bacteria were grown in mineral medium 154 (9) supplemented with carbon and nitrogen sources at a concentration of 20 mM each.
Specific activities are recorded as µmol product formed (mg protein)$^{-1}$ h^{-1}.

DISTRIBUTION OF THE ARGININE CATABOLIC PATHWAYS IN THE BACTERIAL KINGDOM

In order to understand the evolution of arginine catabolic pathways, it is of interest to know what are the pylogenetic relationships in the bacterial world. By determining the degree of homology between 16 S ribosomial RNA sequences, Fox et al. (71) classified the existing bacteria into two major kingdoms : the archaebacteria and the eubacteria (true bacteria). The archaebacterial line includes a group of organisms that occupy very specialized niches such as methanogenes, extreme halophiles such as *Halobacterium salinarium*, and thermoacidophiles. The eubacterial line contains most of the recognized bacterial species and is divided into eight groups : one containing the enterobacteria, *Pseudomonas, Agrobacterium* and *Rhizobium* ; another including all Gram-positive bacteria, namely *Clostridium, Bacillus, Lactobacilli, Streptococcus, Staphylococcus* which also contains *Mycoplasma* and *Spiroplasma*. A third group includes the actinomycetes, a fourth group concerns the genus *Spirocheta* ; a fifth is related to the genus *Cyanobacterium*. The other groups are irrelevant to the topic described here.

The arginine deiminase pathway exists in taxonomic groups that differ markedly in phenotypic respects and in mean DNA base composition. The possession of the deiminase route is undoubtly ancient as attested by its wide distribution among the different bacterial groups. It is also present in the archaebacterial line and in particular in *Halobacteria* (72, 73). The arginine deiminase route appears in four of the eight groups defined by Fox et al. (i) *Pseudomonas* and *Aeromonas*, (ii) the Gram-positive bacteria, *Lactobacillus, Streptococcus, Staphylococcus, Clostridium* as well as *Mycoplasma* (see ref.5) and *Spiroplasma* (74) ; (iii) the cyanobacteria (26,27) and (iv) *Spirocheta* group (75).
The wide distribution of the arginine deiminase pathway is amenable to two opposing interpretations. The first one is that the pathway arises only once in bacterial evolution. Its present biological dispersion could have resulted from vertical transmission of the structural genes, into a series of bacterial groups accompanied by possible horizontal intergroup transmission. Alternatively, the pathway might have developed independently into a number of different bacterial ancestors of the various taxonomic groups recognized today. As discussed below, the large diversity of solutions used for arginine degradation would appear to

indicate that chemical factors have played a rôle in a particular environment. It is apparent that the biochemical system of the arginine deiminase pathway of all modern organisms is essentially a mechanism for anaerobic arginine decomposition. In all bacteria so far examined, the arginine deiminase pathway is most efficient when the oxygen supply limits respiration, in some cases anaerobic stress elicits this pathway even in the absence of the inducer. It is obvious that the course of evolution of the procaryotic cells was determined by the problem of generating metabolic energy under anaerobic conditions. Possession of the arginine deiminase pathway may have been important in that respect as attested by conservation of this pathway in bacterial groups capable of anaerobic metabolism. With the exception of P. *aeruginosa*, the *Pseudomonas* of the fluorescent group and *P. mendocina* are unable to grow in anaerobiosis at the expense of arginine as an energy source. These aerobic bacteria may represent forms having lost their anaerobic metabolism during the course of evolution.

Amidinotransferases are largely distributed among bacteria. The enzymatic hydrolysis of the guanidino compounds such as allantoic acid, creatine, guanidinobutyric acid, guanidinopropionic acid, guanidinosuccinic acid, agmatine, leads to the production of urea. But, in bacteria, the hydrolysis of arginine by arginase seems to be restricted to three groups, the aerobic *Bacilli*, the *Agrobacterium-Rhizobium* group, and *Cyanobacteria*. The fate of ornithine produced in these bacteria is different ; it is largely excreted by *Cyanobacteria*, the *Bacilli* transforms it by ornithine transaminase ; in *Agrobacterium* an ornithine cyclase produces proline and ammonia (Legrain, pers. comm.). In this latter case, it may be assumed that the *Agrobacterium - Rhizobium* group has acquired its arginase from their host plant with which they are living in parasitic interaction or in symbiosis.

Up to now, the arginine decarboxylase pathway is mainly a characteristics of one eubacterial group, the Enterobacteria - Pseudomonads. The way by which they degrade agmatine is however specific for each subgroup. It may be noted that arginine decarboxylase is absent in strains having ornithine as an intermediate in arginine catabolism, such as *Bacilli* or *Agrobacterium*. Putrescine and polyamines synthesis in these latter cases may proceed directly via ornithine decarboxylase, a chemical option used by all bacteria able to synthesize their own ornithine from glutamate. Consequently, the arginine decarboxylase pathway appears to be the means by which bacteria may produce putrescine from arginine when ornithine is not available for the cell or when its synthesis is blocked.

Up to the present date, the study of distribution of the arginine succinyltransferase pathway was limited to organisms having the ability to use arginine as only carbon and nitrogen source. This pathway is also a characteristics of the Enterobacteria - Pseudomonads group only. Succinylation of α nitrogen atom of arginine and the subsequent intermediate catabolites prevent the operation on an energy-wasteful cycle involving citrulline or ornithine, biosynthetic precursors of arginine. When these biosynthetic precursors are common to both the arginine anabolic and catabolic pathways, this leads to the emergence of regulatory elements that function to control ornithine carbamoyltransferase activity as illustrated by *Bacillus subtilis, Agrobacterium tumefaciens, Pseudomonas putida*, and *Pseudomonas* of the fluorescent group.

It is of interest to note that among the five distinct RNA homology groups defined by Palleroni (76), only a few *Pseudomonas* of group I have an "arginine oxidase" activity although other clusters have the entire pathway of 2-keto-arginine utilization, or at least segments of it. It is difficult to presume whether the "oxidase pathway" is an ancient pathway lost during evolution by most of the present day Pseudomonads or is a recent acquisition. Without exception, the oxidase route appears in *Pseudomonas* together with other arginine catabolic routes. Characterization of a variety of regulatory mutations affecting these divergent pathways still provide a greater insight into the nature and significance of the multiple flow routes in *Pseudomonas putida*.

Multiple arginine catabolic pathways frequently appear in the same bacterium : arginine succinyltransferase, arginine deiminase and arginine decarboxylase are common to the *Pseudomonas* of the fluorescent group, to *P. mendocina* and to *Aeromonas formicans*. *Bacillus licheniformis* and the Cyanobacterium *Aphanocapsa* use arginase and arginine deiminase while in *Klebsiella aerogenes*, arginine is decarboxylated or succinylated. With the exception of the *P. putida* oxidase pathway, the rôles of the divergent catabolic pathways are however largely independent and therefore there is little redundancy of the pathways.

The arginine deiminase route serves mainly to generate ATP under conditions of energy depletion. The pathway is also of major importance as a nitrogen-providing function for *Bacillus licheniformis* and *Aeromonas formicans* under anaerobic conditions, when arginine is the sole nitrogen source.

The arginine decarboxylase pathway fulfils mainly the polyamine requirement but also provides the nitrogen for several *Pseudomonas* and likely for *E. coli*.

The other pathways initiated either by arginine succinyl-transferase and arginine decarboxyoxidase are responsible for the dissimilation of the arginine carbon skeleton and supply the nitrogen either in the form of ammonia or glutamate.

In some particular strains, other pathways of arginine utilization have been described. In Clostridia family, arginine may be involved in Stickland reaction, that is to say mutual oxidation-reduction between an amino acid which acts as an electron donor undergoing oxidation while arginine acts as an electron acceptor, undergoing reduction. The result is the deamination of both amino acids (77). *Arthrobacter simplex* has an arginine transaminase producing 2-ketoarginine converted into 2-ketoornithine by a 2-keto-arginase (78). *Streptomyces* strains can also convert arginine into streptomycin (79) or the antibiotic 2-nitro-imidazole (80).

CONCLUDING REMARKS

The arginine catabolic pathways in bacteria are very diverse, linked or unrelated to energy production, oxygen-requiring or oxygen-sensitive, widely distributed among organisms or restricted to a single species within one group or strain. The diverse significant enzymological data lead of course to the question of how evolution may have produced these biochemical pathways and to what extent the circumstances of the emergence of the various pathways may be deciphered. Generalization concerning evolution and evolutionary relationships among species must be obtained from the comparison of the information now available, an understanding of the catabolism of arginine is only superficial, particularly with respect to its biochemical constitution and function, to the regulatory controls that maintain a balanced metabolic flux through the various segments of metabolism, to the structure-function relationships of the individual enzymes and to the diversity displayed by the enzymes from different organisms, to the distribution of the various metabolic options among bacteria. These are just some aspects of much that remains unknown.

ACKNOWLEDGEMENTS

I am grateful to Nicolas Glansdorff, François Hilger and André Piérard for their criticisms on the original manuscript. I would like to thank Paulette Wilquet for her expert secretarial help without which this review could not have been written.
Original research was supported by grant nr.2.4529.79 from the Belgian Fund for Fundamental Research.
V.S. is Research Associate at the National Fund for Scientific Research.

REFERENCES

1. Oparin, A.I. (1953) "Origin of Life". Dover Publications, Inc., New York.
2. Vogel, R.H., McLellan, W.L., Horvonen, A.P. and Vogel, H.J. (1971). The arginine biosynthetic system and its regulation. In "Metabolic Regulation" (Eds. H.J. Vogel) Vol. 5, pp. 463-488. Academic Press, New York.
3. Baumberg, S. (1976). Genetic control of arginine metabolism in prokaryotes. In "Second Symposium on the Genetics of Industrial Microorganisms" (Ed. K.D. Mc Donald) pp.369-388. Academic Press, New York.
4. Umbarger, H.E. (1978). Amino acid biosynthesis and its regulation. *Annual Review of Biochemistry* 47, 533-606.
5. Abdelal, A.T.H. (1979). Arginine catabolism by microorganisms. *Annual Review of Microbiology* 33, 139-168.
6. Cunin, R. (1983). Regulation of arginine biosynthesis in procaryotes. In "Biotechnology - Aminoacids : Biosynthesis and Genetic Regulation" (Eds. Herrman and Somerville), pp. 53-79. Addison Wesley, London, Boston, Toronto.
7. Paulus, H. (1983). Evolutionary history of the ornithine cycle as a determinant of its structure and regulation. *Current Topics in Cellular Regulation* 22, 177-200.
8. Piérard, A. (1983). Evolution des systèmes de synthèse et d'utilisation du carbamoylphosphate. In "L'Evolution des Protéines" (Ed. G. Hervé) pp. 53-66. Masson, Paris.
9. Stalon, V., Ramos, F., Piérard, A. and Wiame, J.M. (1967). The occurrence of a catabolic and an anabolic ornithine carbamoyltransferase in *Pseudomonas*. *Biochimica Biophysica Acta* 139, 91-97.
10. Issac, J.H. and Holloway, B.W. (1972). Control of arginine biosynthesis in *Pseudomonas aeruginosa*. *Journal of General Microbiology* 73, 427-438.

11. Voellmy, R. and Leisinger, T. (1978). Regulation of enzyme synthesis in the arginine biosynthetic pathway of *Pseudomonas aeruginosa*. *Journal of General Microbiology* 109, 25-35.
12. Haas, D., Kurer, V. and Leisinger, T. (1972). N-acetylglutamate synthetase of *Pseudomonas aeruginosa*. An assay *in vitro* and feedback inhibition by arginine. *European Journal of Biochemistry* 31, 290-295.
13. Haas, D. and Leisinger, T. (1974). N-acetylglutamate synthetase from *Pseudomonas aeruginosa* : synergetic inhibition by N-acetylglutamate and polyamines. *Biochemical Biophysical Research Communications* 60, 42-47.
14. Haas, D. and Leisinger, T. (1975). N-acetylglutamate-5-phosphotransferase of *Pseudomonas aeruginosa* : catalytic and regulatory properties. *European Journal of Biochemistry* 52, 377-383.
15. Udaka, S. (1966). Pathway-specific pattern of control of arginine biosynthesis in bacteria. *Journal of Bacteriology* 91, 617-621.
16. Piérard, A., Glansdorff, N., Mergeay, M. and Wiame, J.M. (1965). The control of the biosynthesis of carbamoylphosphate in *Escherichia coli*. *Journal of Molecular Biology* 14, 382-385.
17. Piérard, A. (1966). The control of the activity of *E. coli* carbamoylphosphate synthetase by antagonistic allosteric effectors. *Science* 154, 1572-1573.
18. Gerhart, J.C. and Pardee, A.B. (1962). The enzymology of control by feedback inhibition. *Journal of Biological Chemistry* 237, 891-896.
19. Abdelal, A.T., Bussey, L. and Vickers, L. (1983). Carbamoylphosphate synthetase from *Pseudomonas aeruginosa* : subunit composition, kinetic analysis and regulation. *European Journal of Biochemistry* 129, 697-702.
20. Paulus, J.T. and Switzer, R.L. (1979). Characteristics of pyrimidine-repressible and arginine-repressible carbamoylphosphate synthetases from *Bacillus subtilis*. *Journal of Bacteriology* 137, 82-91.
21. Issaly, I.M. and Issaly, A.S. (1974). Control of ornithine carbamoyltransferase activity by arginase in *Bacillus subtilis*. *European Journal of Biochemistry* 49, 485-495.
22. De Hauwer, G., Lavallé, R. and Wiame, J.M. (1964). Etude de la pyrroline déshydrogénase et la régulation du catabolisme de l'arginine et de la proline chez *Bacillus subtilis*. *Biochimica Biophysica Acta* 81, 257-269.

23. Harwood, C.R. and Baumberg, S. (1977). Arginine hydroxamate-resistant mutants of *Bacillus subtilis* with altered control of arginine metabolism. *Journal of General Microbiology* 100, 177-178.
24. Broman, K., Lauwers, N., Stalon, V. and Wiame, J.M. (1978). Oxygen and nitrate in utilization by *Bacillus licheniformis* of the arginase and arginine deiminase routes of arginine catabolism and other factors affecting their synthesis. *Journal of Bacteriology* 134, 920-927.
25. Vissers, S., Dessaux, Y., Legrain, C. and Wiame, J.M. (1981). Feedback inhibition by arginine on ornithine carbamoyltransferase of *Agrobacterium tumefaciens*. *Archives Internationales de Physiologie et de Biochimie* 89, B83-B84.
26. Weathers, P.S., Chee, H.L. and Allen, M.M. (1978). Arginine catabolism in *Aphanocapsa* 6308. *Archives of Microbiology* 118, 1-6.
27. Gupta, M. and Carr, N.G. (1981). Enzymology of arginine metabolism in heterocyst forming *Cyanobacteria*. *FEMS Microbiology Letters* 12, 179-181.
28. Prozesky, O.W., Grabow, W.I.K., Vander Merwe, S. and Coetzee, J.N. (1973). Arginine gene clusters in *Proteus-Providence* group. *Journal of General Microbiology* 77, 237-240.
29. Ottow, J.C.G. (1974). Arginine dihydrolase activity in species of the genus *Bacillus* revealed by thin-layer chromatography. *Journal of General Microbiology* 84, 209-213.
30. Baumberg, S. and Harwood, C.R. (1979). Carbon and nitrogen repression of arginine catabolic enzymes in *Bacillus subtilis*. *Journal of Bacteriology* 137, 189-196.
31. Schreirer, H.J., Smith, T.M. and Bernlohr, R.W. (1982). Regulation of nitrogen catabolic enzymes in *Bacillus* spp. *Journal of Bacteriology* 151, 971-975.
32. Huston, J.Y. and Dowing, M. (1968). Pyrimidine biosynthesis in *Lactobacillus leichmanii*. *Journal of Bacteriology* 96, 1249-1254.
33. Pendey, V.W. (1980). Interdependence of glucose and arginine in catabolism of *Streptococcus faecalis* ATCC 8043. *Biochemical Biophysical Research Communications* 96, 1480-1487.
34. Simon, J.P., Wargnies, B. and Stalon, V. (1982). Control of enzyme synthesis in the arginine deiminase pathway of *Streptococcus faecalis*. *Journal of Bacteriology* 150, 1085-1090.

35. Stalon, V., Simon, J.P. and Mercenier, A. (1982) Enzymes of arginine utilization and their formation in *Aeromonas formicans* NCIB 9232. *Archives of Microbiology* 133, 295-299.
36. Stalon, V. and Mercenier, A. (1984). L-Arginine utilization by *Pseudomonas* species. *Journal of General Microbiology* 130, 69-76.
37. Ramos, F., Stalon, V., Piérard, A. and Wiame, J.M. (1967). The specialization of the two ornithine carbamoyltransferases of *Pseudomonas*. *Biochimica Biophysica Acta* 139, 98-106.
38. Mercenier, A., Simon, J.P., Vander Wauven, C., Haas, D. and Stalon, V. (1980). Regulation of enzyme synthesis in the arginine deiminase pathway of *Pseudomonas aeruginosa*. *Journal of Bacteriology* 144, 159-163.
39. Vander Wauven, C., Haas, D. and Piérard, A. (1982). Evidence for a gene cluster coding for arginine deiminase pathway in *Pseudomonas aeruginosa*. *Archives Internationales de Physiologie et de Biochimie* 90, B80-B81.
40. Stalon, V., Legrain, C. and Wiame, J.M. (1977). Anabolic ornithine carbamoyltransferase of *Pseudomonas*. *European Journal of Biochemistry* 74, 319-327.
41. Stalon, V., Ramos, F., Piérard, A. and Wiame, J.M. (1972). Regulation of the catabolic ornithine carbamoyltransferase in *Pseudomonas fluorescens*. A comparison with the anabolic transferase and with a mutationally modified catabolic transferase. *European Journal of Biochemistry* 29, 25-35.
42. Stalon, V. (1972). Regulation of the catabolic ornithine carbamoyltransferase of *Pseudomonas fluorescens*. A study of the allostric properties. *European Journal of Biochemistry* 29, 36-46.
43. Haas, D., Evans, R., Mercenier, A., Simon, J.P. and Stalon, V. (1979). Genetic and physiological characterization of *Pseudomonas aeruginosa* mutants affected in the catabolic ornithine carbamoyltransferase. *Journal of Bacteriology* 139, 713-720.
44. Legrain, C., Stalon, V., Glansdorff, N., Gigot, D., Piérard, A. and Crabeel, M. (1976). Structural and regulatory mutations allowing utilization of citrulline or carbamoylaspartate as a source of carbamoylphosphate. in *E. coli* K-12. *Journal of Bacteriology* 128, 39-48.
45. Legrain, C., Halleux, P., Stalon, V. and Glansdorff, N. (1972). The dual genetic control of ornithine carbamoyltransferase in *Escherichia coli* : a case of bacterial hybrid enzymes. *European Journal of Biochemistry* 27, 93-102.

46. Morris, D.R. and Pardee, A.B. (1966). Multiple pathways of putrescine biosynthesis in *E. coli*. *Journal of Biological Chemistry* 241, 3129-3135.
47. Tabor, C.W. and Tabor, H. (1976). 1-4-Diaminobutane (putrescine) spermidine and spermine. *Annual Review of Biochemistry* 45, 285-306.
48. Mercenier, A., Simon, J.P., Haas, D. and Stalon, V. (1980). Catabolism of L-arginine by *Pseudomonas aeruginosa*. *Journal of General Microbiology* 116, 381-389.
49. Roon, R.J. and Barker, H.A. (1972). Fermentation of agmatine in *Streptotoccus faecalis* : occurrence of putrescine transcarbamylase. *Journal of Bacteriology* 109, 44-50.
50. Simon, J.P. and Stalon, V. (1982). Enzymes of agmatine degradation and the control of their synthesis in *Streptococcus faecalis*. *Journal of Bacteriology* 152, 676-681.
51. Wargnies, B., Lauwers, N. and Stalon, V. (1979). Structure and properties of putrescine carbamoyltransferase of *Streptococcus faecalis*. *European Journal of Biochemistry* 101, 145-152.
52. Michaels, R. and Kim, K.M. (1965). Comparative studies of putrescine degradation by microorganisms. *Biochimica Biophysica Acta* 115, 59-64.
53. Brohn, F. and Tchen, T.T. (1971). A single transaminase for 1-4 diaminobutyrate and 4-aminobutyrate in a *Pseudomonas* species. *Biochemical Biophysical Research Communications* 415, 573-582.
54. Voellmy, R. and Leisinger, T. (1976). Rôle of 4-aminobutyrate in the arginine metabolism of *Pseudomonas aeruginosa*. *Journal of Bacteriology* 128, 722-729.
55. Jakoby, W.B. (1962). Succinic semialdehyde dehydrogenase. *In* "Methods in Enzymology" (Eds. S.P. Colowick and N.O. Kaplan) Vol. 5, pp.774-778. Academic Press, New York.
56. Haas, D., Matsumoto, H., Moretti, P., Stalon, V. and Mercenier, A. (1984). Arginine degradation in *Pseudomonas aeruginosa* mutants blocked in two arginine catabolic pathways. *Molecular and General Genetics* 193, 437-444.
57. Friedrich, B. and Magasanik, B. (1979). Enzymes of agmatine degradation and the control of their synthesis in *Klebsiella aerogenes*. *Journal of Bacteriology* 137, 1127-1133.

58. Tabor, H. and Tabor, C.W. (1969). Formation of 1,4-diaminobutane and of spermidine by an ornithine auxotroph of *Escherichia coli* grown on limiting ornithine and arginine. *Journal of Biological Chemistry* 244, 2286-2293.
59. Satishchandran, C. and Boyle, S.M. (1984). Antagonistic transcriptional regulation of the putrescine biosynthetic enzyme agmatine ureohydrolase by cyclic AMP and agmatine in *E. coli*. *Journal of Bacteriology* 157, 552-559.
60. Morris, D.R. and Koffrow, K.L. (1967). Urea production and putrescine biosynthesis by *Escherichia coli*. *Journal of Bacteriology* 94, 1516-1519.
61. Kim, R. (1963). Isolation and properties of a putrescine degradating mutant of *Escherichia coli*. *Journal of Bacteriology* 86, 320-323.
62. Miller, D.L. and Rodwell, V. (1971). Metabolism of basic amino acids in *Pseudomonas putida*. *Journal of Biological Chemistry* 246, 5053-5058.
63. Yorifuji, T., Kobayashi, T., Tabuchi, H., Shiritani, Y. and Yonama, K. (1983). Distribution of amidinohydrolases among *Pseudomonas* and comparative studies of some purified enzymes by one dimensional peptide mapping. *Agricultural and Biological Chemistry* 47, 2825-2830.
64. Van Thoai, N., Thombe-Beau, F. and Olomucki, A. (1966). Induction et spécificité des enzymes de la nouvelle voie catabolique de l'arginine. *Biochimica Biophysica Acta* 115, 73-80.
65. Pistorius, E. and Voss, M. (1982). Presence of an amino acid oxidase photosystem II of *Anacystis nidulans*. *European Journal of Biochemistry* 126, 203-209.
66. Pistorius, E. and Voss, M. (1980). Some properties of a basic L-amino acid oxidase from *Anacystis nidulans*. *Biochimica Biophysica Acta* 611, 227-240.
67. Voellmy, R. and Leisinger, T. (1975). Dual rôle for N-acetylornithine aminotransferase from *Pseudomonas aeruginosa* in arginine biosynthesis and arginine catabolism. *Journal of Bacteriology* 122, 799-809.
68. Rahman, M., Laverack, P.D. and Clarke, P.H. (1980). The catabolism of arginine by *Pseudomonas aeruginosa*. *Journal of General Microbiology* 116, 371-380.
69. Friedrich, B., Friedrich, C.G. and Magasanik, B. (1978) Catabolic N^2-acetylornithine-5-aminotransferase in *Klebsiella aerogenes* : control of synthesis by induction, catabolite repression and activation by glutamine synthetase. *Journal of Bacteriology* 133, 689-691.

70. Friedrich, B. and Magasanik, B. (1978) Utilization of arginine by *Klebsiella aerogenes*. *Journal of Bacteriology* 133, 680-685.
71. Fox, G.E., Stakebrandt, E., Hespell, R.B., Gibson, J., Maniloff, J., Dyer, A., Wolfe, R.S., Balch, W.E., Tanner, R.S., Magrom, L.J., Zablen, L.B., Blakemore, R., Gupta, R., Bonen, L., Lewis, D.A., Stahl, D.A., Luehrsen, K.R., Chen, K.N. and Woese, C.R. (1980). The phylogeny of prokaryotes. *Science* 209, 457-463.
72. Dundas, I.E. and Halvorson, O.H. (1966). Arginine metabolism in *Halobacterium salinarium*, an obligate halophilic bacterium. *Journal of Bacteriology* 91, 113-119.
73. Hartman, R., Sichinger, H.D. and Oesterhelt, D. (1980). Anaerobic growth of halobacteria. *Procedings of the National Academy of Sciences (USA)* 77, 3821-3825.
74. Igwebe, E.C.K. and Thomas, C. (1978). Occurrence of enzymes of arginine dihydrolase pathway in *Spiroplasma citri*. *Journal of General and Applied Microbiology* 24, 261-269.
75. Blakemore, R.P. and Canale-Parola, E. (1976). Arginine catabolism by *Treponema denticola*. *Journal of Bacteriology* 128, 616-622.
76. Palleroni, N.J., Kunisawa, R., Contropoulou, R. and Doudoroff, M. (1974). Nucleic acid homologies in the genus *Pseudomonas*. *International Journal of Systematic Bacteriology* 23, 333-339.
77. Barker, H.A. (1981). Amino acid degradation by anaerobic bacteria. *Annual Review of Biochemistry* 50, 23-40.
78. Tochikura, T., Sugiyama, S., Bunno, M., Matsubara, T., and Tachiki, T. (1980). Degradation of arginine via α-keto-δ-guanidinovalerate and α-ketovalerate in arginine-grown *Arthrobacter simplex*. *Agricultural and Biological Chemistry* 44, 1173-1178.
79. Walker, J.B. and Hndica, V.S. (1964). Developmental changes in arginine amidinotransferase activity in Streptomycin-producing strains of *Streptomyces*. *Biochimica Biophysica Acta* 89, 473-482.
80. Nakane, A., Nakamura, T., and Eguchi, Y. (1977) A novel metabolic fate of arginine in *Streptomyces eurocidicus*. *Journal of Biological Chemistry* 252, 5267-5273.

PHYLOGENY AND PHYLOGENETIC
CLASSIFICATION OF PROKARYOTES

E. Stackebrandt

Institut für Allgemeine Mikrobiologie
Christian-Albrechts-Universität
Kiel, Federal Republic of Germany

1. INTRODUCTION

The comparative analysis of genetic sequence has provided us with such an overwhelming wealth of data and exciting and surprising insights into the genealogical relationships of prokaryotes that to day we stand at the threshold of a new area in microbiology. The emerging phylogenetic tree not only shows the degree of relatedness among the bacterial species but consequently will unravel the evolution of biochemical, physiological, morphological and regulatory features of prokaryotic geno- and phenotypes.

The first glimpse at bacterial phylogenies was achieved by DNA-hybridization, a method developed as early as 1961 (1). When the conserved character of the ribosomal RNA genes was detected (2) it soon became obvious that the large rRNA species, although covering a minor part of the genome only, were the molecules of choice for phylogenetic studies. Their conserved primary structures together with regions of higher variability (3), their ubiquitous distribution, genetic stability (4) and functional constancy (5) make them superior phylogenetic probes, today used in the determination of rRNA cistron similarities and in the comparative analysis of 16S rRNA oligonucleotides. These methods, used alone or in combination (depending on the phylogenetic position of the organism concerned), allow a reliable and objective placement of any prokaryote into the phylogenetic tree. Organisms are exclusively according to their genetic relationships while

phenotypic characters are not considered at all.
Although there is growing excitement, interest and acceptance, one main problem is evident: the phylogenetic structure, highly suitable for description of a system above the strain level, comes up against the traditional classification scheme, e.g. as represented by Bergey's Manual; a comparison of the two hierarchic systems immediately reveals the immense discrepancies in the ranking of taxa. [The situation has been improved in the latest edition of Bergey's Manual, Vol. 1, 1984, in that "preliminary rearrangements of taxa have been made based on phylogenetic information" (6)].

It is the goal of this chapter to highlight two aspects of the situation: Firstly, unless the phylogenetic coherency of taxa has been demonstrated by methods measuring phylogenetic relationships, it has to be assumed that within the traditional classification scheme all ranks above the strain level are artificial and have no phylogenetic meaning. Any ranking and naming of taxa above the genus level should be avoided since the latin-greek nomenclature falsely implies a natural relationship. Secondly, the construction of a synthesis classification is feasible, in that the membership of organisms to a phylogenetic group can be decided on the basis of phylogenetic date, and only then will ranks be determined by common phenotypic characters, e.g. chemotaxonomic, physiological, biochemical and morphological properties.

THE PHYLOGENETIC TREE

The number of bacterial strains investigated by the various phylogenetic approaches how exceeds 2000. Among these, most strains have been investigated by DNA hybridization (see 7) for references between 1974 and 1982). The success of the DNA hybridization technique is probably due to two facts: (i) there exists a salient interest in the determination of relationhips at the strain level, which allows newly described organisms to be allocated to existing taxa,and by which misclassified strains can be detected easily; (ii) the various methods available today are reliable, fast, highly reproducible, they support each other and they are inexpensive. The drawbacks of this approach are its restricted ability to resolve higher phylogenetic units and, something which is true for all hybridization methods, the lack of cumulative data storage. The determination of relationships among higher taxa, i.e. from the intrageneric to the interfamily level, was achieved by rRNA cistron similarity studies. Today taxo-

PHYLOGENY AND CLASSIFICATION OF PROKARYOTES

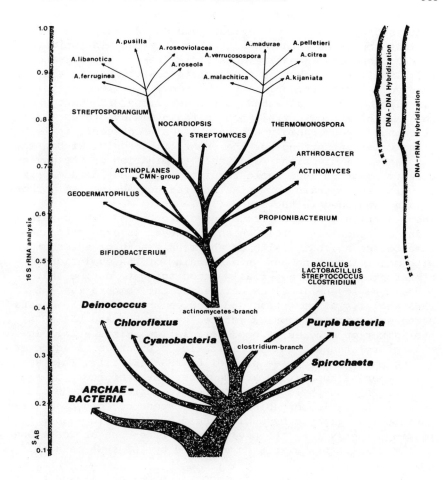

Fig. 1 Schematic phylogenetic tree of 16S rRNA of prokaryotes showing those regions which are covered by DNA-DNA and DNA-rRNA hybridization techniques.

nomists can choose among various methods (8 - 11) and the relationships of more than 450 strains have been investigated, mainly among Gram-negative eubacteria (see 7 for references till 1982). Although this approach is cost effective and has made valuable contributions to taxonomy, e.g. leading to the description of new taxa and to the detection of heterogenous genera, the method suffers from certain disadvantages (12): it cannot detect rapidly evolving lines of descent, it is of no value in sorting out the order of branching within major bacterial divisions, and the determination of the phylogenetic position as a whole requires a complete data matrix, in which each strain serves as a reference against all other strains included in the study (13).

These difficulties are circumvented by the 16S rRNA cataloguing approach (14 - 16), which, granted, is expensive and time-consuming, but yields an enormous wealth of information with major impact on our understanding of the evolution of prokaryotes. It was especially encouraging to notice that certain phylogenetic lines defined by the comparative 16S rRNA analysis of more than 450 species from about 150 genera agree well with categories defined by DNA-DNA and DNA-rRNA hybridization studies. A good correlation exists at all levels of relatedness the hybridization approaches could cover in the phylogenetic tree of the 16S rRNA (Fig. 1). This was furthermore a supporting argument for the usefulness of ribosomal RNAs as phylogenetic probes to represent the evolution of the entire organism. As a consequence, groups of organisms investigated by nucleic acid hybridization can be placed directly into the phylogenetic tree when at least one of the strains has also been analysed by 16S rRNA cataloguing.

In early sequencing studies on ribosomal RNA it was hoped that the 5S rRNA would turn out to be an excellent molecule for phylogenetic studies since it was small enough to be totally sequenced routinously. However it could be shown that this RNA species is a less exact chronometer than the 16S rRNA (17, 18). Likewise, the significance of certain proteins, useful in the determination of relationships among eukaryotes, has been lost in studies on prokaryotes with the introduction of nucleic acid sequencing techniques. Nevertheless, the comparison of trees derived from the analysis of cytochrome c (19) and 16S rRNA (20, 21) of purple non-sulfur bacteria revealed such a high degree of correlation that this result was taken as a supporting argument for the genetic stability of both molecules (20) - a prerequisite for a molecule to serve as a phylogenetic probe.

In general, however, phylogenetic patterns derived from the

analysis of 5S rRNA (18, 22, 23), cytochrome c (19, 24) and ferredoxins (24) are neither congruent among each other, nor does any of them match completely the branching pattern of the 16S rRNA. Most of these discrepancies will probably be explained by differences in the mutation rate, but we will have to await new insights into the mode and speed of the evolution of macromolecules as well as improved methods (e.g. full sequence comparison of large rRNAs and other suitable molecules), which in the near future will not only improve the data analysis (25) but consequently will elucidate the branching points more accurately, expecially those of the largest phylogenetic units.

THE PHYLOGENY OF PROKARYOTES

The concept of the primary kingdoms (26), startling and revolutionary, has lost nothing of its attractiveness and stimulating influence on various fields of biology (27). The genealogical gap between eubacteria, archaebacteria and the eukaryotic cytoplasm, originally discovered by the low degree of simularity in the primary structure of the 16S rRNA of about 80 organisms investigated in 1977, is still valid in 1984 with about 450 organisms investigated. Similarly, the fundamental differences at the genetic and epigenetic level between archaebacteria and eubacteria are not blurred with additional data on hundreds of organisms available (the only exceptions are the finding of a group of peptidoglycanless eubacteria (28) and of a glycerol diether containing eubacterium (29).

The evolutionary branching order of the primary kingdoms, however, is still not settled. The pattern we are familiar with, i.e. all three urkingdoms branch off at the same time in evolution (as expressed by similar S_{AB} values) in tentative only, reflecting the sad fact that at present no better methods are available to decide whether or not any two of the urkingdoms are specifically related to one another, to the exclusion of the third. Although archaebacteria resemble eukaryotes in certain aspects while they resemble eubacteria in others (30) an increasing number of studies reveal a distinct similarity in certain features between archaebacteria and eukaryotes (18, 31- 33). The determination of the branching order of major lines and the extent to which lateral gene transfer may have occurred in early stages in evolution are of high priority in the more accurate elucidation of the relationships of the various ancient lines.

The last comprehensive picture of the phylogenetic structure of prokaryotes was drawn by Stackebrandt und Woese (16). Since then, a number of phylogenetic groupings have been analyzed more thorougly and new lines have been discovered. The following section summarizes these data with emphasis on the recent findings.

Eubacteria

As shown in Figure 2, eubacteria investigated so far fall into 11 main lines of descent which branch off from each other at S_{AB} values between 0.15 and 0.23. Lines defined by *Planctomyces* and *Pirella*, by Haloanaerobiaceae, by *Bacteroides*, *Cytophaga* and others, and by the group defined by *Bdellovibrio*, myxococci and the sulfur and sulfate respiring Gram-negative eubacteria could be added to the 7 major lines known to constitute the eubacterial kingdom in 1981 (16). Table 1 lists the genera defining the lines 1 - 10.

1. The Gram-positive eubacteria. Except for *Deinococcus*, all Gram-positive eubacteria form a phylogenetically coherent cluster, in which organisms with a low DNA G+C content ("Clostridium" subbranch) are clearly separated from organisms with a high DNA G+C content ("Actinomycetes" subbranch). The phenotype of members of the former subbranch is predominantly clostridial, e.g. the anaerobic, rod-shaped and spore-forming bacteria are found in several deep branching lines. Almost every line, however, contain non-spore-forming rods and/or cocci as well e.g. *Peptococcus*, *Sarcina*, *Ruminococcus*, and so on. The facultative and aerobic representatives are members of one subline in which taxa are phylogenetically equivalent e.g. *Lactobacillus*, *Streptococcus*, *Enterococcus*, *Bacillus*, and mycoplasmas containing certain "clostridia", "lactobacilli" and *Erysipelothrix*. Recent studies have focused on lactobacilli (42) and streptococci (40, 61). Within the actinomycetes subbranch most recent work has been done on *Arthrobacter* and relatives (50, 62) and on sporeforming actinomycetes (48, 63) which in most cases revealed the phylogenetic heterogeneity of morphologically defined genera.

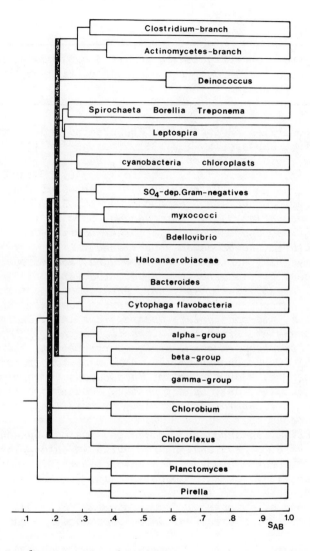

Fig. 2 Dendrogram of relationships among the main lines of descent of the eubacterial kingdom.

TABLE I

List of genera investigated by 16S rRNA cataloguing defining the main branches 1-10 of the eubacterial kingdom. Numbers refer to references.

1. Gram positives ("Clostridium" subbranch) Acetobacterium, Clostridium, Eubacterium (34, 35), Acetogenium (34), Acholeplasma, Mycoplasma, Spiroplasma (36), Bacillus (14, 37), Brochothrix, Listeria (38), Gemella (39, 40), Enterococcus, Streptococcus (40), Kurthia (40, 41), Lactobacillus, Leuconostoc, Pediococcus (42), Peptococcus (15, 35), Planococcus, Staphylococcus (43, 44), Ruminococcus, Sarcina (15), Sporolactobacillus, Sporosarcina (14), Thermoactinomyces (45), Erysipelothrix, Aerococcus (unpublished).
("Actinomycetes" subbranch) Actinomadura, Nocardiopsis (46), Actinomyces, Bifidobacterium, Propionibacterium (45), Actinoplanes, Ampullariella (47, 48), Arthrobacter, Micrococcus (43, 49), Aureobacterium, Brevibacterium, Cellulomonas, Curtobacterium, Microbacterium, Nocardioides (49), Chainia, Elytrosporangium, Kitasatoa, Microellobosporia, Promicromonospora, Streptosporangium, Streptoverticillium, Thermomonospora (48), Corynebacterium (45, 49), Dactylosporangium, Micromonospora, Microbacterium, Rhodococcus, Nocardia, Streptomyces (45, 48), Dermatophilus, Geodermatophilus (50), Oerskovia (47), Stomatococcus (44), Frankia (unpublished).
2. Deinococcus (51)
3. Agmenellum, Aphanocapsa, Anacystis, Nostoc, Synechococcus, Fischerella, Euglena and Porphyridium chloroplasts (52), Prochloron, Chlamydomonas chloroplast (53), Corn chloroplast (54), Lemna chloroplast (15, 53), Spirulina (unpublished).
4. Bacteroides, Cytophaga, Flavobacterium, Flexibacter, Haliscomenobacter, Saprospira, Sporocytophaga, Fusobacterium, nucleatum (55)
5. Spirochaeta, Borellia, Leptospira, Treponema (56)
6. Bdellovibrio (25), Desulfobacter, Desulfobulbus, Desulfococcus, Desulfonema, Desulfovibrio, Desulfuromonas (57), Myxococcus, Cystobacter, Stigmatella, Sorangium, Nannocystis (58)
7. Planctomyces, Pirella (59)
8. Haloanaerobium, Haloanaeromicrobium (60)
9. Chloroflexus (15), Herpetosiphon (unpublished)
10. Chlorobium (15), Chloroherpeton (unpublished)

2. *Deinococcus* (51). This line is so far only defined by the closely related, radiation-resistant cocci, characterized by their atypical "holey" cell wall, which contains ornithin as diaminoacid in the peptidoglycan. Recently, a rod shaped organism (*Ornithinobacter*, Oyaizu, Pohla, Stackebrandt, unpublished) was found to be closely related to Deinococcus.

3. *Cyanobacteria*. Little additional work has been done on cyanobacteria and chlorplasts of higher plants and algae since 1981. It could be shown that *Prochloron*, a chlorophyll a and b containing, unicellular prokaryot, as well as *Spirulina* are phylogenetically placed within the radiation of cyanobacteria, with *Nostoc*, *Fischerella* and *Agmenellum* as their closest relatives (53, and unpublished results). In the light of the evolution of chloroplasts, *Prochloron* was once hailed as the ancestor of those organisms which must have given rise to the chloroplasts of higher plants and green algae. The phylogenetic analysis of *Prochloron* by the standard analysis does not support this conclusion although its position cannot be determined unambigously (64). The 16S rRNA of Chlamydomonas showed a deep branching point within the cyanobacteria/chloroplast line of descent similar to those found for other chloroplasts (53). So far, no heterotrophic representative of this group has been found, although a variety of gliding, nonphototrophic gram-negative eubacteria have been investigated.

4. The group of *Bacteroides*, *Cytophage*, *Flavobacterium* and related taxa (55). This group forms one of the major lines of the eubacteria. The bacteroides form one, and cytophagas, certain low DNA G+C containing flavobacteria, *Haliscomenobacter*, *Flexibacter*, *Sporocytophaga* and *Saprospira* the second subbranch. The latter one is further divided into two subclusters, one containing cytophagas and certain flavobacteria while the second embraces *H.hydrossis*, *F.elegans*, *S.grandis* and some other flavobacteria. *Fusobacterium nucleatum* is a member of *Bacteroides*. The membership of the anaerobic non-gliding bacteroides to the mainly aerobic gliding cytophagas, *Saprospira* and *Haliscomenobacter* and non gliding flavobacteria is so far not supported by other data, but may be rationalized by the relationship of an anaerobic gliding isolate (K.O. Stetter) to bacteroides.

5. The *spirochaetes* (56). While in earlier dendrograms *Leptospira* was shown to be phylogenetically isolated from other spirochaete genera (15, 16) recent improvements in the data analysis ("signature analysis, 25, 36, 55, 56) allowed a more precise grouping of *Leptospira* which is now considered a peripheral member of the main spirochaete cluster, defined by *Spirochaeta*, *Borellia* and *Treponema*. This main cluster comprises three sublines, one embracing the bulk of *Spirochaeta* and *Treponema*, while the second and third are represented by *B.hermsii* and *T.hyodysenteriae*, respectively.

6. The group of *Bdellovibrio*, myxococci and sulfurdependent Gram-negatives. The relationship of these three diverse groups is extremely remote and could not be detected by the traditional signature analysis (14, 25). Almost no common features of taxonomic value support this relationships. The myxococci (*Myxococcus*, *Cystobacter*, *Stigmatella*, *Nannocystis* and *Sorangium*) form a coherent cluster with the first three genera displaying a close relationship (58). The dissimilatory sulfur and sulfate reducing Gram-negatives (*Desulfuromonas*, *Desulfococcus*, *Desulfobulbus*, *Desulfobacter*, *Desulfonema*, *Desulfosarcina*, *Desulfovibrio*) define a separate cluster, with the sulfur reducing *Desufuromonas* forming a sub-branch of the sulfate reducers and in which *Desulfovibrio* is peripherally related only to the other genera (57). The occurence of sulfate reducing bacteria in various parts of the phylogenetic tree (archaebacteria, Gram-positives and Gram-negatives) indicate that this mode of energy yielding process has evolved independently by convergence.

7. The *Planctomyces-Pirella* group (59). These budding, non-prosthecate organisms are unique among the eubacteria in that they lack muramic acid and diaminopimelic acid, possessing a proteinaceous cell wall instead of a murein sacculus, and in that they are resistant to cell wall antibiotics (28). 11 strains have been investigated by oligonucleotide cataloguing and the very low S_{AB} values found with eubacteria (average of 0.15) indicate that members of the *Planctomyces/Pirella* group are descendants of the most ancient group of eubacteria, a group which branched off after the separation of eubacteria from archaebacteria but before the main radiation of eubacteria into the various major lines occurred. Membership of *Planctomyces* and *Pirella* to the ar-

chaebacterial kingdom, whose members are also free of peptidoglycan, could be excluded on the basis of a negative diphtheria toxin reaction (59) and because of the presence of common eubacterial fatty ester based lipids (T. Langworthy, pers. communication). The lack of peptidoglycan in members of such ancient group of eubacteria may be of interest from the viewpoint of the evolution of peptidoglycan, which may have been "invented" at a later stage in the evolution of eubacteria, after the budding, non-prosthecate eubacteria had branched off.

8. The Haloanaerobiaceae (60). This family embraces the genera *Haloanaerobium* and *Halobacteroides* which are defined as anaerobic, moderately halophilic eubacteria with a low DNA G+C content. They are not related to any recognized major group and so far no additional information on these organisms is available which would support their phylogenetically unique position.

9. and 10. *Chloroflexus* and *Chlorobium*: the green sulfur bacteria. With *Herpetosiphon aurantiacus* and a *Herpetosiphon*-type isolate from sewage two non-phototrophic gliding bacteria were found to be remotely related to *Chloroflexus* (Woese, Ludwig, Stackebrandt unpublished). Phylogenetically unrelated to *Chloroflexus* and representing an individual line of descent is *Chlorobium vibrioforme*. The recently described *Chloroherpeton thalassium* (65), a non-filamentous, flexing and gliding green sulfur bacterium with chlorophyll c could be allocated phylogenetically to *Chlorobium* (Gibson and Woese, unpublished, quoted in 65).

11. The purple bacteria and their relatives (Table 2). This group is the most complex collection of different phenotypes and the emerging relationships are hardly in accord with the existing calssification scheme. Hundreds of strain have been investigated by nucleic acid hybridization techniques and 16S rRNA cataloguing to show the true picture of their natural relationships. The group comprises three distinct major subgroups (alpha, beta, gamma) each of them containing phototrophic bacteria. Table 2 lists the genera allocated to the three subgroups by rRNA hybridization and sequencing.

Since it does not seem reasonable that the complex phototrophic apparatus evolved by convergence individually in the three subgroups, it is highly likely that the common an-

TABLE II

List of phylogenetically defined taxa of the branch of the
purple photosynthetic bacteria and related taxa investigated by rRNA cataloguing and hybridization

1. *Alpha group* (15, 21, 66)
Rhodospirillum, *Rhodopseudomonas*, *Rhodomicrobium*, *Rhodopila*,
Rhodobacter (21, 95), *Acetobacter*, *Beijerinckia*, *Gluconobacter*, *Phyllobacterium*, *Xanthobacter*, *Zymomonas* (67), *Agrobacterium* (68), *Azospirillum* (69), *Nitrobacter* (70), *Paracoccus* (15), *Phenylobacterium* (71), *Rhizobium* (68), *Aquaspirillum itersonii* (72), *Pseudomonas diminuta* (73, 74), *Ery-
throbacter* (66), *Bradyrhizobium*, *Hyphomicrobium*, *Prosteco-
microbium* (unpublished).

2. *Beta group* (15, 21, 75)
Rhodocyclus (21, 95), *Aquaspirillum* (72), *Achromobacterium*,
Bortedella (76), *Chromobacterium*, *Janthinobacterium* (83),
Alcaligenes, *Comamonas*, *Nitrosococcus*, *Nitrosolobus*, *Nitro-
somonas*, *Nitrosospira*, *Nitrosovibrio*, *Thiobacillus* (75),
Derxia (69), *Sphaerotilus* (15), *Spirillum* (72), *Pseudomonas
acidovorans* group, *Pseudomonas solanacearum* group (73, 74),
Simonsiella (unpublished).

3. *Gamma group* (15, 21, 77)
Chromatium (21, 78), *Amoebobacter*, *Lamprocystis*, *Thiocapsa*,
Thiocystis, *Thiodyction*, *Thiospirillum* (78), *Ectothiorhodo-
spira* (79), *Acinetobacter* (15), *Beggiatoa*, *Halomonas*, *Leu-
cothrix*, *Lysobacter* (77), *Enterobacteriaceae* (15, 81),
Vibrionaceae (15, 80, 81), *Legionella* (82), *Oceanospirillum*,
Serpens (72), *Alteromonas* (81), *Azomonas*, *Azotobacter* (69),
Frateuria (73), *Xanthomonas*, fluorescent pseudomonas (73,
74), *Alysiella* (unpublished).

cestor of the entire group was phototrophic (as mentioned above, a possible gene transfer as the cause for the presence of the photosynthetic apparatus in the various subgroups can be excluded (20). From this it follows that non-phototrophic members of the subbranches evolved from the phototrophes by conversion as indicated previously (15, 16, 21, 70). This makes it easier to understand that budding organisms, e.g. *Nitrobacter*, *Blastobacter* or *Hyphomicrobium* are phylogenetically related to budding phototrophs, or that spiral Gram-negatives are likewise related to spiral phototrophic bacteria. A closer analysis of some genera of this branch will be dealt with in the forthcoming chapter.

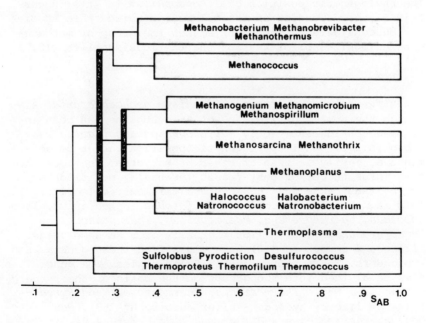

Fig. 3 Dendrogram of relationships among the main lines of descent of the archaebacterial kingdom.

Archaebacteria
The great diversity of the archaebacterial geno- and phenotypes, comprehensively depicted in the book "Archaebacteria" (27) has been confirmed and extended by the investigations on methanogenic, haloalkaliphilic and thermoacidophilic archaebacteria. Within this primary kingdom two deep gaps separate the methanogens and halophiles from the single species *Thermoplasma acidophilum* on the one side and from the sulfur dependent thermoacidophiles on the other side (Figure 3).

1. Methanogenic and halophilic archaebacteria. This branch comprises for main subbranches whose organisms are classified in Methanobacteriales,Methanococcales and Methanomicrobiales (84), together with the halobacteria. The phylogenetic position of the latter group, however, is not yet settled. While 16S rRNA analysis group them peripherally with Methanobacteriales (15), 5S rRNA data indicate a closer relationship to Methanomicrobiales (18). *Natronobacterium* and *Natronococcus*, described to harbor the haloalkaliphilic isolates, are members of the group of halobacteria (85). The phylogenetic position of *T.acidophilum* is still uncertain since, depending on the molecule studied, this species can be considered a member of either the group of methanogens/halophiles or of that of the thermoacidophiles (15, 18, 86).

2. Sulfur-dependent, thermoacidophilic archaebacteria. Although the number of isolates of this subline has been increased considerably (only *Sulfolobus acidocaldarius* was known in 1981) their phylogenetic position cannot be determined as easily as that of other prokaryotes. The presence of a large number of post-transcriptionally modified bases in the rRNA makes the sequence analysis extremely difficult. This main branch now contains *Sulfolobus, Pyrodictium, Thermococcus, Thermoproteus, Thermophilum, Thermodiscus* and *Desulfurococcus* (10, 87) on the basis of rRNA homologies, Zillig and coworkers (10) have classified these genera into two orders, Thermoproteales and Sulfolobales. The genera can also be divided into two groups on the basis of their energy metabolism: While *Thermodiscus, Thermoproteus* and *Pyrodictium* can live by hydrogen-sulfur autotrophy (88), *Desulfurococcus, Thermococcus* and *Thermophilium* exist by dissimilatory sulfate reduction (89). These division do not however correspond to their phylogenetic groupings.

The discovery that some methanogenic bacteria can also respire sulfate (90) has led to the discussion of a possible closer relationship between methanogens and sulfur-dependant archaebacteria than was originally believed.

TOWARDS A PHYLOGENETIC CLASSIFICATION

The comparison of the hierarchic phylogenetic structure with the traditional system immediately reveals considerable discrepancies, especially within and among higher ranks. The 8th edition of Bergey's Manual (1974) lists 5 orders. Of these fives, Myxobacterales and Spirochaetales are phylogenetically coherent units (as far as representatives have been investigated) revealing that the complex life cycle and morphology, respectively, of these organisms are actually phylogenetically valid features. The situation is different with "Rhodospirillales" and "Cytophagales", since their families are members of various lines of descent. Some genera of "Cytophagales" are more closely related to certain taxa of "Rhodospirillales" than they are related to genera of their own "order", *Rhodocyclus tenue-Vitreoscilla*, *Chromatium-Leucothrix* or they are related to representatives of other phenotypically defined groupings, (e.g. *Cytophaga-Bacteroides*, *Cytophaga-Flavobacterium*, *Chloroflexus-Herpetosiphon*, *Lysobacter-Xanthomonas*. More than that, the purple photosynthetic bacteria are highly intermixed with Gram-negative eubacteria which are today listed in a dozen or so sections or parts of Bergey's Manual (Table 2).
The fifth order, Actionomycetales, contains a few genera which would have to be excluded to make this order phylogenetically coherent. Others, not considered to be members according to the present scheme, ought to be included. The internal structure of this order also needs major revisions. At the intrafamily level a similar situation is encountered. Prime examples of phylogenetically heterogenous families are, among others, Pseudomonadaceae, Simonsiellaceae, Cytophagaceae, Micrococcaceae, Dermatophilaceae, Micromonosporaceae and Actinoplanaceae. Hardly any family investigated at all intensively can maintain its status in a phylogenetic system. Heterogenous genera are predominantly those defined by simple morphological and physiological characteristics, e.g. *Flavobacterium*, *Pseudomonas*, *Peptococcus*, *Clostridium*, *Eubacterium* and *Corynebacterium*.

With the knowledge that most of the traditional taxa are phylogenetically poorly defined the question of a future

treatment of the phylogenetic data has become a point of
controversy. Especially within the eubacterial kingdom, the
situation is difficult, with the 100 year old tradition of
eubacterial classification, with the fixed idea that morphological characteristics actually reflect natural relationships, and with a certain degree of frustration that a long
sought goal has been reached within ten years only by methods not previously used in microbial systematics. (The situation is different with the archaebacteria. The discovery
of this kingdom enabled taxonomists to classify its members
right from the beginning according to their natural relationship by the consequent use of the hierarchic structure
emerging from the analysis of the 16S rRNA.)
While some taxonomists favour the continued existance of
two systems, the traditional system (to work with) and the
phylogenetic one (which is nice to have), others believe
that a <u>single</u>, phylogenetic system would not only be highly
advantageous over the traditional one, but is actually currently feasible. The main differences lie in one's estimation of priorities in handling phylogenetic data for taxonomic purposes. For a taxonomist the classification of a
new isolate at the strain and species level has absolute
priority, while relationships of higher taxa are of secondary importance, if not actually unnecessary for classification. The phylogenetic approach is thus a priori useless
for this description level of taxonomy. But are phenetic
and phylogenetic approaches really mutually exclusive? In
the opinion of some taxonomists a combination of the phylogenetic with the phenetic data is the most promising approach.

It is too early to propose a formal system of classification of the eubacteria that is consistent with their phylogeny. However, a strategy can be devised for handling the
data to lead to one synthesis classification scheme: 1. during the delineation of a hierarchic structure outlining
the phylogenetic diversity of prokaryotes, it will be possible to sort out which of the traditional phenotypic characters are homologous, which evolved by convergence, which
are modern, which ancient, which are phylogenetically valid,
which are meaningless. 2. The phylogenetic skeleton obtained will serve as the basis for the definition of taxonomic categories; taxa will not be defined by phylogenetic
parameters alone (e.g. homology, $T_{m(e)}$, or S_{AB} values), but
phylogenetically coherent units at all levels of relationships will be described in terms of appropriate phenotypic
characters useful for distinguishing related taxa at the
same level of relationship, to allow for future classifi-

cation and recognition. The characters will vary from taxon, they may span the whole scale of phenotype and the usefulness of a character will be decided on the basis of its distribution within the taxon concerned. Schleifer and Stackebrandt (7) have summarized some chemotaxonomic markers which have already been found useful in the characterization of phylogenetic groupings. In many cases these characters can be investigated routinely (DNA G+C content, the presence or absence of isomers of diaminopimelic acid, Gram-staining, pigments, spore formation, morphology), while others are more difficult to determine (e.g. peptidoglycan type (91), isoprenoid quinone composition (92), fatty acid composition (93), enzyme patterning (13, 94),phage typing and so on). This strategy is already in use, especially at high levels of relationships,and it is encouraging to see that Volume 1 of the 9th edition of Bergey's Manual has already used phylogenetic data to improve the systematics. Most progress has been achieved among the Gram-positive eubacteria, with several chemotaxonomically valuable markers available (7). Although the traditional classification of Gram-positives is in better accord with their phylogenetic relationships than that seen among Gram-negatives (although things are changing, too (96)) even here taxonomists have had to become accustomed to the idea that mycelium formation is not restricted to actinomycetes (e.g. *Thermoactinomyces/Bacillus*) or that in certain cases the formation of lactic acid is a more reliable marker than morphology (e.g. *Lactobacillus, Pediococcus, Leuconostoc*). Among the phototrophic purple bacteria and their non phototrophic relatives, most phylogenetically-defined families would comprise genera of various morphological and physiological types with almost no common features available which at present would rationalize the groupings. The capacity to reproduce by budding is as unreliable for the definition of higher taxa as chemolithoautotrophy, nitrogen fication, spiral morphology, sulfate respiration or the posession of bacteriochlorophyll a and/or b. Many predjudices must be thrown out if certain apparently dissimilar genera are to be encompassed with one phylogenetically-defined taxon, e.g. *Nitrobacter, Rhodopseudomonas palustris* and *Bradyrhizobium*, or *Rhodocyclus* tenue, *Chromobacterium lividum* and *Alcaligenes eutrophus*. But whatever the decision about the rank of individual taxa, in no case should any more taxa be defined which are inconsistent with the phylogenetic data.

As Staley and Krieg (6) pointed out "a classification (of bacteria) that is of little use to microbiologists, no matter how fine a scheme and who devised it, will soon be ignored or significantly modified". Since schemes have been significantly modified during the last century these schemes must have been poor as we today know they were - Although no classification scheme can claim to be definite, the newly proposed strategy will provide a scheme in which a measure of phylogenetic relationship serves as a framework for a hierarchic structure which is considerably less susceptible than formerly to the subjective judgements of taxonomists about the validity of taxonomic markers. While the framework remains stable, the decisions concerning the ranks of the individual phylogenetic clusters will still be up to taxonomists, keeping this discipline lively, interesting, and demanding.

REFERENCES

1 Schildkraut, C.L., Marmur, J. and Doty, P. (1961). The formation of hybrid DNA molecules and their use in studies of DNA homologies. *Journal of Molecular Biology* 3, 595-617.
2 Moore, R.L. and McCarthy, B. (1967). Comparative study of ribosomal ribonucleic acid cistrons in enterobacteria and myxobacteria. *Journal of Bacteriology* 94, 1066-1074.
3 Woese, C.R., Fox, G.E., Zablen, L., Uchida, T., Bonen, L., Pechman, K., Lewis, B.J. and Woese, C.R. (1975). Conservation of primary structure in 16S ribosomal RNA. *Nature* 254, 83-86.
4 Woese, C.R., Gibson, J. and Fox, G.E. (1980). Do genealogical patterns in purple photosynthetic bacteria reflect interspecific gene transfer? *Nature* 283, 212-214.
5 Nomura, M., Traub, P. and Bechmann, H. (1968). Hybrid 30S ribosomal particles reconstituted from components of different bacterial origins. *Nature* 219, 793-799.
6 Staley, J.T. and Krieg, N.R. (1984). Classification of prokaryotic organisms: an overview. *In*: "Bergey's Manual of Determinative Bacteriology" (Eds. N.R. Krieg and J.G. Holt) Vol. 1, pp. 1-4. Williams and Wilkins, Baltimore, London.
7 Schleifer, K.H. and Stackebrandt, E. (1983). Molecular systematics of prokaryotes. *Annual Review of Microbiology* 37, 143-187.
8 De Ley, J. and De Smedt, J. (1975). Improvements of the membrane filter method for DNA:rRNA hybridization *Antonie van Leeuwenhoek, Journal of Microbiology and Serology* 41, 287-307.

9 Palleroni, N.J., Kunisawa, R., Contopoulou, R. and Douodoroff, M. (1973). Nucleic acid homologies in the genus. *Pseudomonas*. *International Journal of Systematic Bacteriology* 23, 333-339.
10 Tu, J., Prangishvilli, D., Huber, H., Wildgruber, G., Zillig, W. and Stetter, K.O. (1982). Taxonomic relations between archaebacteria including 6 novel genera examined by cross-hybridization of DNAs and 16S rRNAs. *Journal of Molecular Evolution* 18, 109-114.
11 Baharaeen, S., Melcher, U. and Vishniac, H.S. (1983). Complementary DNA-25S ribosomal RNA hybridization: an improved method for phylogenetic studies. *Canadian Journal of Microbiology* 29, 546-551.
12 Stackebrandt, E. and Woese, C.R. (1984). The phylogeny of prokaryotes. *Microbiological Sciences* 1, 117-122.
13 Byng, G.S., Johnson, J.L., Whitaker, R.J., Gherna, R.L. and Jensen, R.A. (1983). The evolutionary pattern of aromatic amino acid biosynthesis and the emerging phylogeny of pseudomonas bacteria. *Journal of Molecular Evolution* 19, 272-282.
14 Fox, G.E., Pechman, K.J. and Woese, C.R. (1977). Comparative cataloging of 16S ribosomal ribonucleic acid: molecular Approach to prokaryotic systematics. *International Journal of Systematic Bacteriology* 27, 44-57.
15 Fox, G.E., Stackebrandt, E., Hespell, R.B., Gibson, J., Maniloff, J., Dyer, T.A., Wolfe, R.S., Balch, W.E., Tanner, R.S., Magrum, L.J., Zablen, L.B., Blakemore, R., Gupta, R., Bonen, L., Lewis, B.J., Stahl, D.A., Luehrsen, K.R., Chen, K.N. and Woese, C.R. (1980). The phylogeny of prokaryotes. *Science* 209, 457-463.
16 Stackebrandt, E. and Woese, C.R. (1981). The evolution of prokaryotes. *In*: "Molecular and Cellular Aspects of Microbial Evolution " (Eds. M.J. Carlile, J.F. Collins and B.E.B. Moseley), pp. 1-31. Cambridge University Press, Cambridge.
17 Woese, C.R., Luehrsen, K.R., Pribula, C.D. and Fox, G.E. (1976). Sequence characterization of 5S ribosomal RNA from eight Gram-positive procaryotes. *Journal of Molecular Evolution* 8, 143-153.
18 Fox, G.E., Luehrsen, K.R. and Woese, C.R. (1982). Archaebacterial 5S ribosomal RNA. *In:* "Archaebacteria" (Ed. O. Kandler), pp. 330-345. G. Fischer, Stuttgart.
19 Dickerson, R.E. (1980). Evolution and gene transfer in purple photosynthetic bacteria. *Nature* 283, 210-212.
20 Woese, C.R., Gibson, J. and Fox, G.E. (1980). Do genealogical patterns in purple photosynthetic bacteria reflect interspecific gene transfer? *Nature* 283, 212-214.

21 Gibson, J., Stackebrandt, E., Zablen, L.B., Gupta, R. and Woese, C.R. (1980). A genealogical analysis of the purple photosynthetic bacteria. *Current Microbiology* 3, 59-66.
22 Hori, H. and Osawa, S. (1979). Evolutionary change in 5S RNA secondary structure and a phylogenetic tree of 54 5S RNA species. *Proceedings of the National Academy of Sciences USA* 76, 381-385.
23 Hori, H., Itoh, T. and Osawa, S. (1982). The phylogenetic structure of the metabacteria. *In:* "Archaebacteria" (Ed. O. Kandler), pp. 18-30. G. Fischer, Stuttgart.
24 Schwartz, R.M. and Dayhoff, M.O. (1978). Origins of prokaryotes, eukaryotes mitochondria and chloroplasts. *Science* 199, 395-403.
25 Hespell, R.B., Paster, B.J., Macke, T.J. and Woese, C.R. The Origin and Phylogeny of the Bdellovibrios. *Applied and Systematic Microbiology* 5, 196-203.
26 Woese, C.R. and Fox, G.E. (1977). Phylogenetic structure of the prokaryotic domain: The primary kingdoms. *Proceedings of the National Academy of Sciences USA* 74, 5088-5090.
27 "Archaebacteria" (1982). (Ed. O. Kandler) G. Fischer, Stuttgart.
28 König, H., Schlesner, H. and Hirsch, P. (1984). Cell wall studies on budding bacteria of the *Planctomyces/Pasteuria* group and on a *Prosthecomicrobium* sp. *Archives of Microbiology* 138, 200-205.
29 Langworthy, T.A., Holzer, G., Zeikus, J.G., and Tornabene, T.G. (1983). Iso- and anteiso-branched glycerol diethers of the thermophilic anaerobe *Thermodesulfotobacteriae commune*. *Systematic and Applied Microbiology* 4, 1-17.
30 Woese, C.R. (1981). Archaebacteria. *In:* "Scientific American" 244, 98-122.
31 Zillig, W., Schnabel, R., Tu, J. and Stetter, K.O. (1982). The phylogeny of archaebacteria, including novel anaerobic thermoacidophiles in the light of RNA polymerase structure. *Naturwissenschaften* 69, 197-204.
32 Lake, J.A., Henderson, G., Michael, W. and Matheson, A.T. (1982). Mapping evolution with ribosome structure: Intralineage constancy and interlineage variation. *Proceedings of the National Academy of Sciences (Washington)* 79, 5948-5952.
33 Schmidt, G. and Böck, A. (1984). Immunoblotting analysis of ribosomal proteins from archaebacteria. *Systematic and Applied Microbiology* 5, 1-10.
34 Tanner, R.S., Stackebrandt, E., Fox, G.E. and Woese, C.R.

(1981). A phylogenetic analysis of *Acetobacterium woodii*, *Clostridium barkeri*, *Clostridium butyricum*, *Clostridium lituseburense*, *Eubacterium limosum* and *Eubacterium tenue*. *Current Microbiology* 5, 35-38.
35 Tanner, R.S., Stackebrandt, E., Fox, G.E. and Woese, C.R. (1982). A phylogenetic analysis of anaerobic eubacteria capable of synthesizing acetate from CO_2. *Current Microbiology* 7, 127-132.
36 Woese, C.R., Maniloff, J. and Zablen, L.B. (1980). Phylogenetic analysis of the mycoplasmas. *Proceeding of the National Academy of Sciences USA* 77, 494-498.
37 Pechman, K.J., Lewis, B.J. and Woese, C.R. (1976). Phylogenetic status of *Sporosarcina urea*. *International Journal of Systematic Bacteriology* 26, 305-310.
38 Ludwig, W., Schleifer, K.H. and Stackebrandt, E. (1984). 16S rRNA analysis of *Listeria monocytogenes* and *Brochothrix thermosphacta*. *FEMS Microbiology Letters* (in press).
39 Stackebrandt, E., Wittek, B., Seewaldt, E. and Schleifer, K.H. (1982). Physiological, biochemical and phylogenetic studies on *Gemella haemolysans*. *FEMS Microbiology Letters* 13, 361-365.
40 Ludwig, W., Seewaldt, E., Kilpper-Bälz, R., Schleifer, K.H., Magrum, L.J., Woese, C.R., Fox, G.E. and Stackebrandt, E. (1985). The phylogenetic position of *Streptococcus* and *Enterococcus*. *Journal of General Microbiology* (in press).
41 Ludwig, W., Seewaldt, E., Schleifer, K.H. and Stackebrandt, E. (1981). The phylogenetic status of *Kurthia zopfii*. *FEMS Microbiology Letters* 10, 193-197.
42 Stackebrandt, E., Fowler, V.J. and Woese, C.R. (1983). A phylogenetic analysis of lactobacilli, *Pediococcus pentosaceus* and *Leuconostoc mesenteroides*. *Systematic and Applied Microbiology* 4, 326-337.
43 Stackebrandt, E. and Woese, C.R. (1979). A phylogenetic dissection of the family Micrococcaceae. *Current Microbiology* 2, 317-322.
44 Ludwig, W., Schleifer, K.H., Fox, G.E., Seewaldt, E. and Stackebrandt, E. (1981). A phylogenetic analysis of Staphylococci, *Peptococcus saccharolyticus* and *Micrococcus mucilaginosus*. *Journal of General Microbiology* 125, 357-366.
45 Stackebrandt, E. and Woese, C.R. (1981). Towards a phylogeny of actinomycetes and related organisms. *Current Microbiology* 5, 131-136.
46 Fowler, V.J., Ludwig, W. and Stackebrandt, E. (1984). The phylogenetic structuring of the genus *Actinomadura*. Academic Press, London (in press).

47 Stackebrandt, E., Ludwig, W., Schleifer, K.H. and Gross, H.J. (1981). Rapid cataloguing of ribonuclease T1 resistant oligonucleotides from ribosomal RNAs for phylogenetic studies. *Journal of Molecular Evolution* 17, 227-236.
48 Stackebrandt, E., Ludwig, W., Seewaldt, E. and K.H. Schleifer (1983). Phylogeny of sporeforming members of the order Actinomycetales. *International Journal of Systematic Bacteriology* 33, 173-180.
49 Stackebrandt, E., Lewis, B.J. and Woese, C.R. (1980). The phylogenetic structure of the coryneform group of bacteria. *Zentralblatt für Bakteriologie und Hygiene, I. Abteilung, Originale* C2, 137-149.
50 Stackebrandt, E. Kroppenstedt, R.M. and Fowler, V.J. (1983). A phylogenetic analysis of the family Dermatophilaceae. *Journal of General Microbiology* 129, 1831-1838-
51 Brooks, B.W., Murray, R.G.E., Johnson, J.L., Stackebrandt, E., Woese, C.R. and Fox, G.E. (1981). A study of the red-pigmented micrococci as a basis for taxonomy. *International Journal of Systematic Bacteriology* 30, 627-646.
52 Bonen, L., Doolittle, W.F. and Fox, G.E. (1979). Cyanobacterial evolution: results of 16S ribosomal ribonucleic acid sequence analysis. *Canadian Journal of Biochemistry* 57, 879-888.
53 Seewaldt, E. and Stackebrandt, E. (1982). Partial sequence of 16S ribosomal RNA and the phylogeny of Prochloron. *Nature* 295, 618-620.
54 Schwarz, A. and Kössel, H. (1980). The primary structure of 16S rDNA from *Zea mays* chloroplast is homologoues to E. coli 16S rRNA. *Nature* 283, 739-742.
55 Paster, B.J., Ludwig, W., Weisburg, W.G., Stackebrandt, E., Reichenbach, H., Hespell, R.B., Hahn, C.M., Gibson, J., Stetter, K.O. and Woese, C.R. (1984). A phylogenetic grouping of the bacteroides, cytophagas and certain flavobacteria. *Systematic and Applied Microbiology* (in press).
56 Paster, B.J., Stackebrandt, E., Hespell, R.B., Hahn, C.M. and Woese, C.R. (1984). The phylogeny of the spirochetas. *Systematic and Applied Microbiology* (in press).
57 Fowler, V.J. (1984). Phylogenetic analysis of the phototrophic purple sulfur bacteria and the dissimilatory sulfate-reducing bacteria. pH. D. thesis, Technical University, Munich, F.R.G.
58 Ludwig, W., Schleifer, K.H., Reichenbach, H. and Stackebrandt, E. (1983). A phylogenetic analysis of the myxobacteria *Myxococcus fulvus, Stigmatella aurantiaca, Cystobacter fuscus, Sorangium cellulosum* and *Nannocystis exedens. Archives of Microbiology* 135, 58-62.

59 Stackebrandt, E., Ludwig, W., Schubert, W., Klink, F., Schlesner, H., Roggentin, T. and Hirsch, P. (1984). Molecular genetic evidence for early evolutionary origin of budding peptidoglycan-less eubacteria. *Nature* 307, 735-737.
60 Oren, A., Paster, B.J. and Woese, C.R. (1984). Haloanaerobiaceae: a new family of moderately halophilic, obligatory anaerobic bacteria. *Systematic and Applied Microbiology* 5, 71-80.
61 Kilpper-Bälz, R., Fischer, G. and Schleifer, K.H. (1982). Nucleic acid hybridization of group N and group D streptococci. *Current Microbiology* 7, 245-250.
62 Stackebrandt, E., Scheuerlein, C. and Schleifer, K.H. (1983). Phylogenetic and biochemical studies on *Stomatococcus mucilaginosus*. *Systematic and Applied Microbiology* 4, 207-217.
63 Fischer, A., Kroppenstedt, R.M. and Stackebrandt, E. (1983). Molecular genetic and chemotaxonomic studies on *Actinomadura* and *Nocardiopsis*. *Journal of General Microbiology* 129, 3433-3446.
64 Van Valen (1982). Phylogenies in molecular evolution: *Prochloron*. *Nature* 298, 493-494.
65 Gibson, J., Pfennig, N. and Waterbury, J.B. (1984). *Chloroherpeton thalassium* gen. nov et spec. nov., a non-filamentous, flexing and gliding green sulfur bacterium. *Archives of Microbiology* 138, 96-101.
66 Woese, C.R., Stackebrandt, E., Weisburg, W.G., Paster, B.J., Madigan, M.T., Blanz, P., Fowler, V.J., Hahn, C.M. and Fox, G.E. (1984). The phylogeny of purple bacteria: the alpha group. *Systematic and Applied Microbiology* (in press).
67 Gillis, M. and De Ley, J. (1980). Intra- and intergeneric similarities of the ribosomal ribonucleic acid cistrons of *Acetobacter* and *Gluconobacter*. *International Journal of Systematic Bacteriology* 30, 7-27.
68 De Smedt, J. and De Ley, J. (1977). Intra- and intergeneric similarities of *Agrobacterium* ribosomal Ribonucleic acid cistron. *International Journal of Systematic Bacteriology* 27, 222-240.
69 De Smedt, J., Bauwens, M., Tytgat, R. and De Ley, J. (1980). Intra- und intergeneric similarities of ribosomal ribonucleic acid cistrons of free-living, nitrogen fixing bacteria. *International Journal of Systematic Bacteriology* 30, 106-122.
70 Seewaldt, E., Schleifer, K.H., Bock, E. and Stackebrandt, E. (1982). The close phylogenetic relationship of *Nitrobacter* and *Rhodopseudomonas palustris*. *Archives of Mic-*

robiology 131,287-290.
71 Ludwig, W., Eberspächer, J., Lingens, F. and Stackebrandt, E. (1984). 16S ribosomal RNA studies on the relationship of a chloridazon-degrading Gram-negative eubacterium. *Applied and Systematic Microbiology* 5, 241-246.
72 Woese, C.R., Blanz, P., Hespell, R.B. and Hahn, C.M. (1982). Phylogenetic relationships among various helical bacteria. *Current Microbiology* 7, 119-124.
73 De Vos, P. and De Ley, J. (1983). Intra- and intergeneric similarities of *Pseudomonas* and *Xanthomonas* ribosomal ribonucleic acid cistrons. *International of Systematic Bacteriology* 33, 487-509.
74 Woese, C.R., Blanz, P. and Hahn, C.M. (1984). What isn't a Pseudomonad: The importance of nomenclature in bacterial classification. *Systematic and Applied Microbiology* 5, 179-195.
75 Woese, C.R., Weisburg, W.G., Paster, B.J., Hahn, C.M., Koops, H.P., Harms, H., Stackebrandt, E. and Woese, C.R. (1984). The phylogeny of purple bacteria: the beta group. *Systematic and Applied Microbiology* (in press).
76 Kersters, K., Hinz, K.H., Hertle, A., Segers, P., Lievens, A., Siegmann, O. and De Ley, J. (1984). *Bortedella avium* sp. nov., isolated from the respiratory tracts of turkeys and other birds. *International Journal of Systematic Bacteriology* 34, 56-70.
77 Woese, C.R., Fowler, V.J., Fox, G.E. and Stackebrandt, E. (1985). The phylogeny of purple bacteria: the gamma group (in preparation).
78 Fowler, V.J., Pfennig, N., Schubert, W. and Stackebrandt, E. (1984). Towards a phylogeny of phototrophic purple sulfur bacteris - 16S rRNA oligonucleotide cataloguing of 11 species of Chromatiaceae. *Archives of Microbiology* (in press).
79 Stackebrandt, E., Fowler, V.J., Schubert, W. and Imhoff, J.F. (1984). Towards a phylogeny of phototrophic purple bacteria - the genus *Ectothiorhodospira*. *Archives of Microbiology* 137, 366-370.
80 Baumann, L. and Baumann, P. (1976). Study of relationship among marine and terrestrial enterobacteria by means of in vitroDNA/ribosomal RNA hybridization. *Microbios Letters* 3, 11-20.
81 Baumann, P. and Baumann, L. (1981). The marine Gram-negative eubacteria: genera *Photobacterium, Beneckea, Alteromonas, Pseudomonas* and *Alcaligenes*. *In*: "The Prokaryotes" (Eds. M.P. Starr, H. Stolp, H.G. Trüper, A. Balows and H.G. Schlegel), Vol. II, pp. 1302-1331. Springer,

Berlin, Heidelberg, New York.
82 Ludwig, W. and Stackebrandt, E. (1983). A phylogenetic analysis of *Legionella*. *Archives of Microbiology* 135, 45-50.
83 De Ley, J., Segers, P. and Gillis, M. (1978). Intra and intergeneric similarities of *Chromobacterium* and *Janthinobacterium* ribosomal ribonucleic Acid Cistrons. *International Journal of Systematic Bacteriology* 28, 154-168.
84 Balch, W.E., Fox, G.E., Magrum, L.J., Woese, C.R. and Wolfe, R.S. (1979).Methanogens: Reevaluation of a unique biological group. *Microbiological Reviews* 43, 260-296.
85 Tindall, B.J., Ross, H.N.M. and Grant, W.D. (1984). *Natronobacterium* gen. nov. and *Natronococcus* gen. nov., two new genera of haloalkaliphilic archaebacteria. *Systematic and Applied Microbiology* 5, 41-57.
86 Schnabel, R., Thomm, M., Gerardy-Schahn, R., Zillig, W., Stetter, K.O. and Huet, J. (1983). Structural homology between different archaebacterial DNA-dependent RNA polymerases analyzed by immunological comparison of their components. *EMBO Journal* 2, 751-755.
87 Woese, C.R., Gupta, R., Hahn, C.M., Zillig, W. and Tu, J. (1984). The phylogenetic relationship of three sulfur dependent archaebacteria. *Systematic and Applied Microbiology* 5, 97-105.
88 Fischer, F., Zillig, W., Stetter, K.O. and Schreiber, G. (1983). Chemolithoautotrophic metabolism of anaerobic extremely thermophilic archaebacteria. *Nature* 301, 511-513.
89 Zillig, W., Schnabel, R., Tu, J. and Stetter, K.O. (1982). The phylogeny of archaebacteria, including novel anaerobic thermoacidophiles in the light of RNA polymerase structure. *Naturwissenschaften* 69, 197-204.
90 Stetter, K.O. and Gaag, G. (1983). Reduction of molecular sulphur by methanogenic bacteria. *Nature* 305, 309-311.
91 Schleifer, K.H. and Kandler, O. (1972). Peptidoglycan types of bacterial cell walls and their taxonomic implications. *Bacteriological Reviews* 36, 407-477.
92 Collins, D.M. and Jones, D. (1981). Distribution of isoprenoid quinone structural types in bacteria and their taxonomic implications. *Microbiological Reviews* 45, 316-354.
93 Minnikin, D.E., Goodfellow, M. and Collins, M.D. (1978). Lipid composition in the classification and identification of coryneform and related taxa. *In*: "Coryneform Bacteria". (Eds. I.J. Bousfield and A.G. Callely) pp. 85-160. Academic Press Inc., London.

94 Harford, S., Jones, D. and Weitzman, P.D. (1976). The mode of regulation of bacterial citrate synthese as a taxonomic tool. *Journal of Applied Bacteriology* 44, 465-471.
95 Imhoff, J.F., Trüper, H.G. and Pfennig, N. (1984). Rearrangement of the species and genera of the phototrophic "purple nonsulfur bacteria". *International Journal of Systematic Bacteriology* 34, 340-343.

ACKNOWLEDGMENT

Research work described here that was done in the author's laboratory at the Technical University Munich was supported by grants from the Deutsche Forschungsgemeinschaft and the Gesellschaft für Biotechnologische Forschung m.b.H. I would like to thank Valerie Joan Fowler for reading the manuscript and Gudrun Gentzen for typing the manuscript.

EVOLUTION OF THE SYSTEMATICS OF BACTERIA

Otto Kandler

Botanisches Institut der Universität München
Munich, Federal Republic of Germany

The numerous new results and excellent reviews presented at this workshop demonstrate the impressive progress recently achieved in our understanding of evolution and our attempts towards phylogenetically oriented systematics. Having been asked to choose the subject of this evening's closing talk myself, I decided to devote my lecture to the "Evolution of Systematics of Bacteria", so as to remind us where our ideas came from. Since bacteriology branched off from botany I shall concentrate, when talking on the early history of our subject, on plants.

You are all aware that systematics has two roots:
1) a practical one, namely the necessity to reliably recognize plants suitable for food, medicine etc., and to communicate about them.
2) a philosophical one, derived from our immanent desire to recognize and understand the world around us, which requires us to bring the recognized items into a logical order.

THE DAWN OF SYSTEMATICS

Primitive man, through trial and error, learned to use more and more plants and products of spontaneous microbial fermentation for food and other purposes. Some day, he invented names for ready reference. Reports on serviceable plants are found among the earlist writings in any antique civilisation. Our European tradition goes back to the Greek Philosophers. Among them, Theophrastus (370 - 285 B.C.) is considered as the Father of Botany. He was a pupil of the great Plato and, later on, of his older fellow-student Aristotle, from whom he probably learned the principles of classification.

In his voluminous writings, of which only fragments are still extant, he listed about 500 names of plants used by farmers, charcoal-burners, apothecaries etc. Because his books were so widely used, these names gained general recognition. Some of them are still in use as generic names, e.g. <u>daucon</u> (Daucus), <u>aspharagos</u> (Asparagus). The writings of Theophrastus were heavily laden with Greek speculative philosophy, they did contain many descriptions but no illustrations.

The first collection of names accompanied by illustrations of the plants was probably compiled by Krateuas (120 - 63 B.C.). However, his books are lost, and only copies of some of his illustrations are to be found in later Roman and Byzantine books.

When the Greeks lost their freedom, their scholarship faded away. However, it fertilized, as the Hellenic culture, the whole Mediterranean area. The Romans, being more practically minded, for the most part compiled and repeated Greek philosophical ideas. However, they did improve considerably books on serviceable plants, especially those used in medicine.

The MATERIA MEDICA, written by the Sicilian born Greek Pedanios Dioscorides (first century A.D.), is said to have been the alpha and omega of European botany. In fact, this book was copied in many variations for more than 1000 years, and E.L. Core (1955) states: "During the Middle Ages, no drug plant was recognized as genuine unless the 'MATERIA MEDICA' had been used for its identification."

The original 'MATERIA MEDICA' was divided into 5 books in accordance with the different uses of the plants or plant products described. The common names of about 600 plants were given in Greek together with, in some cases, the synonyma in Latin and in other tongues. Most important was that not only the usage was given but also a description of the plants, sometimes even including seeds and fruits. While the first edition was probably not accompanied by illustrations, the numerous later copies, produced over the centuries, did contain illustrations, partly taken from various older books which have been lost, for instance from Krateuas. The order of arrangement of the plants in the original MATERIA MEDICA on the basis of practical use - later copies show an alphabetical order - has no systematic significance as compared to modern systems based on relationships. However, there are several series of related plants recognized even in present systematics as belonging to distinct taxa, e.g. the <u>Labiatae</u> and the <u>Umbelliferae</u>. Such groupings were based on the fact that the plants contained similar essential oils important for their medical usage, and these arrangements may thus be considered as an example of a primitive, still unconscious approach towards chemotaxonomy.

THE "LAG PHASE" OF SYSTEMATICS: THE EUROPEAN MIDDLE AGES

The decline and fall of the Roman Empire was paralleled by a loss of interest in literature and science. For almost 1000 years the European Middle Ages were characterized by slavish adherance to classical authorities. The numerous copyings of Dioscorides' book became more and more degenerate, descriptions of plants were almost completely dropped, and fantastic recipes for medical and mystical use became more and more dominant. The authors of these copies did not even realize that many of the Central European plants were different from the Mediterranean plants. They tried to fit together details of Central European plants they had really seen with those of Mediterranean plants they knew only from copies of copies of old books. Hence, illustrations often became primitive combinations of different plants overcrowded with mysticism. For example, melons were presented as growing on trees (Fig. 1B), and the split root of Mandragora (Mandrake), the mystic plant believed to have grown on Mount Calvary, was depicted as either a woman or a man (Fig. 1A).

The only important exception in this period was Albert von Bollstädt (1193 - 1287), the highly esteemed philosopher of the early Renaissance, who received the title Albertus Magnus by general acclamation. In his book "DE VEGETABILIS" he describes many cultivated plants and gives methods for cultivation. Trained in Aristotle's philosophy, he proposed a general system suitable for all plants (Tab. 1). Most of his groups contain mainly plants which are still recognized as belonging to one and the same taxon in present systems.

TABLE 1

System of plants according to Albertus Magnus

A) Leafless plants (mostly Cryptogams)
B) Leafy plants (mostly Phanerogams)
  1) Corticatae (mostly Monocotyledons)
  2) Tunicatae (mostly Dicotyledons)
    a) Herbaceous plants
    b) Woody plants

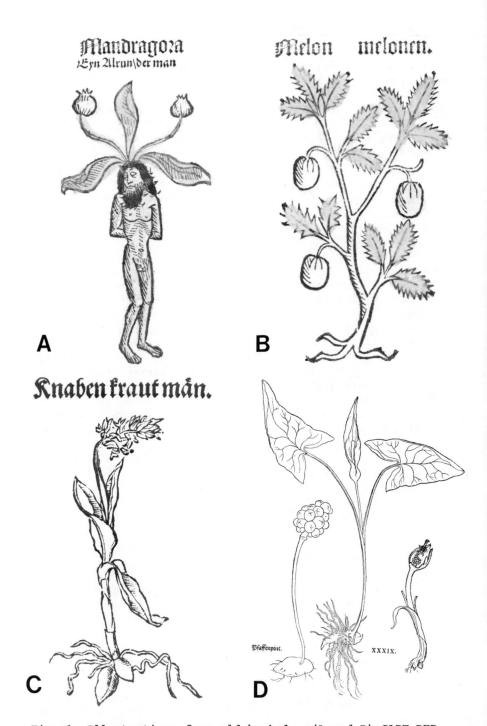

Fig. 1 Illustrations from old herbals. (A and B) GART DER GESUNDHEIT. Published by Renatus Beck, Straßburg (1515).
(C) CONTAFAYT KRÄUTERBUCH by Otto Brunnfelß, Straßburg (1532).
(D) NEW KRÄUTERBUCH by Leonhard Fuchs, Basel (1543).

Albertus Magnus' systematic approach found no resonance in his time. Even at the end of the 15th and the beginning of the 16th century, when plants were depicted very realistically in the paintings of the great European masters, and movable-type printing had been invented, the most widespread herbal "HORTUS SANITATIS (GART DER GESUNDHEIT)", edited in many variations from 1485 to ca. 1540, still contained mostly primitive unrealistic illustrations like the ones mentioned above (Fig. 1A, B), and the text was still dominated by medieval superstition. However, some of the illustrations were obviously drawn according to nature, indicating the influence of the Renaissance spirit.

THE ACCELERATION PHASE OF SYSTEMATICS: THE TIME OF THE GREAT HERBALS

Three great herbals issued within a short period at the beginning of the 16th century by Otto Brunfels (1530), Hieronymus Bock (1539) and Leonhard Fuchs (1542), the three so-called German Fathers of Botany, are the link between medieval copyings of classical books and "modern" systematics. While the text of Brunfels' herbal is still concerned with medieval medicine without much description of the plants themselves, the illustrations, done by Weidenitz, a disciple of the great Renaissance painter Albrecht Dürer, are highly naturalistic (Fig. 1C). They show the whole plants including the roots and thus enable us to recognize the species even today. Bock, on the other hand, greatly improved the description of the plants, but did not use illustrations in his first edition. Fuchs applied real species descriptions to 500 plants, accompanied by high quality realistic drawings which, in a few cases, even included details of fruits and flowers (Fig. 1D). The break with the medieval tradition of copying imaginary illustrations from older books is documented by the last page of his herbal where the artists are depicted preparing drawings and woodcarvings from living flowers (Fig. 2).
    Neither of the three authors achieved significant progress in arranging plants in a systematic order. Brunfels and Fuchs, both used alphabetical order, while Bock stuck to the classical subdivision into herbs, shrubs and trees. However, no rational order is seen within the three groups. How far those great men were still away from our present way of thinking may be demonstrated by citing the arguments given by Bock to explain why he started his herbal with the Stinging Nettle (Fig. 3).

Fig. 2 Artists preparing the illustrations of Fuchs' herbal

Fig. 3 Text from KREUTERBUCH by Hieronymus Bock, Straßburg, enlarged edition (1572)

English version of Fig. 3: "....Further I should like to explain why the Stinging Nettles take the first place in my herbal. It is well known that the Stinging Nettles are the most tender and the purest of all plants, for they cannot be used for all kinds of unclean purposes like the other herbs. In fact, they are completely protected from use by the common

herd when they relieve themselves. It is true, the Stinging
Nettles may sometimes be polluted by dogs and other ignorant
beasts, close to fences, their favourite places. There the
Nettles may be besmirched and bespattered. However, this
happens because of ignorance and lack of judgement. Hence,
even the noble Nettles have to suffer, but soon they will be
washed and cleaned by rain and dew from the heavens and, thus,
such dirt cannot do them any harm. It is, therefore,
justifiable to consider them the purest and cleanest herbs
man can find. ... "

Such argumentation is not only light-years away from the one
we heard the other day from Carl Woese, when he described
the progenotes as being the first creatures in the hierarchy
of organisms, but Bock's argumentation was anachronistic
even a few years later by the second half of the 16th century.
The younger generation of herbalists developed a strong
feeling for morphological similarity and tried to arrange at
least certain groups of plants according to such "natural"
relationships. The enthusiasm for the new approach to
elucidate common features among plants is expressed in the
preface to Lobel's herbal (1570) as quoted by Arber: "For
thus, in an order, than which nothing more beautiful exists
in the heavens or in the mind of a wise man, things which
are far and widely different become, as it were, one thing."
The end and at the same time the culmination of the age of
the herbals is marked by Caspar Bauhin (1560 - 1624) who
finally became completely independent of medicine. His work
is devoted only to the description and systematics of plants.
It is Bauhin's merit to have fully recognized the distinction
between species and genus, to have applied a consistent
nomenclature with a generic and a specific name for each
plant - sometimes still tri- or quadrinominals - and to have
arranged the plants on the basis of morphological relationships of the vegetative organs - mainly leaves - and the
habitus. However, such a more intuitive judgement on
morphological relationships did not result in a coherent
system, but led to the description of several groups of
related plants, which had to be arranged into a more or less
arbitrary order.
Unlike his contemporaries, the great herbalists, Andrea
Cesalpino (1519 - 1603) was not a medical doctor but a
philosopher following the school of Aristotle. He developed
a system of classification of plants based on reasoning and
not on a utilitarian approach or on emotionally recognized
morphological similarities as used by the other herbalists.

He considered the fruits and seeds to be most important for classification. In his main opus "DE PLANTI LIBERI" he designed an artificial but coherent system of plants based for the most part on a priori characteristics, mainly the morphology of seeds and fruits. His work found no resonance among his contemporaries, but Linné called him the first of the systematists.

THE "LOG PHASE" OF SYSTEMATICS: THE LINNEAN PERIOD

As a consequence of the exploration of new continents, the 17th and 18th centuries were characterized by an enormous increase in the number of known plant species collected all over the world. There was also a growing public eagerness to learn about plants, which led to the foundation of many botanical gardens and publications called "HORTUS" or "FLORA", books containing elaborate engravings of plants in the baroque style, which were, however, of limited systematic value.

The number of described species rose from the 600 listed in Bauhin's "PINAX" (1624) to 7000, when Joseph Pitton de Tournefort (1656 - 1708) wrote his "INSITUTIONES REI HERBARIAE" in 1700, the most widespread scientific book on systematics in those days. Although the distinction between genera and species had already been made clear by Bauhin, as we have seen, Tournefort was the first to attach descriptions to each of the 600 recognized genera, most of them established and named by himself. However, he did not include descriptions of species, but listed species names under the respective generic names. He arranged the numerous genera on the basis of Cesalpino's principles, but the characteristics used were extended from the morphology of fruits and seeds to that of other parts of the flowers, in order to approach more closely a "natural" relationship. However, his neglect of the morphology of the habitus and that of most of the vegetative organs as used by the herbalists, led to a system even more artificial than that previously published by John Ray ("HISTORIA PLANTARUM", 1686) which already contained groups corresponding to our present cryptogams, phanerogams, monocotyledons and dicotyledons.

The dominant figure of the 18th century was, of course, Carl von Linné (1707 - 1778). His most important opus "SPECIES PLANTARUM" (1753) contains not only the outline of his "sexual system", but lists all plants using the generic name, the trivial name, a polynomial descriptive phrase (intended by Linné as the specific name), reference to previous publications or his own herbarium specimen, and the location where the plant was found (Fig. 4).

LUNARIA.

1. **LUNARIA** filiculis oblongis. *rediviva.*
Lunaria foliis cordatis. *Hort. cliff.* 333; *Fl. fuec.* 529.
Roy. lugdb. 332.
Viola lunaria major, filiqua oblonga. *Bauh. pin.* 203.
Viola latifolia. Lunaria odorata. *Cluf. hift.* 1. *p.* 297.
Habitat in Europa *feptentrionaliore.* ♃

2. **LUNARIA** filiculis fubrotundis. *annua.*
Lunaria major, filiqua rotundiore. *Bauh. hift.* 2. *p.* 881.
Viola latifolia. *Dod. pempt.* 161. *Dalech. hift.* 805.
Habitat in Germania.
Ita affinis præcedenti, ut etiamnum dubium utrum vere diftincta.

Fig. 4 Example of a species description from "SPECIES PLANTARUM", Linné (1753).

The combination of the generic name and the trivial name came into general use, resulting in the establishment of a binomial nomenclature, which had not originally been thought by Linné to be a major aspect of his book. Later on, the book "SPECIES PLANTARUM" and the date 1753 were chosen as the starting point (priority rule !) of the modern nomenclature of plants and microorganisms. In the case of bacteria, the reference book "SPECIES PLANTARUM" was replaced by the "Approved Lists of Bacterial Names" published on January 1st, 1980, as mentioned below. Linné certainly had a feeling for morphological relationships, but recognized that the overwhelming number of species can only be handled by an artificial system based on a limited number of characteristics. He chose the easily countable stamina and the overall organization (monoecious, dioecious) of the flowers, or the lack of flowers, to divide the plant kingdom into 24 classes (Fig. 5).

Although Linné's work was considered by his contemporaries as the beginning of a new epoque in systematics, it was really the end of a development starting with Albertus Magnus and continued by Cesalpino, Tournefort and others. Never again has such a purely artifical system been created, since the time was then ripe for the new ideas of evolution, the basis of a truly natural system. Basic facts on which such ideas could be founded were just being discovered: the sexuality of plants (Camerarius (1665 - 1721)) and the

Class		
1. Monandria	Stamens one, *Canna, Salicornia*
2. Diandria	Stamens two, *Olea, Veronica*
3. Triandria	Stamens three, many grasses
4. Tetrandria	Stamens four, *Protea, Galium*
5. Pentandria	Stamens five, *Ipomoea, Campanula*
6. Hexandria	Stamens six, *Narcissus, Lilium*
7. Heptandria	Stamens seven, *Trientalis, Aesculus*
8. Octandria	Stamens eight, *Vaccinium, Dirca*
9. Enneandria	Stamens nine, *Laurus, Butomus*
10. Decandria	Stamens ten, *Rhododendron, Oxalis*
11. Dodecandria	Stamens 11-19, *Asarum, Euphorbia*
12. Icosandria	Stamens 20 or more, on the calyx, *Cactus, Mesembryanthemum*
13. Polyandria	Stamens 20 or more, on the receptacle, *Tilia, Ranunculus*
14. Didynamia	Stamens didynamous (in two pairs of different lengths), many mints
15. Tetradynamia	Stamens tetradynamous (with four long stamens and two shorter), mustards
16. Monadelphia	Stamens monadelphous (united in one group), mallows, etc.
17. Diadelphia	Stamens diadelphous (united in two groups), legumes, etc.
18. Polyadelphia	Stamens polyadelphous (united in three or more groups), *Theobroma, Hypericum*
19. Syngenesia	Stamens syngenesious (united by their anthers), many composites
20. Gynandria	Stamens united to the gynoecium, orchids, etc.
21. Monoecia	Plants monoecious, *Carex, Morus*
22. Dioecia	Plants dioecious, *Salix, Juniperus*
23. Polygamia	Plants polygamous, *Acer, Nyssa*
24. Cryptogamia	Flowerless plants, *Pteris, Agaricus, Lichen*

Fig. 5 Outline of Linné's artificial system of plants (1753)

formation of hybrids between species (Koelreuther (1753 - 1806)); the first publication on plant fossils (Scheuchzer, 1723).

However, the term "relationship" in those days meant no more than morphological similarity, and a system in which plants were grouped according to phenetic similarities was called a "natural" system. Linné himself also had such a natural system mind, but he never went very far and only fragments remained when he died.

In recognizing ever increasing numbers of horticultural varieties and closely related species exhibiting intermediate characteristics, Linné may have felt that his doctrine of the constancy of species was not so firmly founded as he had originally thought when he stated that there were as many species as different forms had been created by the Lord

in the beginning (ab initio). Later he switched from the
realistic temporal expression ab initio to the idealistic
expression in principio which was more in accordance with
his philosophy. In the 6th edition of his "GENERA PLANTARUM"
he gives more details about his opinion on the origin of
species by saying that plants were created by the Lord at all
levels of the hierarchy at the same time. In his view, the
mixing of the type plants of the classes by the Lord resulted
in the different plant forms at the generic level, while the
mixing of the type plants of the genera by nature gave rise
to the species; varieties arose from species simply by random
deviations. Mixing according to Linné, was the combination
of the "wood substance" of one form which produced the pollen,
with the "pith substance" of the other form, whose pistil
became fertilized by the pollen.

Some people interpreted these ideas of Linné as early
approaches to a theory of descent. However, as is pointed
out by Sachs (1875): " ... this theory of Linné is no
precursor of our theory of evolution, but is most distinctly
opposed to it; it is utterly and entirely the fruit of
scholasticism."

TOWARDS A PHYLOGENETIC SYSTEM OF PLANTS

The fascination of Linné's personality and the convenience
of his system were so great that many of his followers may
have become distracted from further developing the principles
of systematics. They devoted themselves instead to polishing
up the rough edges in Linné's system by verbose discussions
and to enlarging it by describing numerous new species. Thus,
Sachs (1875) complains that in the early 19th century some
botanical text books looked more like a German-Latin
dictionary than a book on natural science. However, several
ingenious men, especially Antonie Laurent de Jussieu (1748 -
1836) and Augustin Pyrame de Candoll (1778 - 1841) - the
latter introduced the term taxonomy to designate the theory
of plant classification - continued to work towards a
natural system based on the comparative morphology not only
of the generative but also of the vegetative organs and of
the whole habitus of the plants. To some extent, they
followed Bauhin's principles more closely than Linné's,
while holding to the latter's doctrine of the constancy
of species. In their systems, they distinguished, respectively,
100 or 161 orders, most of them still recognized as families
even today. They also formed subdivisions for vascular plants
and cryptogams and created classes for dicotyledons and
monocotyledons.

The last great work on plant systematics under the doctrine of the constancy of species, containing not less than 97 205 species grouped into 200 orders, was published by Bentham and Hooker in their "GENERA PLANTARUM" issued 1862 - 1883. By the time of its publication, it was rather anachronistic in respect of its philosophical background. Not only had Jean Lamarck's idea of a continuous evolution of organisms caused by their interaction with the environment been well known since the publication of his "PHILOSOPHIE ZOOLOGIQUE" in 1809, but even Darwin's theory of evolution had already been made public a few years before. This event took place on July 1, 1858, when the joint paper by Ch. Darwin and A.R. Wallace "On the tendency of species to form varieties and on the perpetuation of the varieties and species by natural means of selection" was read by A.R. Wallace to the Linnean Society. One year later, in 1859, Darwin's "ORIGIN OF SPECIES" was published. The period of the belief in the doctrine of the constancy of species was definitely closed, and a new era had opened.

The plant kingdom had so far been regarded as a great puzzle, composed of a fixed number of building blocks which had to be arranged by the systematists in such a way as to obtain the picture the Creator had in mind when he created the plants. Darwin replaced this static concept by a dynamic one of living beings continuously changing their genetic make up, and, directed by the selective forces of the environment, evolving new forms which matched ever better the environmental requirements.

The way was paved towards a real natural system, i.e. a system based on genealogical relationship not merely on morphological or other phenetical similarities. The new concept was a tremendous challenge to the systematists, and gradually they changed their systems of classification so as to express actual relationships between organisms from an evolutionary point of view. Surprisingly, Darwin himself never drew a phylogenetic tree of any group of organisms, but he designed a theoretical scheme suggesting how new species might arise from existing species by successive small changes in the genetic make up over a number of generations (Fig. 6).

Ernst Heinrich Haeckel, in 1866, was the first to design a phylogenetic tree (Fig. 7). He also invented the term phylogeny to designate the science of the genealogical development of organisms in the various taxa. In his tree, the world of organisms was subdivided into three <u>Archephyla</u>: the <u>Protista</u>, the <u>Plantae</u> and the <u>Animalia</u>, which arose from early divergences of the <u>Radix communis</u> formed by <u>Moneres autogamum</u>,

EVOLUTION OF THE SYSTEMATICS OF BACTERIA   347

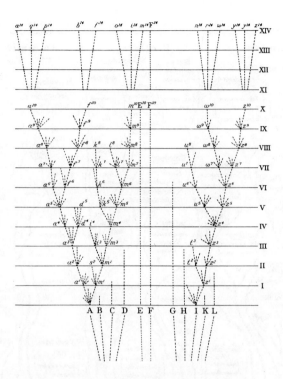

Fig. 6 Charles Darwin's scheme demonstrating the evolution of new species from existing species.

the most primitive organism which, he thought, consisted only of a droplet of proteinacous slime. Haeckel suggested that <u>Moneres autogamum</u> may have a complicated molecular fine-structure which cannot be seen under the microscope. He further assumed that <u>Moneres autogamum</u> arose by spontaneous generation eons ago. In his view, some <u>Moneres</u> did not develop their internal structure any further and thus remained primitive organisms down to his day forming the existing group of <u>Moneres</u> which multiplied by fission or budding, but not longer arose by spontaneous generation. Although most of Haeckel's <u>Moneres</u> were protists, mainly amoeba, he also mentioned Vibrio, a name given by earlier authors to a characteristic form of bacteria. Thus, it so happened that bacteria were recognized just in time to have at least one representative listed in the first phylogenetic tree ever designed.

At this point we shall leave the evolution of the systematics of plants and focus on the systematics of bacteria.

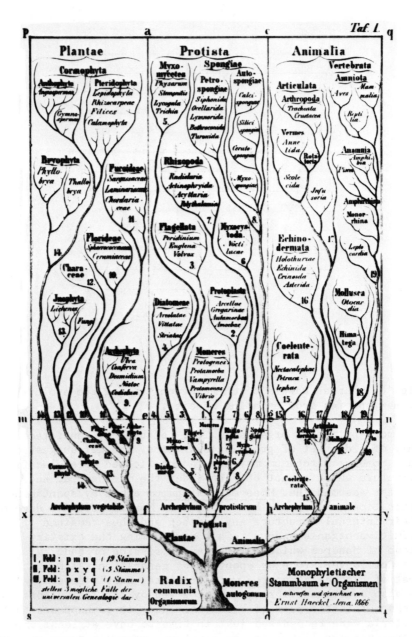

Fig. 7 Ernst Haeckel's first phylogenetic tree of organisms designed 1866.

SYSTEMATICS OF BACTERIA

When bacteria were first observed under the microscope and documented in instructive drawings by Antonie van Leeuwenhook (1675), he called them "animalcules" (little animals). In Linné's system, Leeuwenhook's "animalcules" are found as "specia dubia" among the worms. "Vibrio" and "Monas" were the first names attached to characteristic forms of bacteria by O.F. Müller (1786). In his well known publication "Die Infusionstierchen als vollkommene Organismen", Ehrenberg (1838) described, among many genuine infusorians, several bacterial forms and named them, e.g. Bacterium, Spirillum, Spirochaeta etc.. He stated that they were simply too small for him to see, under his poor microscope, details of their internal organs, organs which he believed he had identified in the larger infusorians, and which he knew from planctonic animals.

The confusion of bacteria with infusorians went on into the fourth quarter of the 19th century. However, when Carl von Nägeli (1878) found cellulose in the cell walls of Acetobacter xylinum and Sarcina ventriculi, he concluded that bacteria belonged to the plant kingdom and called them Schizomycetes. Very soon afterwards, it turned out that the two organisms were unique in possessing cellulose. However, the presence of a rigid cell wall consisting of an unknown nitrogen-containing polymer was confirmed for all the other bacteria and thus they continued to be considered as belonging to the plant kingdom rather than to the animal kingdom.

THE PLEOMORPHISM HYPOTHESIS

The main motivation to grow and study bacteria came from fermentation and food technology, and from hygiene and medicine with Louis Pasteur (1822 - 1895) and Robert Koch (1843 - 1910) respectively being the two outstanding personalities in these fields. Since at first no sterile techniques were available, mixed cultures and even infections with yeasts and fungi were very common. Thus, great confusion arose about the constancy of bacterial forms. In 1848, Carl von Nägeli suggested that the species concept could not be applied to bacteria, since they arose by spontaneous generation, and their forms varied depending on the substrate and other environmental conditions. Even after the de novo formation of bacteria had been disproved by Pasteur, Cohn, Tyndal and others, many different versions of the pleomorphism hypothesis were proposed. Theodor Billroth (1874) believed that there was only one bacterial species, Coccobacillus septica, which could occur in any form, depending on the growth conditions. Ernst Hallier (1866 - 1871) as well as Joseph Lister (1872) suggested that bacteria originated from the spores of different species of fungi and thus, exhibited differing morphologies. Zopf (1884) and Lankester (1873) recognized different species of bacteria, but, in their view, each species could appear in very different forms. Zopf (1884) even published a system based on pleomorphic species. Finally, the pleomorphism hypothesis was disproved by Ferdinand Cohn (1828 - 1898) and Koch who demonstrated the life cycle of spore forming bacilli, and by Koch (1881) who obtained pure cultures using gelatine plates. However, the discussion went on beyond the end of the century, and there are still echoes even today.

COHN'S SPECIES CONCEPT AND EARLY SYSTEMS OF BACTERIA

Being a trained botanist and a pupil of Ehrenberg, Cohn was convinced that the species concept must also be applied to bacteria. Based on careful microscopical studies, he recognized the morphological similarity between bacteria and blue-green algae, for instance the presence of a cell wall and the identical mode of multiplication by binary fission in both groups of organisms. Therefore he considered the bacteria to be plants closely related to blue-green algae. In his first system (1872), the bacteria form the first family in the order Schizosporeae, whereas the following families contain the blue-green algae. The subdivision of the Bacteriaceae into genera is exclusively based on cell morphology (Table 2).

TABLE 2

First system of Bacteria (Cohn 1872)

Division : Thallophyta
  Order : Schizosporeae
    Familia 1: Bacteriaceae (Schizomycetes)
      Tribus   I: Sphaerobacteria (Kugelbakterien)
               Genus 1. Micrococcus
      Tribus II: Microbacteria (Stäbchenbakterien)
               Genus 2. Bacterium
      Tribus III: Desmobacteria (Fadenbakterien)
               Genus 3. Bacillus
               Genus 4. Vibrio

Tribus IV: <u>Spirobacteria</u> (Schraubenbakterien)
Genus 5. <u>Spirillum</u>
Genus 6. <u>Spirochaete</u>

In his second system (1875), Cohn integrated the known groups of bacteria with the known groups of blue-green algae in one class, the <u>Schizophyta</u>.

As a result of Koch's invention of the pure culture technique employing gelatine and, later, agar plates, a vast number of new bacterial species were discovered.

The numerous systems designed during the last quarter of the 19th century were for the most part extended versions of Cohn's system, but physiological characteristics, mainly growth requirements, were included in the descriptions of species. Very original systems, based almost exclusively on physiological characteristics such as aerobic/anaerobic, pathogenous/zymogenous/saprogenous, hydrolysis of gelatine or starch, etc., were designed by Miquel (1891) and Ward (1892).

The most widely accepted system, strictly based on morphology, and summarizing all the species described by the end of the 19th century, was published by Migula (1897) and included in Engler-Prantl's "DIE NATÜRLICHEN PFLANZEN-FAMILIEN" (1900).

At the same time, Lehmann and Neumann (1896) published their "BAKTERIOLOGISCHE DIAGNOSTIK" which came into international use for the identification of bacteria in the first quarter of the 20th century (last edition 1927). It contained mainly bacteria important in clinical and other applied fields, and was arranged in a very handy system based mainly on morphology and cultural characteristics. The unique feature of this book was its brilliant illustration by coloured drawings of the cells and the colonies on various substrates. However, when the number of known species grew in an almost logarithmic manner at the beginning of the 20th century, the book, with its expensive illustrations could not keep up, and its position as the internationally leading handbook for identifying bacteria was taken over by the well known "BERGEY'S MANUAL OF DETERMINATIVE BACTERIOLOGY" first printed in 1923 as a result of an initiative of the American Society of Bacteriologists.

The nomenclature employed in all bacterial systems from the very beginning up to the the middle of the 20th century followed more or less the rules for botanical or zoological nomenclature. Only in 1948, was an "INTERNATIONAL CODE OF NOMENCLATURE OF BACTERIA" published by the "Commission of Nomenclature and Taxonomy of Bacteria", constituted at the

"First International Microbiological Congress" in Paris in 1930. With minor adaptations to bacteriological needs, its rules still follow those of botanical nomenclature even in the most recent revised edition of 1976. However, a decisive change was made in order to remove confusion caused by names given to incompletely described bacterial taxa in the early days of bacteriology. Thus the starting date for nomenclature was shifted from 1753 to January 1st 1980, the day when the "APPROVED LISTS OF BACTERIAL NAMES" were published in the International Journal of Systematic Bacteriology. This publication replaces Linné's "SPECIES PLANTARUM", 1753, as the reference publication for bacterial names.

SYSTEMS BASED ON COMPARATIVE MORPHOLOGY AND PHYSIOLOGY

To some extent, the fate of bacterial systematics in the 20th century up to 1974 is reflected by the systems used in the eight editions of BERGEY'S MANUAL published with in this time span. Thanks to the progress in physiology, biochemistry, cytology etc., in the 20th century, an overwhelming number of characteristics became available to classify bacteria. However, as a result of the lack of fossil record and the scarcity of morphological details, it was not possible to decide which end of a morphological or physiological series of characteristics had to be considered as the primitive one and which as the deduced one. Thus, it was up to the intuitive judgement of the individual authors as to how to arrange the lower ranks of taxa, which often represented naturally related groups still recognized today. There was no rationale for the construction of a system based on genealogical relationship or for designing a phylogenetic tree of bacteria. Thus, we find all gradations between systems almost exclusively based on comparative physiology, such as the one proposed by Orla Jensen (1909), who considered the autotrophic bacteria to be the most primitive ones, and systems based on comparative morphology such as the one proposed by Kluyver and van Niel (1936), who considered the cocci to be the most primitive bacteria, from which several lines of descent were deduced. Prevot (1935) even claimed in his first "taxonomic law": "morphology overrules physiology".

Numerical taxonomy, developed in the late 50s, together with the invention of computers, has facilitated the handling of large numbers of strains and characteristics, and led to a refined clustering of strains at different levels of phenetic similarity. However, numerical taxonomy is not a reasonable way of improving our knowledge of phylogenetic

relationships or for constructing satisfactory groups at
taxonomic levels above the genus.

The general difficulty in defining bacterial taxa at the
higher ranks of the systematic hierarchy (above family or
order) is reflected in the 8 th edition of BERGEY'S MANUAL
OF DETERMINATIVE BACTERIOLOGY" (1974) and the first edition
of "BERGEY'S MANUAL OF SYSTEMATIC BACTERIOLOGY" (1984).
Because of the lack of evidence for phylogenetic relationships of taxa at higher ranks, all attempts to approach a
phylogenetic system were abandoned and the bacteria were
subdivided into sections under vernacular headings for the
purpose of recognition and identification; i.e., it resulted
in a practical, artificial system.

TOWARDS A PHYLOGENETIC SYSTEM

A first break-through towards a deeper understanding of
phylogenetic relationships at the highest level of the
hierarchy has come from the progress in cytology and cytochemistry as a result of advances in electron microscopy
and molecular biology in the late 50s and early 60s.

The essential findings were:

1) The procaryotic organisation of the bacterial and
   cyanobacterial cell, as opposed to the eucaryotic
   organisation of the animal cell and the plant cell.
2) All bacteria and cyanobacteria have a common cell
   wall chemistry (peptidoglycan, murein).
3) Ribosomes of procaryotes and cytoplasmic ribosomes
   of eucaryotes are different in size.
4) Ribosomes of procaryotes, and ribosomes of the mitochondria and the chloroplasts of eucaryotes exhibit
   similar size, similar membrane lipids and similar
   sensitivity towards antibiotics.
5) Mitochondria and chloroplasts of eucaryotes contain
   circular procaryotic DNA.

4) and 5) indicate a symbiotic origin of eucaryotic mitochondria and chloroplasts from procaryotic ancestors.

Based on these findings, Murray (1968) proposed a formal
kingdom, the <u>Procaryotae</u>, subdivided into 4 divisions based
on cell wall structure. An emended system (Table 3) based
on cell wall structure and physiological characteristics
was published by Gibbons and Murray (1978):

TABLE 3

System of bacteria according to Gibbons and Murray 1978

Kingdom: <u>Procaryotae</u>
Division   I: <u>Gracilicutes</u>
Class   I: <u>Photobacteria</u>
Subclass I: <u>Oxyphotobacteria</u>
Order I: <u>Cyanobacteriales</u>
Order II: <u>Prochlorales</u>
Subclass II: <u>Anoxyphotobacteriae</u>
Order I: <u>Rhodospirillales</u>
Order II: <u>Chlorbiales</u>
Class II: <u>Scotobacteria</u>
Division II: <u>Firmacutes</u>
Division III: <u>Mollicutes</u>

In this system, Gibbons and Murray integrate bacteria and blue-green algae (cyanobacteria) in a way similar to that of Cohn, working on an intuitive, purely morphological basis exactly 100 years earlier. Their system also includes the unique procaryotic alga <u>Prochloron</u>, whose pigment composition is very similar to that of the eucaryotic green-algae. Thus, <u>Prochloron</u> rather than cyanobacteria has recently been assumed to be the ancestor of the chloroplasts in green-algae and higher plants.

    The procaryotes had generally been considered to be the more primitive organisms from which the eucaryotes might have developed. This view was further supported, when praecambrian microfossils up to 3.5 billion years old were discovered. Thus, Whittaker (1969) placed the procaryotes at the basis of his phylogenetic tree of organisms which he subdivided into five kingdoms. The most recent version of Whittaker's five kingdoms concept, indicating the supposed symbiotic origin of eucaryotic organelles, was designed by L. Margulis and K.V. Schwartz (1982; Fig. 8). In addition, these authors suggested a rather detailed phylogenetic tree of the bacterial kingdom, (Fig. 9) utilizing data on comparative morphology, cytology and biochemistry accumulated by the middle of the 70s. When their book, containing these trees, was issued in 1982, the first papers of Carl Woese and his colleagues on the partial sequencing of ribosomal RNAs, which let us have a first glance at the outline of

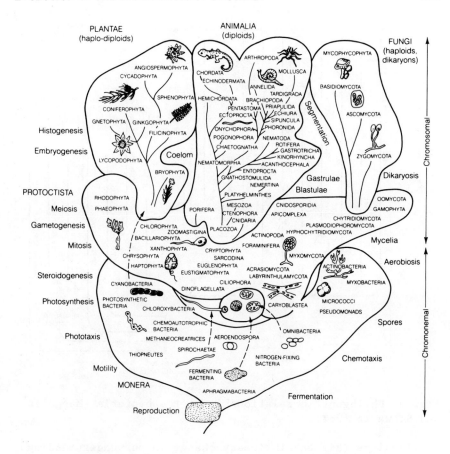

Fig. 8 Phylogenetic tree of organisms designed by Whittacker, modified by Margulis and Schwartz (1982).

bacterial phylogeny, had already been published (1977).

The publications of Woese and his colleagues closed the period of phylogenetic speculation based on comparative morphology and biochemistry. They demonstrated the efficiency of nucleid acid sequencing in unravelling the phylogenetic record contained in highly conserved macromolecules, a principle first discussed by Zuckerkandl and Pauling in 1965.

During this symposium, we have seen the phylogenetic trees derived from this unrabelling presented by Carl Woese and Erko Stackebrandt, demonstrating the new concept of three kingdoms, the Archaebacteria, the Eubacteria and the Eucaryotes, which are thought to have evolved in separate lines from a common ancestor, the Progenote.

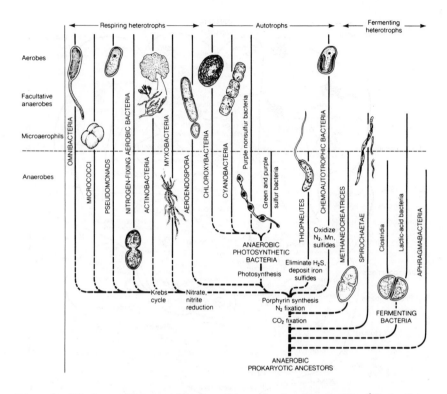

Fig. 9 Phylogenetic relationship of the bacteria (Margulis and Schwartz 1982).

To visualize this revolutionary change in our understanding of the phylogeny of organisms, I prefer to use not a tree, but a disc with the common ancestor occupying the centre and the lineages of descent extending to the periphery (Fig. 10). In addition to the S_{ab}-values given by the rings, the cell wall chemistry is indicated in Fig. 10 by different signatures, as an example of one of the early biochemical arguments which "put meat on the bones of the phylogenetic skeleton", as Carl Woese expressed it. Of course, the phylogenetic relations between the organisms are much more interwoven than we can ever depict by any type of drawing. Horizontal transfer of genetic information at any level up to the transfer of complete genomes by endosymbiosis may confuse part of the picture but such transfers are neglected in this presentation. The translation of phylogenetic relationships into systematics is a real challenge, and some people feel that there will be a struggle between phylogenetically and phenetically oriented systematists. Thus, I should like to end this talk on the evolution of systematics by a short look at how these matters may be resolved.

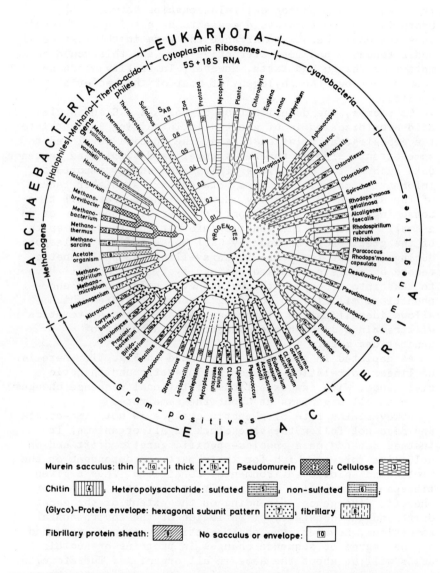

Fig. 10 Diagramatic presentation of the phylogenetic relatedness between main lines of organisms based on Woese's S_{ab}-values. Main components of cell walls are indicated by different signatures (Kandler 1982).

PHYLOGENETIC RELATIONSHIP VERSUS PHENETIC SIMILARITY

If we knew the genealogical relationships between all bacteria, systematics would cease to be a science. The only question would be how to arrange the taxa in the most easily understandable and most attractive way, and this would be a matter of design and taste. However, in reality we do not know much about the phylogeny of most of the bacterial species.

Of course, the determination of mol % G + C and nucleic acid hybridization is of great help in gaining insight into the genealogical relationships at the species level and to some extent also at the genus level, and simplified sequence data such as the S_{ab}-values of 16S rRNA supply important information on relationships between taxa at higher ranks of the systematic hierarchy. However, these are still time-consuming procedures based only on overall properties or on a small portion of the genetic information. Thus, in spite of the great progeess made by Carl Woese and his colleagues during the last few years in unravelling the phylogenetic record by nucleic acid sequencing, systematists still have to take into account both phylogenetic relationship and phenetic similarity in building a coherent system which allows identification of any new isolate. Thus systematics will remain a Janus-headed science for a couple of generations more.

As depicted in the scheme presented in Fig. 11, there is no linear correlation between phylogenetic and phenetic divergence. With few exceptions, mutational sequence changes go on in all organisms at about the same rate.

However, the rate of phenetic changes is not time-constant and does not follow the same clock in all organisms. It instead depends on a species-specific genetic drift and on selective forces, which for their part are dependent on the biotic and abiotic environmental conditions. Thus a given timespan of evolution will result in a greater or smaller phenetic divergence, depending on the extent of the genetic drift and the constancy or variation of the environmental conditions. However, the phylogenetic divergence, measured by the degree of sequence changes in proteins or nucleic acids will be about the same in all organisms. Therefore, phenetic and phylogenetic grouping yields different groups of organism at all levels of the systematic hierarchy as indicated in Fig. 11.

Convergent evolution leads to a grouping of phylogenetically very diverse lines in one and the same phenetic group (Fig. 11, Nos. 1-7), while divergent evolution causes the

EVOLUTION OF THE SYSTEMATICS OF BACTERIA

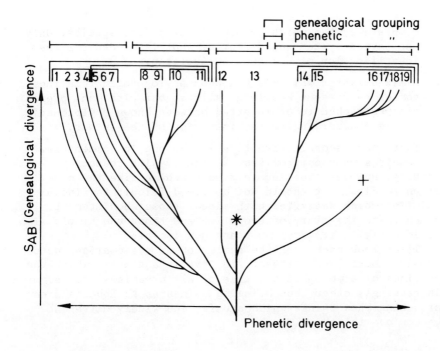

Fig. 11 Diagramatic presentation of the correlation between genealogical and phenetical divergence. Nos. 1-19 = extant organisms; + and * = extinct organisms. (modified after Ehrendorfer, 1983)

grouping of genealogically closely related forms into different phenetic groups (Fig.11, Nos. 14 + 15, and 16 + 19). With respect to the ranking of the groups, even greater discrepancies are encountered, for instance the groups formed by Nos. 1 - 4 and Nos. 16 - 19 show about the same span of phenetic divergence and thus should be ranked at the same level, i.e. as genera. However, with respect to their phylogenetic depth, they differ considerably. While the group containing Nos. 16 - 19 may be ranked as a genus, that formed by Nos. 1 - 4 must be ranked as an order.

Since we do not know the phylogenetic relationships between most of the bacteria, a systematist may approach classification in two different ways:

1. Construction of two independent systems:
   A truly phylogenetic but incomplete system which contains only those organisms between which the phylogenetic relationships are known, and a purely artificial but

complete system which covers all described species. Only the latter system would provide what is needed for the identification of unknown new isolates.

2. A mixed system which is based on phylogenetic relationships as far as present knowledge allows, but with many phenetic groups attached to or inserted between known phylogenetic lineages according to the intuition of the author.

In fact, the second alternative is the one in common use. It requires many compromises especially with respect to ranking. However, bearing in mind a scheme like the one shown in Fig.11 it should not be too difficult to find a solution which satisfies both sides: those who intend to demonstrate the glory of evolution and those who simply wish to identify their very important new isolate.

If we look back over the evolution of systematics, we can consider ourselves fortunate to live in a time in which the outlines of a monophyletic tree of all organisms are becoming increasingly clear thus bringing us nearer to Lobel's dream of seeing "how things which are far and widely different become one thing".

REFERENCES

References to particular chapters may be found in the following reviews and books.

I   Early botanical systematics:

   Sachs, J. (1975). Geschichte der Botanik vom 16. Jahrhundert bis 1860. R. Oldenburg, München
   Core, E.L. (1955). Plant Taxonomy. Prentice Hall, Inc.

II Early bacterial systematics up to 1900:

   Migula, W. (1897). System der Bakterien. Gustav Fischer, Jena
   Bergey's Manual of Determinative Bacteriology. First (1923), second (1925) and third (1930) editions.

III Bacterial systematics after 1900:

   Archaebacteria. Proceedings of the First International Workshop on Archaebacteria (1982). Kandler, O. ed. Fischer Verlag, Stuttgart, New York
   Bergey's Manual of Determinative Bacteriology, Sixth edition (1949)
   Bergey's Manual of systematic Bacteriology, Vol. 1. Krieg, N.R. and Holt, J.G. eds. First edition 1984

Ehrendorfer, F. Samenpflanzen. In: Lehrbuch der Botanik für Hochschulen. Begründet von: Strasburger, E., Noll, F., Schenk, H., Schimper, A.F.W. Gustav Fischer, Stuttgart, New York, 1983.
Kandler, O. (1982). Cell wall structures and their phylogenetic implications. Zbl. Bakt. Hyg. I. Orig. C 3, 149-160.
Margulis, L. and Schwartz, K. (1982). Five Kingdoms. Freemann and Company, San Francisco.
Woese, C.R. and Fox, G.E. (1977). Phylogenetic structure of the prokaryotic domain: The primary kingdoms. Proc. Nat. Acad.Sci.Wash.) 74, 5088-5090.

Index

A

Acetate, assimilation, 243
Acetyl CoA
  assimilation, 244
  carbon assimilation, 236
  carbon dioxide fixation, 245, 246
  incorporation, 243
  synthesis, 240, 245, 247
  synthesizing multienzyme complex, 243
Acetylornithine transaminase, action of, 279
Aeromas formicans, catabolism of arginine, 286
Agmatinase, of *Klebsiella aerogenes*, 291
Agmatine, deiminase pathway, 290
Ammonia monooxygenase, oxidation of ammonia, 209
AMP, *see* Citrate synthase
Antibiotic sensitivity, of protein synthesis, 84
Archaebacteria
  acetogenic, 240
  ATP phosphorylase, 176
  difference from eubacteria, 60
  discovery, 37-38
  halophilic, 322
  homologies between corresponding components of *Eucyta*, 61
  line of, 298
  methanogens and their relatives, 20, 240, 241, 322
  sensitivity to protein synthesis inhibitors, 81, 82
  sequences of ribosomal proteins, 80
  signatures defining various groups, 18
  sulfur-dependent, 20, 74, 75, 240, 241, 245, 322
  transcriptional, translational or posttranslational splits, 60
  translational apparatus, 85
  usage of the citric acid cycle, 264
Arginine
  biosynthesis, 278-281
  catabolism of, 282-291
  catabolism via 2-ketoarginine formation, 291, 292
  decarboxylase pathway, 288, 289, 299
  decarboxyoxidase, 292, 293
  deiminase pathway, 284-287, 298, 299
  oxidase, 292, 293
  succinyltransferase pathway, 293-295, 300

B

Bacillus licheniformis, arginase and arginine deiminase, 283, 285, 286
Bacillus subtilis, induced arginase pathway, 283
Bergey's Manual of Determinative Bacteriology, 351, 352

C

Calvin cycle, *see* Carbon dioxide fixation
Carbamoylphosphate synthetase, 281
Carbon dioxide fixation
  Calvin-Bassham-Benson cycle, 237, 238, 245, 246
  mechanism, 236
  pathways in prokaryotes, 245
  via a reductive carboxylic acid pathway, 245
  via the reductive citric acid cycle, 239
  via the reductive pentosephosphate cycle, 237, 238
Carbon isotope fractionation, in the oldest sediments, 247

Carbon monoxide
  dehydrogenase, 213, 214, 243
  incorporated into the carboxyl of acetate, 242
Carboxylase, of ribulose 1,5-biphosphate, 238
Carboxylation, of phosphoenolpyruvate (PEP)
    by PEP carboxylase, 240
Carboxyl group of acetate, origin of, 242
Chemolithoautotrophy, classification of
    bacteria, 206
Chemotoxonomic markers, in the
    characterization of phylogenetic
    groupings, 325
Chloroplasts, origin, 164
Chromatophores, for isolation of pure reaction
    centers, 146.
Citric acid cycle
  enzymes of, 268, 269
  evolution of, 256
  glyoxylate bypass, 255
  nearly complete, 257
  retrograde evolution, 255
  split in anaerobic organisms, 259
Citrate synthase
  control by α-oxoglutarate, 267
  control by AMP, 266
  control by NADH, 267
  control by succinyl-CoA, 267
Citrullin
  phosphorolysis of, 284
  synthesis of, 281
Cohn, system of bacteria, 350, 351
Cyanobacteria, oxygenic photosynthesis, 257
Cytochrome b, second membrane-bound
    complex, 155
Cytochrome c
  ability to synthesize, 185, 186
  usage for phylogenetic measurement, 4

D

Darwin, origin of species, 346

E

Electron transfer
  cyclic, in blue-green bacteria, 186, 187
  light-dependent, 184, 185
  non-cyclic, oxidization of reduced sulfur
      compounds, 188
Electron transport
  in chloroplasts, 151

  light-induced, from donor to acceptor, 146
  photosynthetic, in green bacteria, 145
  in photosynthetic bacteria and plants, 149,
      157, 158
  respiratory and photosynthetic chains, 163
Electrophosphorylation, transition from
    heterotrophs to phototrophs, 161
Energy conservation, protonmotive,
    force-dependent, 187
Eubacteria
  Chloroflexus group, 16, 319
  chloroplasts, 15
  cyanobacteria, 15, 317
  Deinococcus, 317
  Gram-positive, 13, 314–316
  green, sulfur, 15, 319
  group of *Bacteroides*, cytophage,
      Flavobacterium, 15, 317
  line, 298
  molecular biological differences between
      eukaryotes and prokaryotes, 33
  myxococci, 318
  Planctomyces–Pasteuria group, 17
  Planctomyces–Pirella group, 318
  purple bacteria and their relatives, 14, 319
  sulfate-reducing bacteria and their relatives,
      16
  sulfur-dependent, Gram-negative, 15, 317
Eukaryote
  cytoplasmic ribosomes, 75
  nuclear molecular biology, 35
Eukaryote–prokaryote
  concept, 3
  discontinuity, 32
Evolution
  branching order of the primary kingdoms,
      313
  cellular, 177
  energy-transducing systems, 159
  eukaryotic algae, 165
  fermentation reactions, 178
  fumarate respiration, 183
  in the *Mycoplasma* lines of descent, 22
  metazoan, 21
  molecular biology, 31, 32
  of a new gene, 106
  oxygenic, photosynthetic prokaryotes, 165
  prokaryote, classical opinion, 136
  relationships among the three primary
      kingdoms, 23, 24
  translation, 73–90
  transposon family, 98

F

Fermentation, accessory oxidant-dependent, 180, 181
Formaldehyde, assimilation in aerobic, methylotrophic bacteria, 238
Fumarate reductase
  link between fermentation and anaerobic respiration, 182
  scalar, 184

G

Gene
  activation by excisive deletion, 108
  Antirrhinum majus, integration of plant transposable elements, 108, 109
  integration of SPM-I8, 110
Guanidinobutyrase, split, 291

H

Haloanaerobiaceae, anaerobic, moderately halophilic eubacteria, 319
Halobacterium halobium, light energy conversion, 164
Hybridization, technique, DNA, 310
Hydrogen
  via ATP hydrolysis, 180
  ejection via end-product efflux, 179
Hydrogenase
  from bacterial sources, 216
  structure and function, 215
Hydrogen autotrophy, genetic transfer of, 220
Hydroxylamine oxidoreductase, location in *Nitrosomonas* cells, 209

I

Insertion element, IS1 and IS2, constituents of the *E. coli* chromosome, 105, 106
Intervening sequences
  antiquity, 40
  in eukaryotic protein-coding genes, 31
  introduction, 34
Intron, origin, 34

K

Kreuterbuch, see Stinging nettles

L

Light-harvesting complex, antenna–pigment complex, 148
Linné, *Species Plantarum*, 343–345
Lithoautotrophic bacteria
  carboxydotrophic, 207
  hydrogenotrophic, 207
Lithoautotrophic genes, pHG1-encoded, 222–225

M

Materia Medica, alpha and omega of European botany, 336
Methyl group of acetate, origin, 241
Morphology, development
  comparative, of the whole habitus of plants, 345
  of seeds and fruit, 342
  similiarity in plants, 341
  summary of all species, Migula, 351

N

NADH, *see* Citrate synthase
Nitrate
  denitrification, 200, 201
  respiration, 200, 201
Nitrification, evolution, 199
Nitrite
  membrane-bound oxidoreductase, 210
  oxireductase in anaerobes, 201
Nitrobacter, phylogenetic relationship with *Rhodopseudomonas*, 210
Nitrosomonas, hydroxylamine oxidoreductase, 210
Nomenclature
  for component patterns of different RNA polymerase types, 54, 57, 58
  nucleic acid sequencing, 358
  publication of *International Code*, 351, 352

O

Oligonucleotide
  cataloging, *see*, Ribosomal RNA
  defining the three primary kingdoms, 8
  signatures defining the eubacterial phyla, 10–12
Ornithine carbamoyltransferase, for the synthesis of citrullin, 281

Oxidation
  ammonia, 209
  nitrite, 210
  sulfur, 211

P

Paracoccus denitrificans,
  thiosulfate-oxidizing, 218
Photosynthesis
  apparatus, 150, 152
  artificial, 166
  green bacteria, 150
  oxygen-evolving, 150
  purple bacteria, 146
Photosynthetic organism
  bacteria, 145
  characteristics, 144
  electron transport, 157
Photosystem
  in the chloroplast membranes, 156
  mechanism of water oxidation, 152, 154
Phylogenetic classification, of archaebacteria and eubacteria, 323–324
Phylogenetic disc concept, 356, 357
Phylogenetic reconstruction, 125
Phylogenetic relationships, in bacteria, 4
Phylogenetic tree, 346–348, 355
  corresponding to DNA-dependent RNA polymerase types, 64
  from progenote, 355
  relating the three kingdoms, 38
Phylogeny
  bacterial, 2, 309
  pleomorphism hypothesis, 349, 350
  of prokaryotes, 313
Plant transposable element
  evolution of new functions, 112
  excision–integration process, 112
Plasmid
  Hox-encoding, 221, 222
  in lithoautotrophic bacteria, 218, 219
Progenote, concept, 31, 36
Promoter
  archaebacterial, 40
  eukaryotic, 40
Pseudomonas, derepression of arginine deiminase 286, 287
Putrescine
  diversion into polyamines, 289
  production from arginine, 288
Pyruvate, cyclic scheme of reactions, 264

R

Respiration
  aerobic, 176
  anaerobic, 176
  chemoheterotrophs, 195
  Paracoccus denitrificans, 193, 194
  sulfate, 188, 189
  Thiobacillus, 197, 198
Respiratory chain
  from fumarate-respiring bacteria, 190, 191
  phosphorylation, 176
  principal function in chemotrophic bacteria, 175
Rhodospirillaceae, non-cyclic electron transfer, 186
Ribonucleic acid
  DNA-dependent polymerases, 48, 50, 51
  polymerase, common origin, 60
  SDS polyacrylamide gel electrophoresis patterns, 49
Ribosomal ribonucleic acid
  chronometer, 5, 6
  clock, 7
  genes, 39
  moiety, 74
  as phylogenetic probe, 312
  sequences, by oligonucleotide cataloging, 7
  5 S
    evolution, 125
    gene sequence, 76
    isolation and sequencing, 116–119
    phenogram, 120, 121
    secondary structure, 122–124
    sequence alignment, 121
  16 S, sequence, 76, 77
Ribosome
  archaebacterial, physical properties, 73, 74
  DNA gene organization in archaebacteria, 75
  protein acidity, 80
  proteins
    from methanogens and halophiles, 78
    from sulfur-dependent archaebacteria, 79
  reverse electron flow, 207
  subunits from *Methanococcus*, 74
  subunits from *Sulfolobus*, 74

S

Stinging nettles, and Hieronymus Bock, 340, 341

Streptococcus faecalis, oxidative phosphorylation, 285
Succinate thiokinase
  in the citric acid cycle, 269, 270
  pathway in cyanobacteria, 257
Succinylarginine, transferase pathway, 297
Succinyl CoA, *see* Citrate synthase
Sulfite oxidase, integral part of the multienzyme complex, 212
Systematics
  development, 335–340
  Linnéan period, 342–345
  phylogenetic system of plants, 345–348

T

Thiobacillus versutus, thiosulfate-oxidizing enzyme complex, 212

Transfer RNA gene, with intervening sequences, 41
Transposon
  DNA rearrangements and evolution, 97
  duplication, 106, 107
  genetic and physical maps, 96
  insertion, 96
  promoted rearrangements, 97
  structure and gene organization, 92
  transposition, 93–97, 100
Tricarboxylic acid cycle, *see* Citric acid cycle

W

Western blotting, immune chemical cross-reactivity, 53, 54, 59